GEOMETRY

The Easy Way

THIRD EDITION

Lawrence S. Leff
Assistant Principal,
Mathematics Supervision,
Franklin D. Roosevelt High School
Brooklyn, New York

BARRON'S

All inquiries should be addressed to:
Barron's Educational Series, Inc.
250 Wireless Boulevard
Hauppauge, New York 11788

Library of Congress Catalog Card No. 97-482

International Standard Book No. 0-7641-0110-2

Library of Congress Cataloging-in-Publication Data
Leff, Lawrence S.
 Geometry the easy way / Lawrence S. Leff.—3rd ed.
 p. cm. — (Easy way)
 Rev. ed. of: Barron's geometry the easy way. 2nd ed. ©1990.
 Includes index.
 Summary: Explains the principles of plane geometry and includes practice exercises and model problems.
 ISBN 0-7641-0110-2
 1. Geometry, Plane. [1. Geometry.] I. Leff, Lawrence S. Barron's geometry the easy way. II. Title. III. Series.
QA455.L35 1997
516.22—dc21 97-482
 CIP
 AC

CONTENTS

PREFACE

Many geometry-help books tend to resemble factual handbooks that reduce the study of geometry to a list of formulas followed by a continuous stream of exercises. *Geometry the Easy Way* is different. This book "teaches" a typical introductory course in geometry, using a friendly writing style that incorporates many of the same strategies that an experienced classroom teacher would use.

This book has several special features that make it ideal for self-study and rapid learning:

- Most concepts are *not* presented in "finished form," but instead are carefully developed using a clear writing style with many helpful examples and diagrams. Emphasis is on *understanding* geometric principles as well as on knowing how to apply them in new problem situations.
- The discussions, numerous examples, and convenient summaries anticipate and answer the "why" types of questions students might ask if the material were being explained by a classroom teacher.
- Each chapter concludes with a comprehensive set of original exercises that are designed to clinch understanding of key ideas while building skill and confidence in solving problems. Also, three sets of cumulative review exercises, strategically placed in the book, serve as progress checkpoints and ensure long-term learning. Answers to all exercises are provided at the end of the book.
- Key geometric formulas and relationships have been organized in easy-to-read tables at the end of the book.
- Important geometric terms have been collected and placed in a glossary at the end of the book.

LAWRENCE S. LEFF
January 1997

BUILDING A GEOMETRY VOCABULARY

WHAT YOU WILL LEARN

Studying geometry is very different from studying elementary algebra. In geometry we are concerned with developing a logical structure in which mathematical relationships are *proved* as well as *applied*. In this chapter you will learn:

- some basic terms and assumptions of geometry;
- the nature of geometric proofs;
- the type of mathematical reasoning that forms the basis of this course.

SECTIONS IN THIS CHAPTER

The Building Blocks of Geometry

Definitions and Postulates

Inductive Versus Deductive Reasoning

The IF . . . THEN . . . Sentence Structure

The Building Blocks of Geometry

Studying geometry is, in a sense, like building a house. Cement and bricks are often used to give a house a strong foundation. Instead of bricks, geometry uses the following types of building blocks:

- UNDEFINED TERMS. These words are so fundamental that they cannot be defined using simpler terms. *Point, line,* and *plane* are undefined terms. Although these terms cannot be defined, they can be described.
- DEFINED TERMS. These words help to create a common vocabulary, making it easier to refer to geometric figures and relationships.
- POSTULATES. These are basic assumptions that are accepted as true.
- THEOREMS. A theorem is a generalization that can be proved. A familiar theorem is that the sum of the measures of the three angles of a triangle is 180 degrees. Much of our work in geometry will be devoted to building on our current knowledge by suggesting and then proving theorems. These theorems, in turn, will be used to prove additional theorems. The result will be an expanding knowledge of a variety of geometrical relationships.

Geometry is an example of a *postulational system* in which a beginning set of assumptions and undefined terms is used as a starting point in developing new relationships that are expressed as theorems. These theorems, together with the postulates and defined and undefined terms, are used to *prove* other theorems. Here, "to prove" simply means to use a logical chain of reasoning to show how undefined terms, definitions, postulates, and previously established theorems lead to a new generalization.

UNDEFINED TERMS

Table 1.1 lists some undefined terms.

TABLE 1.1

Undefined Term	Description	Notation
Point	A *point* indicates position; it has no length, width, or depth.	• A A point is named by a single capital letter.
Line	A *line* is a set of continuous points that extend indefinitely in either direction.	← • A • B → ℓ A line is identified by naming two points on the line and drawing a line over the letters: $\overleftrightarrow{AB}$ Alternatively, a line may be named by using a single lowercase letter. line ℓ

(Continued)

TABLE 1.1 *Continued*

Undefined Term	Description	Notation
Plane	A *plane* is a set of points that forms a flat surface that has no depth and that extends indefinitely in all directions.	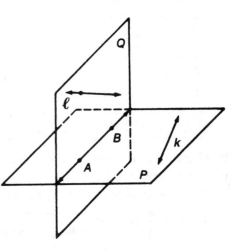 A plane is usually represented as a closed four-sided figure and is named by placing a capital letter at one of the corners.

Figure 1.1 illustrates that lines may lie in different planes or in the same plane. Line ℓ and line *AB* both lie in plane *Q*. Line *k* and line *AB* lie in plane *P*. Lines ℓ and *k* are contained in different planes, while line *AB* (the intersection of the two planes) is common to both planes. To simplify our discussions, we will always assume that we are working with figures that lie in the same plane. This branch of geometry takes a "flat," two-dimensional view of figures and is referred to as *plane geometry. Solid geometry* is concerned with figures and their spatial relationships as they actually exist in the world around us.

FIGURE 1.1

DEFINED TERMS

Table 1.2 lists some geometric terms and their definitions.

TABLE 1.2

Term	Definition	Illustration
1. Line segment	A *line segment* is a part of a line consisting of two points, called *end points*, and the set of all points between them.	NOTATION: $\overline{AB}$
2. Ray	A *ray* is a part of a line consisting of a given point, called the *end point*, and the set of all points on one side of the end point.	NOTATION: $\overrightarrow{LM}$ A ray is always named by using two points, the first of which must be the end point. The arrow drawn above always points to the right.

(Continued)

TABLE 1.2 *Continued*

Term	Definition	Illustration
3. Opposite rays	*Opposite rays* are rays that have the same end point and that form a line.	$\overrightarrow{KX}$ and $\overrightarrow{KB}$ are opposite rays.
4. Angle	An *angle* is the union of two rays having the same end point. The end point is called the *vertex* of the angle, and the rays are called the *sides* of the angle.	VERTEX: K SIDES: $\overrightarrow{KJ}$ and $\overrightarrow{KL}$

An angle may be named in one of three ways:

1. Using three letters, the center letter corresponding to the vertex of the angle and the other letters representing points on the sides of the angle. For example, in Figure 1.2, the name of the angle whose vertex is T can be angle RTB ($\angle RTB$) or angle BTR ($\angle BTR$).
2. Placing a number at the vertex and in the *interior* of the angle. The angle may then be referred to by the number. For example, in Figure 1.3, the name of the angle whose vertex is T can be $\angle 1$ or $\angle RTB$ or $\angle BTR$.

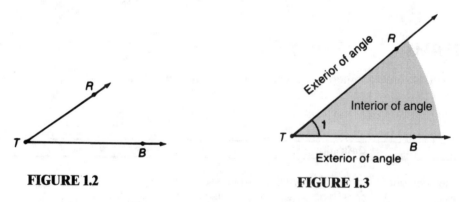

FIGURE 1.2 **FIGURE 1.3**

3. Using a single letter that corresponds to the vertex, provided that this does not cause any confusion. There is no question which angle on the diagram corresponds to angle A in Figure 1.4, but which angle on the diagram is angle D? Actually three angles are formed at vertex D:

- Angle ADB
- Angle CDB
- Angle ADC

FIGURE 1.4

In order to uniquely identify the angle having *D* as its vertex, we must either name the angle using three letters or introduce a number into the diagram.

a. Name the accompanying line in three different ways.
b. Name three different segments.
c. Name four different rays.
d. Name a pair of opposite rays.

SOLUTION

a. $\overleftrightarrow{JW}$, $\overleftrightarrow{WR}$, and $\overleftrightarrow{JR}$

b. $\overline{JW}$, $\overline{WR}$ and $\overline{JR}$

c. $\overrightarrow{JR}$, $\overrightarrow{WR}$, $\overrightarrow{RJ}$ and $\overrightarrow{WJ}$

d. $\overrightarrow{WJ}$ and $\overrightarrow{WR}$

Use three letters to name each of the numbered angles in the accompanying diagram.

SOLUTION

$\angle 1 = \angle BAD$ or $\angle DAB$ or $\angle LAB$

$\angle 2 = \angle CDB$ or $\angle BDC$ (NOTE: Letter *E* may be used instead of letter *B*.)

$\angle 3 = \angle BEM$ or $\angle MEB$

$\angle 4 = \angle DLM$ or $\angle MLD$ (NOTE: Letter *E* may be used instead of letter *M*.)

Definitions and Postulates

DEFINITIONS

The purpose of a definition is to make the meaning of a term clear. A good definition must:

- Clearly identify the word (or expression) that is being defined.
- State the distinguishing characteristics of the term being defined, using only words that are commonly understood or that have been previously defined.
- Be expressed in a grammatically correct sentence.

As an example, consider the term *collinear*. In Figure 1.5, points *A*, *B*, and *C* are collinear. In Figure 1.6, points *R*, *S*, and *T* are *not* collinear.

FIGURE 1.5 **FIGURE 1.6**

DEFINITIONS OF COLLINEAR AND NONCOLLINEAR POINTS
- *Collinear points* are points that lie on the same line.
- *Noncollinear points* are points that do not lie on the same line.

Notice that a definition begins by identifying the term being defined. The definition of collinear points uses only geometric terms (*points* and *line*) that have been previously discussed. Contrast this definition with the following one:

An *apothem* is a line segment drawn from the center of a regular polygon perpendicular to a side of the polygon.

Is this a good definition? No, it is not clear what an apothem is since several terms, including *regular polygon* and *perpendicular*, which have not been explained, are used in the definition.

Much of geometry involves building on previously discussed ideas. For example, we can use our current knowledge of geometric terms to arrive at a definition of a triangle. How would you draw a triangle? If you start with three noncollinear points (Figure 1.7a) and connect them with line segments (Figure 1.7b), a *triangle* is formed.

(a) Points *A, B, C* are noncollinear.

(b) Triangle *ABC* (NOTATION: △*ABC*)

FIGURE 1.7

DEFINITION OF TRIANGLE

A *triangle* is a figure formed by connecting three noncollinear points with three different line segments each of which has two of these points as end points.

Notice that the definition uses the term *noncollinear*, which has been defined. Is it necessary to include that the three noncollinear points are connected by line segments? Yes; observe (in Figure 1.8) that it is possible to join three noncollinear points without using line segments.

FIGURE 1.8

A good definition must be reversible as shown in the following table.

Definition	Reverse of the Definition
Collinear points are points that lie on the same line.	Points that lie on the same line are collinear points
A *right angle* is an angle whose measure is 90 degrees.	An angle whose measure is 90 degrees is a right angle.
A *line segment* is a set of points.	A set of points is a line segment.

The first two definitions are reversible since
the reverse of the definition is a true statement.
The reverse of the third "definition" is false since
the points may be scattered as in Figure 1.9.

A set of points

FIGURE 1.9

The Reversibility Test

The reverse of a definition must be true. If the reverse of a statement which is
being offered as a definition is false, then the statement is not a good definition.

The reverse of a definition will prove useful in our later work when attempting to
establish geometric properties of lines, segments, angles, and figures. For example, a
midpoint of a segment may be defined as a point that divides a segment into two seg-
ments of equal length. In Figure 1.10, how can we prove that point M is the midpoint
of $\overline{AB}$? We must appeal to the *reverse* of the definition of a midpoint: a point that
divides a segment into two segments of equal length is the midpoint of the segment. In
other words, we must first show that $AM = MB$. Once this is accomplished, we are
entitled to conclude that point M is the midpoint of $\overline{AB}$.

FIGURE 1.10

As another illustration, we may define an *even* integer as an integer that leaves a
remainder of 0 when divided by 2. How can we prove that a particular integer is an
even number? Simple—we use the reverse of the definition to show that, when the
integer is divided by 2, the remainder is 0. If this is true, then the integer must be an
even number.

INITIAL POSTULATES

In building a geometric system, not everything can be proved since there must be some
basic assumptions, called *postulates* (or *axioms*), that are needed as a beginning. Here
are our first two postulates.

POSTULATE 1.1

Two points *determine* a line.

A second *line* cannot be drawn through
points A and B.

Three noncollinear points *determine* a plane.

Postulate 1.1 implies that through two points exactly one line may be drawn while Postulate 1.2 asserts that a plane is defined when a third point not on this line is given.

More than one plane may contain *two* points.

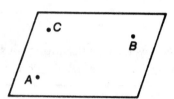

Exactly one plane contains three noncollinear points.

Inductive Versus Deductive Reasoning

Consider the result of accumulating consecutive odd integers beginning with 1.

String of Odd Integers	Sum
1 + 3	4
1 + 3 + 5	9
1 + 3 + 5 + 7	16
1 + 3 + 5 + 7 + 9	25

Do you notice a pattern? It appears that the sum of consecutive odd integers, beginning with 1, will always be a *perfect square*. (A perfect square is a number that can be expressed as the product of two identical numbers.) If, on the basis of this evidence, we now conclude that this relationship will always be true, regardless of how many terms are added, we have engaged in *inductive reasoning*. Inductive reasoning involves examining a few examples, observing a pattern, and then assuming that the pattern will never end. Inductive reasoning is *not* a valid method of proof, although it often suggests statements that can be proved by other methods.

Deductive reasoning may be considered to be the opposite of inductive reasoning. Rather than begin with a few specific instances as is common with inductive processes, deductive reasoning uses accepted *facts* (i.e., undefined terms, defined terms, postulates, and previously established theorems) to reason in a step-by-step fashion until a desired conclusion is reached.

EXAMPLE 1.3

Assume the following two postulates are true. (1) All last names that have seven letters with no vowels are the names of Martians. (2) All Martians are 3 feet tall. Prove that Mr. Xhzftlr is 3 feet tall.

SOLUTION

For illustrative purposes we will use the two-column proof format that will be explained more fully in subsequent chapters.

PROOF:	Statements	Reasons
	1. The name is Mr. Xhzftlr.	1. Given.
	2. Mr. Xhzftlr is a Martian.	2. All last names that have seven letters with no vowels are the names of Martians. (See Postulate 1.)
	3. Mr. Xhzftlr is 3 feet tall.	3. All Martians are 3 feet tall. (See Postulate 2.)

Notice that each statement has a corresponding justification.

The IF . . . THEN . . . Sentence Structure

"If I graduate from high school with a 90 or better grade average, then my parents will buy me a car." Will the student receive a car as a graduation present? Only if his or her average is 90 or better will the student receive a car. Notice that the statement in the "if clause" of the sentence represents a *condition*; if the condition is true, then the "then clause" specifies the action to be taken.

Statements in mathematics often take the "If . . . , then . . ." form. For example,

If <u>two lines are parallel</u>, then <u><u>they never meet</u></u>.

The statement in the if clause (the single-underlined clause) represents the hypothesis or Given. The statement in the then clause (the double-underlined clause) represents the conclusion. If we assume the lines are parallel, then we may conclude that they never meet.

It is common for theorems to be expressed in this conditional form. When a *proposed theorem* is to be proved, the statement in the if clause represents what we are given or entitled to assume; the statement in the then clause corresponds to what has to be proved. For example, a theorem that will be proved in a later chapter is "If two sides of a triangle are equal in length, then the angles opposite them are equal." What is the Given? Equal sides. What is to be proved? Equal angles. Here is the diagram that would be used in this proof, together with a summary of the Given and the Prove.

GIVEN: $AB = CB$.
PROVE: Angle A = Angle C.

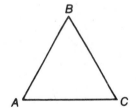

PROOF: Statements Reasons

1. $AB = CB$.	1. Given.
2.	2. ____ ⎧ Undefined terms
3.	3. ____ ⎪ and previously
4.	4. ____ ⎨ established
	⎪ definitions,
	⎩ postulates, and
	theorems may be
	entered.

Once this proof is complete, the theorem is taken as fact and can then be used to prove other theorems.

REVIEW EXERCISES FOR CHAPTER 1

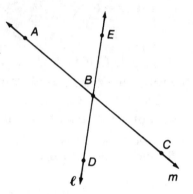

1. For the accompanying diagram:
 (a) Name four rays each of which has point B as an end point.
 (b) Name line ℓ in three different ways.
 (c) Name line m in three different ways.
 (d) Name four angles that have the same vertex.
 (e) Name two pairs of opposite rays.

Use the following diagram for Exercises 2 and 3.

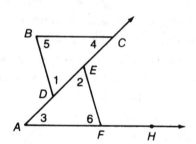

2. Name the vertex of each angle: (a) 1 (b) 3 (c) 5.

3. Use three letters to name each angle: (a) 2 (b) 4 (c) 6.

For Exercises 4 to 11, use the following diagram.

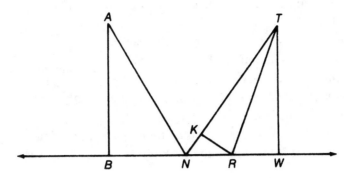

4. Name four collinear points.

5. If point N is the midpoint of $\overline{BW}$, name two segments that have the same length.

6. Name the different triangles that appear in the diagram.

7. Name each angle that has point R as its vertex.

8. Name an angle that is not an angle of a triangle.

9. Name two pairs of opposite rays.

10. Name a segment that is a side of two different triangles.

11. To prove R is the midpoint of $\overline{WN}$, which two segments must be demonstrated to have the same length?

12. Write the reverse of each of the following definitions:
 (a) An acute angle is an angle whose measure is less than 90.
 (b) An equilateral triangle is a triangle having three sides equal in length.
 (c) A bisector of an angle is the ray (or segment) which divides the angle into two congruent angles.

13. Identify each of the following as an example of inductive or deductive reasoning.
 (a) The sum of 1 and 3 is an even number; the sum of 3 and 5 is an even number; the sum of 5 and 7 is an even number; the sum of 7 and 31 is an even number; the sum of 19 and 29 is an even number. *Conclusion:* The sum of any two odd numbers is an even number.
 (b) All students in Mr. Euclid's geometry class are 15 years old. John is a member of Mr. Euclid's geometry class. *Conclusion:* John is 15 years old.
 (c) It has rained on Monday, Tuesday, Wednesday, Thursday, and Friday. *Conclusion:* It will rain on Saturday.
 (d) The sum of the measures of a pair of complementary angles is 90. Angle A and angle B are complementary. The measure of angle A is 50. *Conclusion:* The measure of angle B is 40.

14. A *median* of a triangle is a segment drawn from a vertex of the triangle to the midpoint of the opposite side of the triangle. Draw several large *right* (90-degree) triangles. See the diagram. For each triangle, locate the midpoint of the hypotenuse (the side opposite the 90-degree angle). Draw the median to the hypotenuse. Using a ruler, compare the lengths of the median and the hypotenuse in each triangle drawn. Use inductive reasoning to draw an appropriate conclusion. Note in the diagram that M is the midpoint of AB if $\overline{AM}$ and $\overline{BM}$ measure the same length.

15. Draw several large triangles (not necessarily right triangles). In each triangle locate the midpoint of each side. Draw the three medians of each triangle. Use inductive reasoning to draw a conclusion related to where the medians intersect.

16. Use deductive reasoning to arrive at a conclusion based on the assumptions given.
 (a) ASSUMPTIONS 1. All Martians have green eyes.
 2. Henry is a Martian.
 (b) ASSUMPTIONS 1. The sum of the measures of the angles of a triangle is 180.
 2. In a particular triangle, the sum of the measures of two angles is 100.

17. A prime number is any whole number that is divisible only by itself and 1. For example, 7, 11, and 13 are prime numbers. Evaluate the formula $n^2 + n + 17$ using all integer values of n from 0 to 9, inclusive. Do you notice a pattern?
 (a) Using inductive reasoning, draw a conclusion.
 (b) Is your conclusion true for all values of n? Test $n = 16$.

MEASURE AND CONGRUENCE

WHAT YOU WILL LEARN

This chapter focuses on measuring angles and segments. In this chapter you will learn:

- definitions and terms related to segments and angles that have the same measures;
- the properties of equality that are useful in working with the lengths of line segments and the degree measures of angles;
- the way to draw conclusions using the properties of equality and congruence;
- the format of a formal two-column geometric proof.

SECTIONS IN THIS CHAPTER

Measuring Segments and Angles

Betweenness of Points and Rays

Congruence

Midpoint and Bisector

Diagrams and Drawing Conclusions

Properties of Equality and Congruence

Additional Properties of Equality

Measuring Segments and Angles

We often describe the size of something by comparing it to another thing we are already familiar with. "She is as thin as a rail" creates the image of an underweight person, but it is not very exact. In geometry we must be precise. How could we determine the exact weight of a person? We might use a measurement instrument that is specifically designed for this purpose—the scale. To determine the length of a segment or the measure of an angle, we must also use special measurement instruments—the *ruler* for measuring the length of a segment and the *protractor* for measuring an angle.

The units of measurement that we choose to express the length of a segment are not important, although they should be convenient. It would not be wise, for example, to try to measure and express the length of a postage stamp in terms of kilometers or miles. As illustrated in Figure 2.1, a segment is measured by lining up the end points of the segment with convenient markings of a ruler. In this example, the length or measure of line segment AB is 2 inches. We abbreviate this by writing $m\overline{AB} = 2$, read as "The measure of line segment AB is 2." Alternatively, we could write $AB = 2$, read as "The distance between points A and B is 2." It is customary to use the expressions $m\overline{AB}$ and AB (no bar over the letters A and B) interchangeably and to interpret each as the length of line segment $\overline{AB}$. *Caution:* It is incorrect to write $\overline{AB} = 2$, which implies that the infinite set of points that make up segment $\overline{AB}$ is equal to 2.

FIGURE 2.1

USING A PROTRACTOR

To measure an angle we use a protractor (see Figure 2.2), where the customary unit of measure is the degree. In our example, the measure of angle ABC is 60 degrees. We abbreviate this by writing $m\angle ABC = 60$, read as "The measure of angle ABC is 60." It is customary to omit the degree symbol (°). Thus, we *never* write $m\angle ABC = 60°$ or $\angle ABC = 60$ (omitting the "m").

FIGURE 2.2

In this course, however, we will assume that the measure of an angle corresponds to some number on the protractor that is greater than 0 and less than *or* equal to 180.

Find the measures of these angles:
a. $\angle APZ$ 50°
b. $\angle FPZ$ 130°
c. $\angle WPB$ 20°
d. $\angle ZPB$ 40°
e. $\angle SPZ$ 80

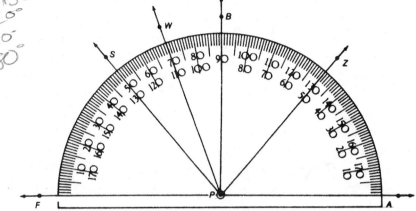

SOLUTION

a. $m\angle APZ$ = 50 (read lower scale when the measure of the angle is less than 90)

b. $m\angle FPZ$ = 130 (read upper scale when the measure of the angle is greater than 90)

c. $m\angle WPB$ = 110 – 90 = 20

d. $m\angle ZPB$ = 90 – 50 = 40

e. $m\angle SPZ$ = 130 – 50 = 80

Find $m\overline{RS}$.

SOLUTION

$m\overline{RS}$ = 6.5 – 3 = 3.5 cm. Since length is a positive quantity, we must always subtract the smaller reading on the ruler (called a *coordinate*) from the larger coordinate.

CLASSIFYING ANGLES

Angles may be classified by comparing their measures to the number 90. An angle whose measure is less than 90 (but greater than 0) is called an *acute angle*. An L-shaped angle is called a *right angle* and its measure is exactly equal to 90. An angle whose measure is greater than 90 (but less than 180) is called an *obtuse angle*. See Figure 2.3.

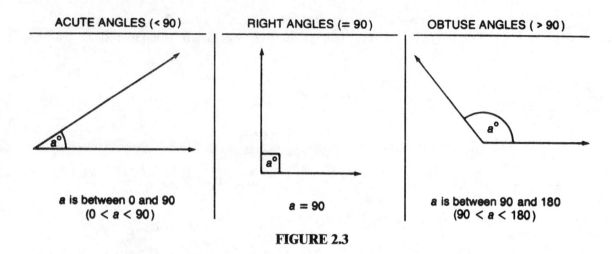

ACUTE ANGLES (< 90)	RIGHT ANGLES (= 90)	OBTUSE ANGLES (> 90)

FIGURE 2.3

Betweenness of Points and Rays

Paul is standing on a line for theater tickets; he is standing *between* his friends Allan and Barbara. We represent this situation geometrically in Figure 2.4. We would like to be able to define this notion formally. The phrasing of the definition should eliminate the possibility that Paul may be standing "off" the ticket line or both Allan and Barbara are in front of Paul, or behind him, on the ticket line. (See Figure 2.5.)

FIGURE 2.4

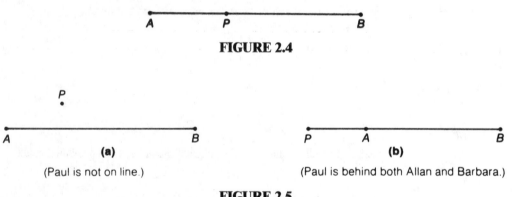

(a)
(Paul is not on line.)

(b)
(Paul is behind both Allan and Barbara.)

FIGURE 2.5

DEFINITION OF BETWEENNESS
Point P is *between* points A and B if both of the following conditions are met:
1. Points A, P, and B are three different collinear points.
2. $AB = AP + PB$.

Condition 1 of the definition of betweenness eliminates Figure 2.5a as a possibility, while condition 2 eliminates the possibility of Figure 2.5b.

 EXAMPLE 2.3 Point Q is between points W and H.
If $WQ = 2$ and $QH = 7$, find WH.

SOLUTION
$WH = 2 + 7 = 9$

 EXAMPLE 2.4 $RT = 2$, $RS = 3$ and $ST = 1$, and points R, S, and T are collinear. Which of the points is between the other two?

SOLUTION

Point T is between points R and S.

The analogous situation with angles occurs when a ray, say $\overrightarrow{OP}$, lies in the interior of an angle, say $\angle AOB$, between its sides. The sum of the measures of the component angles of $\angle AOB$ must equal the measure of the original angle. See Figure 2.6.

NOTE: If $m\angle AOP = 40$ and $m\angle POB = 10$ then $m\angle AOB = 50$. This somewhat obvious relationship is given a special name: the Angle Addition Postulate.

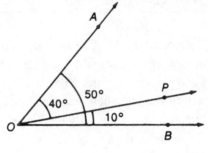

FIGURE 2.6

POSTULATE

ANGLE ADDITION POSTULATE

If ray $\overrightarrow{OP}$ in the interior of angle AOB, then
$$m\angle AOB = m\angle AOP + m\angle POB$$

The Angle Addition Postulate may be expressed in two equivalent forms:
$$m\angle AOP = m\angle AOB - m\angle POB$$
and
$$m\angle POB = m\angle AOB - m\angle AOP$$

 EXAMPLE 2.5 $\overrightarrow{BG}$ lies in the interior of $\angle ABC$. If $m\angle ABG = 25$ and $m\angle CBG = 35$, find $m\angle ABC$.

SOLUTION

$m\angle ABC = 25 + 35 = 60$

 EXAMPLE 2.6 $\overrightarrow{KM}$ lies in the interior of $\angle JKL$. If $m\angle JKM = 20$ and $m\angle LKJ = 50$, find $m\angle MKL$.

SOLUTION

$m\angle MKL = 50 - 20 = 30$

EXAMPLE 2.7

In the accompanying figure the Angle Addition Postulate is contradicted; the measure of the largest angle is *not* equal to the sum of the measures of the two smaller angles. Explain.

SOLUTION

$\overrightarrow{BG}$ is not in the interior of angle *ABC*, thus violating the assumption (hypothesis) of the Angle Addition Postulate.

Congruence

If the windshield of a car shatters or a computer disk drive is not working, we usually do not buy a new car or computer. We replace the broken parts. How do we know that the replacement parts will exactly fit where the old parts were removed? The new parts will fit because they have been designed to be interchangeable; they have been manufactured to have exactly the *same size and shape*. Figures that have the same size and shape are said to be *congruent*.

Figures may agree in one or more dimensions, yet *not* be congruent. Diagrams *ABCD* and *JKLM* (Figure 2.7) each have four sides that are identical in length, but the figures are not congruent since their corresponding angles are not identical in measure. Figures are congruent only if they agree in *all* their dimensions.

FIGURE 2.7

A line segment has a single dimension—its length. Two segments are congruent, therefore, if they have the same length. If line segments $\overline{AB}$ and $\overline{RS}$ have the same length, then they are congruent. We show that these segments are congruent by using the notation $\overline{AB} \cong \overline{RS}$. Similarly, if two angles have the same measure, then they are congruent. If angle *X* has the same measure as angle *Y*, we write $\angle X \cong \angle Y$. See Figure 2.8.

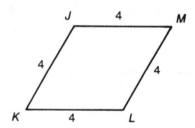

FIGURE 2.8

Congruence is one of the fundamental concepts of geometry. The problem of establishing that two triangles are congruent will be considered in Chapters 6 and 7.

DEFINITION OF CONGRUENT SEGMENTS OR ANGLES
Segments (or angles) are *congruent* if they have the same measure.

The notation $\overline{AB} \cong \overline{RS}$ is read as "Line segment AB is congruent to line segment RS."

Midpoint and Bisector

Consider Figures 2.9 and 2.10. In Figure 2.9, $AM = MB = 3$. Since point M divides $\overline{AB}$ into two congruent segments ($\overline{AM} \cong \overline{MB}$), M is said to be the *midpoint* of $\overline{AB}$. Observe that the measure of each of the congruent segments is one-half the measure of the original segment, $\overline{AB}$. In Figure 2.10, $\overleftrightarrow{XY}$ intersects AB at point M, the midpoint of $\overline{AB}$. $\overleftrightarrow{XY}$ is said to *bisect AB*; a line, ray, or segment that bisects a segment is called a *bisector*.

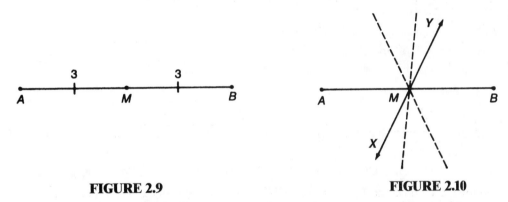

FIGURE 2.9 **FIGURE 2.10**

Since an infinite number of lines, rays, or segments may be drawn through the midpoint of a segment, a line segment possesses an infinite number of bisectors. The terms *midpoint* and *segment bisector* may be formally defined as follows.

DEFINITION OF MIDPOINT
Point M is the *midpoint* of $\overline{AB}$ if
1. M is between A and B and 2. $AM = MB$.

If M is the midpoint of $\overline{AB}$, then M bisects $\overline{AB}$ and the following relationships involving the lengths of the segments thus formed are true:

* $AM = MB$
* $AM = \frac{1}{2}AB$ or $AB = 2AM$
* $MB = \frac{1}{2}AB$ or $AB = 2MB$

DEFINITION OF A SEGMENT BISECTOR
A *bisector* of a line segment AB is *any* line, ray, or segment that passes through the midpoint of $\overline{AB}$. Thus, a segment bisector divides a segment into two congruent segments.

EXAMPLE 2.8

$\overrightarrow{RS}$ bisects $\overline{EF}$ at point P.

a. If $EF = 12$, find PF.
b. If $EP = 4$, find EF.
c. If $EP = 4x - 3$ and $PF = 2x + 15$, find EF.

SOLUTION
a. $PF = \frac{1}{2}EF = \frac{1}{2}(12) = 6$.
b. $EF = 2EP = 2(4) = 8$.
c. Since $EP = PF$,
$$4x - 3 = 2x + 15$$
$$4x = 2x + 18$$
$$2x = 18 \text{ and } x = 9$$
$$EP = 4x - 3 = 4(9) - 3 = 36 - 3 = 33$$
$$EF = 2(EP) = 2(33) = 66$$

Similarly, any ray that lies in the interior of an angle in such a way that it divides the original angle into two congruent angles is the *bisector* of the angle.

DEFINITION OF ANGLE BISECTOR
$\overrightarrow{BM}$ is the *bisector* of $\angle ABC$ if M lies in the interior of $\angle ABC$ and $\angle ABM \cong \angle CBM$.
Thus, an angle bisector divides an angle into two congruent angles.

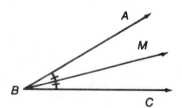

The measure of each of the angles formed by the bisector is one-half the measure of the original angles:
$$m\angle ABM = m\angle CBM \qquad m\angle ABM = \frac{1}{2}m\angle ABC \qquad m\angle CBM = \frac{1}{2}m\angle ABC$$

Diagrams and Drawing Conclusions

Which line segment is longer, $\overline{AB}$ or $\overline{CD}$?

Actually both have the same length, although $\overline{AB}$ may give the illusion of being longer than $\overline{CD}$. When given a geometric diagram, we must exercise extreme caution in drawing conclusions based on the diagram—pictures can be deceiving! In general,

we may assume only collinearity and betweenness of points. We may not make any assumptions regarding the measures of segments or angles unless they are given to us. In Figure 2.11*a*, although segments $\overline{AD}$ and $\overline{DC}$ appear to be equal in length, we may *not* conclude that $AD = DC$. The only assumption that we are entitled to make is that point *D* lies *between* points *A* and *C*. If, however, we are told that $AD = DC$, then we write this *given* information next to the figure. (See Figure 2.11*b*.)

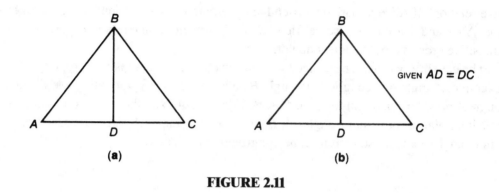

FIGURE 2.11

We may then indicate the equal segments on the diagram by drawing a single vertical bar through each segment. (See Figure 2.12.) An angle may be assumed from the diagram to be a right angle only if the angle contains the "corner" marking (⌐).

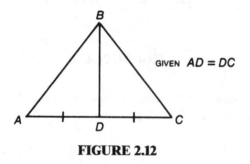

FIGURE 2.12

Properties of Equality and Congruence

John is taller than Kevin and Kevin is taller than Louis. How do the heights of John and Louis compare? We can analyze the situation with the aid of a simple diagram.
(See Figure 2.13.) This leads us to conclude that John must be taller than Louis.

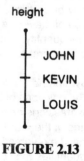

FIGURE 2.13

Using the mathematical symbol for "greater than," >, we can represent the height relationships by the following series of inequality statements:

$$\text{if } J > K$$
$$\text{and } K > L$$
$$\text{then } J > L$$

Without directly comparing John with Louis, we have used a *transitive property* to conclude that John's height is greater than Louis's height. The "greater than" relation is an example of a relation that possesses the transitive property. Is friendship a transitive relation? If Alice is Barbara's friend and Barbara is Carol's friend, does that mean that Alice and Carol must also be friends? Obviously, no. Some relations possess the transitive property, while others do not.

The equality (=) and congruence (≅) relations possess the transitive property. For example, if angle A is congruent to angle B and angle B is congruent to angle C, then angle A must be congruent to angle C. (See Figure 2.14.) Another way of looking at this interrelationship among angles A, B, and C is that angles A and C are each congruent to angle B and must, therefore, be congruent to each other.

FIGURE 2.14

The equality and congruence relations also enjoy some additional properties. These are displayed and illustrated in Table 2.1.

TABLE 2.1

Property	Equality Example	Congruence Example
Reflexive The identical expression may be written on either side of the = or ≅ symbol. Any quantity is equal (congruent) to itself.	1. $9 = 9$. 2. $AB = AB$.	$\angle ABC \cong \angle ABC$.
Symmetric The positions of the expressions on either side of the = or ≅ symbol may be reversed. Quantities may be "flip-flopped" on either side of an = or ≅ sign.	1. If $4 = x$, then $x = 4$. 2. If $AB = CD$, then $CD = AB$	If $\angle ABC \cong \angle XYZ$, then $\angle XYZ \cong \angle ABC$.

(Continued)

TABLE 2.1 *Continued*

Property	Equality Example	Congruence Example
Transitive If two quantities are equal (congruent) to the same quantity, then they are equal (congruent) to each other.	If $AB = CD$ and $CD = PQ$, then $AB = PQ$.	If $\angle X \cong \angle Y$ and $\angle Y \cong \angle Z$, then $\angle X \cong \angle Z$.

Another useful property of the equality relation is the *substitution* property. If $AB = 2 + 3$, then an equivalent number may be substituted in place of the numerical expression on the right side of the equation. We may substitute 5 for $2 + 3$, and write $AB = 5$. Here is a geometric illustration of this often-used property:

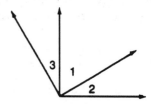

GIVEN: $m\angle 1 + m\angle 2 = 90$,

 $m\angle 2 = m\angle 3$.

CONCLUSION: $m\angle 1 + m\angle 3 = 90$.

REASON: Substitution property. The $m\angle 3$ replaces its equal ($m\angle 2$) in the first equation stated in the Given.

In each of the following examples, the purpose is to justify the conclusion drawn by identifying the property used to draw the conclusion as either the transitive or substitution property.

EXAMPLE 2.9

GIVEN: $\angle 1 \cong \angle 2$,

 $\angle 2 \cong \angle 3$.

CONCLUSION: $\angle 1 \cong \angle 3$.

REASON: ?

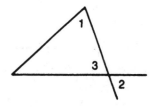

SOLUTION

Since both angles 1 and 3 are congruent to the same angle, angle 2, they must be congruent to each other. This is the *transitive property of congruence*. Since we may only substitute equals in equations, we do *not* have a substitution property of *congruence*.

EXAMPLE 2.10

GIVEN: $m\angle 1 = m\angle 4$,

 $m\angle 3 = m\angle 5$,

 $m\angle 4 + m\angle 2 + m\angle 5 = 180$.

CONCLUSION: $m\angle 1 + m\angle 2 + m\angle 3 = 180$.

REASON: ?

SOLUTION

This is the *substitution property*. In the last equation stated in the Given, the measures of angles 4 and 5 are replaced by their equals, the measures of angles 1 and 3, respectively.

GIVEN: $RS = SM.$ (1)

 $TW = SM.$ (2)

CONCLUSION: $RS = TW.$

REASON: ?

SOLUTION

Since RS and TW are both equal to the same quantity, SM, they must be equal to each other. This is the *transitive property*.

<div align="center">or</div>

In Equation (1), SM may be replaced by its equal, TW. We are using the information in Equation (2) to make a substitution in Equation (1). Hence, the conclusion can be justified also by using the *substitution property*.

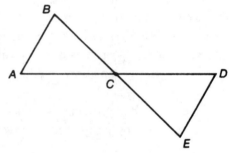

GIVEN: C is the midpoint of $\overline{AD}$,

 $AC = CE.$

CONCLUSION: $CD = CE.$

REASON: ?

SOLUTION

$AC = CD$, since point C is the midpoint of $\overline{AD}$.
We now have the set of relationships:

$$AC = CD \qquad (1)$$
$$AC = CE \qquad (2)$$

Since CD and CE are both equal to the same quantity (AC) they must be equal to each other. Hence, $CD = CE$ by the *transitive property of equality*.

<div align="center">or</div>

We may replace AC by CE in Equation (1), also reaching the desired conclusion.

Examples 2.11 and 2.12 illustrate that the transitive and substitution properties of equality, in certain situations, may be used interchangeably. In each of these examples, *two equations* state that two quantities are each equal to the same quantity, thus leading to either the substitution or transitive property of *equality*.

Additional Properties of Equality

Several properties of equality encountered in elementary algebra prove useful when working with measures of segments and angles. Table 2.2 reviews these properties in their algebraic context.

TABLE 2.2

Property	Algebraic Example	Formal Statement
Addition (+) The same (or =) quantities may be added to both sides of an equation.	Solve for *x:* $x - 3 = 12$ $x - 3 + 3 = 12 + 3$ same $x = 15$	If equals are added to equals, their sums are equal. or If $a = b$, then $a + c = b + c$.
Subtraction (−) The same (or =) quantities may be subtracted from both sides of an equation.	Solve for *n:* $n + 5 = 11$ $n + 5 - 5 = 11 - 5$ same $n = 6$	If equals are subtracted from equals, their differences are equal. or If $a = b$, then $a - c = b - c$.
Multiplication (×) The same quantity may be used to multiply both sides of an equation.	Solve for *y:* $\dfrac{y}{3} = 7$ same $3\left(\dfrac{y}{3}\right) = 3(7)$ $y = 21$	If equals are multiplied by equals, their products are equal. or If $a = b$, then $ac = bc$.

These equality properties may be summarized as follows: "Whatever you do to one side of an *equation*, do the same thing to the other side." The addition, subtraction, and multiplication properties may be applied also to geometric situations. The following examples illustrate how these properties of equality can be used to draw conclusions about the measures of segments and angles. In each of the first four examples, the numerical value in the conclusion is based on the measurements given in the accompanying diagram.

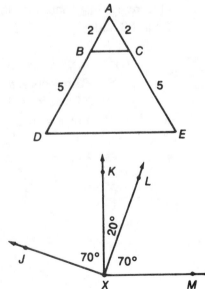

- USING THE ADDITION PROPERTY

 a. GIVEN:

$$AB = AC$$
$$+\ BD = +\ CE$$
$$\overline{AB + BD = AC + CE}$$

 CONCLUSION:

$$AD = AE$$
$$(7 = 7)$$

 b. GIVEN:

$$\text{m}\angle JXK = \text{m}\angle MXL$$
$$+\ \text{m}\angle KXL = +\ \text{m}\angle KXL$$
$$\overline{\text{m}\angle JXL = \text{m}\angle KXM}$$

 CONCLUSION:
$$(90 = 90)$$

- ## USING THE SUBTRACTION PROPERTY

 a. GIVEN: $m\angle BAD = m\angle DCB$

 $\underline{- m\angle PAD = - m\angle BCQ}$

 CONCLUSION: $m\angle BAP = m\angle DCQ$

 $(40 = 40)$

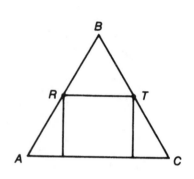

 b. GIVEN: $VI = NE$

 $\underline{\quad - EI = - EI}$

 CONCLUSION: $\underbrace{VI - EI} = \underbrace{NE - EI}$

 $VE = NI$

 $(3 = 3)$

- ## USING THE MULTIPLICATION PROPERTY

 GIVEN: $AB = CB,$

 $AR = \frac{1}{2}AB,$

 $CT = \frac{1}{2}CB.$

 CONCLUSION: $AR = CT.$ Why?

 REASONING: Since we are multiplying equals
 $(AB = CB)$ by the same number
 $(\frac{1}{2})$, their products must be equal:

 $\frac{1}{2}AB = \frac{1}{2}CB$

 By substitution,

 $AR = CT$

This chain of reasoning, in which the multiplying factor is $\frac{1}{2}$, is used so often that we give it a special name, *"halves of equals are equal."*

To help organize and fill in all the necessary logical steps in moving from the Given to an appropriate conclusion, a two-column format, illustrated in Example 2.13, is used. The first column of this proof lists the steps, while the second column justifies each step. Only the Given—definitions, postulates, properties of equality and congruence, and theorems—may appear in the second column.

Table 2.3 outlines the sequence of steps and summarizes the types of justification.

EXAMPLE 2.13

 GIVEN: $m\angle RST = m\angle WTS,$

 $\overline{PS}$ bisects $\angle RST,$

 $\overline{PT}$ bisects $\angle WTS.$

 CONCLUSION: $m\angle 1 = m\angle 2.$

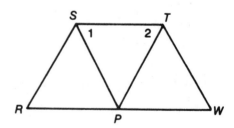

SOLUTION

PROOF:

Statements	Reasons
1. $m \angle RST = m \angle WTS$.	1. Given.
2. $\overline{PS}$ bisects $\angle RST$, $\overline{PT}$ bisects $\angle WTS$.	2. Given.
3. $m \angle 1 = \frac{1}{2} m \angle RST$. $m \angle 2 = \frac{1}{2} m \angle WTS$.	3. Definition of angle bisector.
4. $m \angle 1 = m \angle 2$.	4. Halves of equals are equal.

TABLE 2.3

Pattern of . . .	*Types of . . .*
Statements	Reasons
• Start with the given information supplied in the original problem statement *or* with a conclusion that can be made based on the diagram (for example, betweenness of points). Label this statement number 1, and continue to number statements sequentially. The corresponding reason for each statement is assigned the same number.	• Given (information is included in the statement of the problem.)
	• Definitions
	• Postulates
• Develop a chain of reasoning, writing each logical step as a separate statement and writing its justification across the page in the adjacent "Reasons" column.	• Previously proved theorems
	• Algebraic properties
• Continue until you are able to write a statement that corresponds to the expression or equation in the conclusion (or Prove) of the original problem statement.	

REVIEW EXERCISES FOR CHAPTER 2

1. In the accompanying diagram, classify each of the following angles as acute, right, obtuse, or straight:

 (a) $\angle TOM$
 (b) $\angle LOM$
 (c) $\angle SOM$
 (d) $\angle LOR$
 (e) $\angle ROT$
 (f) $\angle LOT$
 (g) $\angle ROS$
 (h) $\angle MOR$

2. Point P is between points H and G. If $HP = 3$ and $PG = 5$, find HG.

3. If points M, I, and Z are collinear and $IZ = 8$. $MI = 11$, and $MZ = 3$, which point is between the other two?

4. $\overrightarrow{PL}$ lies in the interior of angle RPH; $m \angle RPL = x - 5$ and $m \angle LPH = 2x + 18$. If $m \angle HPR = 58$, find the measure of the smallest angle formed.

5. $\overline{XY}$ bisects $\overline{RS}$ at point M. If $RM = 6$, find the length of $\overline{RS}$.

6. $\overrightarrow{PQ}$ bisects $\angle HPJ$. If $m \angle HPJ = 84$, find $m \angle QPJ$.

7. $\overrightarrow{BP}$ bisects $\angle ABC$. If $m \angle ABP = 4x + 5$ and $m \angle CBP = 3x + 15$, classify angle ABC as acute, right, or obtuse.

8. If R is the midpoint of $\overline{XY}$, $XR = 3a + 1$, and $YR = 16 - 2a$, find the length of $\overline{XY}$.

9. In the accompanying diagram, pairs of angles and segments are indicated as congruent. Use the letters in the diagram to write the appropriate congruence relation.

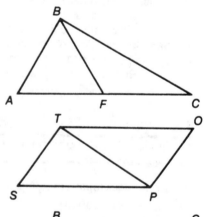

In Exercises 10 to 12, use the diagram and any information given to mark the diagram with the Given and to draw the appropriate conclusion.

10. GIVEN: $\overline{BF}$ bisects $\overline{AC}$.
 CONCLUSION: ?

11. GIVEN: $\overline{PT}$ bisects $\angle STO$.
 CONCLUSION: ?

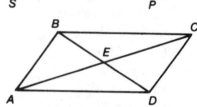

12. GIVEN: $\overline{AC}$ bisects $\overline{BD}$,
 $\overline{BD}$ bisects $\angle ADC$.
 CONCLUSION: ?

In Exercises 13 to 16, justify the conclusion drawn by identifying the property used to draw the conclusion as reflexive, transitive, symmetric, or substitution.

13. GIVEN: $\overline{LM} \cong \overline{GH}$,
 $\overline{GH} \cong \overline{FV}$.
 CONCLUSION: $\overline{LM} \cong \overline{FV}$.

14. GIVEN: Quadrilateral *ABCD*.
 CONCLUSION: $\overline{AC} \cong \overline{AC}$.

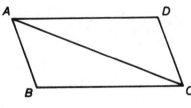

15. GIVEN: $\overline{TW}$ bisects $\angle STV$,
 $\angle 1 \cong \angle 3$.
 CONCLUSION: $\angle 2 \cong \angle 3$.

16. GIVEN: $m\angle 1 + m\angle 2 = 90$,
 $m\angle 1 = m\angle 3$.
 CONCLUSION: $m\angle 3 + m\angle 2 = 90$.

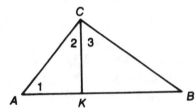

In Exercises 17 to 24, indicate on the diagrams the corresponding pairs of equal or congruent parts. Then state the property of equality that can be used to justify each conclusion.

17. GIVEN: $AC = BT$.
 CONCLUSION: $AT = BC$.
 PROPERTY: ?

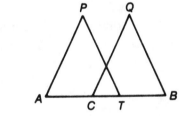

18. GIVEN: $m\angle KPN = m\angle LPM$.
 CONCLUSION: $m\angle KPM = m\angle LPN$.
 PROPERTY: ?

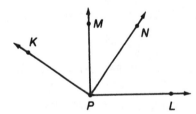

19. GIVEN: $AE = BE$,
 $CE = DE$.
 CONCLUSION: $AC = BD$.
 PROPERTY: ?

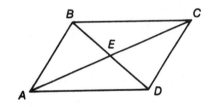

20. GIVEN: m∡1 = m∡3,
 m∡2 = m∡4.
 CONCLUSION: m∡STA = m∡ARS.
 PROPERTY: ?

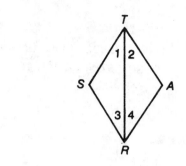

21. GIVEN: m∡WXY = m∡ZYX,
 $\overline{HX}$ bisects ∡WXY,
 $\overline{HY}$ bisects ∡ZYX.
 CONCLUSION: m∡1 = m∡2.
 PROPERTY: ?

22. GIVEN: JQ = LP.
 CONCLUSION: JP = LQ.
 PROPERTY: ?

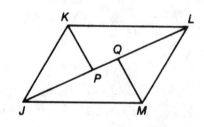

23. GIVEN: m∡SBO = m∡TBH.
 CONCLUSION: m∡SBH = m∡TBO.
 PROPERTY: ?

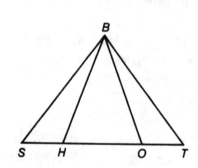

24. GIVEN: AX = YS,
 XB = RY.
 CONCLUSION: AB = RS.
 PROPERTY: ?

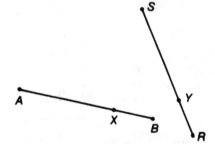

In Exercises 25 to 28, supply the missing reasons.

25. GIVEN: $\overline{PJ} \cong \overline{LR}$.
 CONCLUSION: $\overline{PR} \cong \overline{LJ}$.

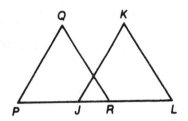

PROOF: Statements	Reasons
1. $\overline{PJ} \cong \overline{LR}$.	1. Given.
2. $PJ = LR$.	2. If two segments are congruent, they are equal in length.
3. $JR = JR$.	3. ?
4. $PJ + JR = LR + JR$.	4. ?
5. $PR = LJ$.	5. Substitution property of equality.
6. $\overline{PR} \cong \overline{LJ}$.	6. If two segments are equal in measure, they are congruent.

26. GIVEN: $\angle RLM \cong \angle ALM$,
 $\angle 1 \cong \angle 2$.
 CONCLUSION: $\angle 3 \cong \angle 4$.

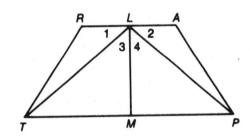

PROOF: Statements	Reasons
1. $\angle RLM \cong \angle ALM$, $\angle 1 \cong \angle 2$.	1. ? given
2. $m\angle RLM = m\angle ALM$, $m\angle 1 = m\angle 2$.	2. ? Congruent lengths are equal
3. $m\angle RLM - m\angle 1 = m\angle ALM - m\angle 2$.	3. ?
4. $m\angle 3 = m\angle 4$.	4. ?
5. $\angle 3 \cong \angle 4$.	5. ?

27. GIVEN: $m\angle TOB = m\angle WOM$,
 $\overline{TB} \cong \overline{WM}$.
 PROVE: (a) $m\angle TOM = m\angle WOB$.
 (b) $\overline{TM} \cong \overline{WB}$.

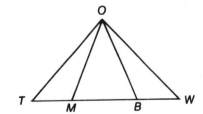

PROOF: Statements Reasons

PROOF OF PART A:

1. $m\angle TOB = m\angle WOM$.	1. ?
2. $m\angle MOB = m\angle MOB$.	2. ?
*3. $m\angle TOB - m\angle MOB =$ $m\angle WOM - m\angle MOB$.	3. ?
*4. $m\angle TOM = m\angle WOB$.	4. Angle addition postulate (and substitution property of equality).

PROOF OF PART B:

5. $\overline{TB} \cong \overline{WM}$	5. ?
6. $TB = WM$.	6. ?
7. $MB = MB$.	7. ?
*8. $TB - MB = WM - MB$.	8. ?
*9. $TM = WB$.	9. Definition of betweenness (and substitution property of equality).
10. $\overline{TM} \cong \overline{WB}$.	10. ?

*In later work, these steps are sometimes consolidated.

28. GIVEN: $\overline{KB}$ bisects $\angle SBF$,
 $\overline{KB}$ bisects $\angle SKF$,
 $\angle SKF \cong \angle SBF$.

 PROVE: $\angle 1 \cong \angle 2$.

PROOF: Statements Reasons

1. $\angle SKF \cong \angle SBF$.	1. Given.
2. $m\angle SKF = m\angle SBF$.	2. ?
3. $\overline{KB}$ bisects $\angle SBF$, $\overline{KB}$ bisects $\angle SKF$.	3. ?
4. $m\angle 1 = \frac{1}{2}m\angle SKF$. $m\angle 2 = \frac{1}{2}m\angle SBF$.	4. ?
5. $m\angle 1 = m\angle 2$.	5. ?
6. $\angle 1 \cong \angle 2$.	6. ?

ANGLE PAIRS AND PERPENDICULAR LINES

WHAT YOU WILL LEARN

This chapter looks at special pairs of angles. In this chapter you will learn:

- definitions and theorems involving complementary angles, supplementary angles, and vertical angles;
- definitions and theorems involving lines that intersect at 90° angles;
- alternative forms of a proof.

SECTIONS IN THIS CHAPTER

Supplementary and Complementary Angle Pairs

Adjacent and Vertical Angle Pairs

Theorems Relating to Supplementary, Complementary, and Vertical Angles

Definitions and Theorems Relating to Right Angles and Perpendiculars

Other Types of Proofs in Mathematics

Supplementary and Complementary Angle Pairs

Pairs of angles may be related in special ways. Supplementary and complementary angle pairs are of particular importance in the study of geometry.

> ### DEFINITIONS OF SUPPLEMENTARY AND COMPLEMENTARY ANGLES
> - Two angles are *supplementary* if the sum of their measures is 180. If angle A is supplementary to angle B, then $m\angle A + m\angle B = 180$, and each angle is called the *supplement* of the other angle.
> - Two angles are *complementary* if the sum of their measures is 90. If angle A is complementary to angle B, then $m\angle A + m\angle B = 90$, and each angle is called the *complement* of the other angle.

GIVEN:

a. Determine the measure of the supplement of angle A.
b. Determine the measure of the complement of angle A.

SOLUTION

a. The supplement of angle A has measure 140.

b. The complement of angle A has measure 50.

In triangle ABC, angle A is complementary to angle B. Find the measures of angles A and B.

SOLUTION

$$2x + 3x = 90$$
$$5x = 90$$
$$x = 18$$
$$m\angle A = 2x = 2(18) = 36$$
$$m\angle B = 3x = 3(18) = 54$$

EXAMPLE 3.3

The measures of an angle and its supplement are in the ratio of 1:8 Find the measure of the angle.

SOLUTION

Method 1: Let x = measure of angle.
Then $180 - x$ = measure of supplement of angle.

$$\frac{x}{180 - x} = \frac{1}{8}$$
$$180 - x = 8x$$
$$9x = 180$$
$$x = 20$$

Method 2: Let x = measure of angle.
Then $8x$ = measure of supplement of angle.

$$x + 8x = 180$$
$$9x = 180$$
$$x = 20$$

Therefore measure of angle = 20.

EXAMPLE 3.4

Determine the measure of an angle if it exceeds twice the measure of its complement by 30.

SOLUTION

Let x = measure of angle.
Then $90 - x$ = measure of complement of angle

$$\begin{aligned} x &= 2(90 - x) + 30 \\ &= 180 - 2x + 30 \\ &= 210 - 2x \\ 3x &= 210 \\ x &= 70 \end{aligned}$$

Therefore, measure of angle = 70.

Adjacent and Vertical Angle Pairs

Adjacent means "next to." But how close do two angles have to be in order to be considered adjacent? Figure 3.1 contrasts four pairs of angles; only the first pair is considered to be adjacent.

(a) Angles are adjacent.

(b) Angles do not share the same vertex.

(c) Angles do not share a common side.

(d) Angles "overlap"—they have interior points in common.

FIGURE 3.1 Adjacent versus nonadjacent angle pairs.

Figure 3.1 suggests that we make the following definition:

DEFINITION OF ADJACENT ANGLE PAIRS

Two angles are an *adjacent pair* if they:
- Have the same vertex.
- Share a common side.
- Have no interior points in common.

If two angles are adjacent, the two sides that are not shared are sometimes referred to as the *exterior* sides of the adjacent angles.

EXAMPLE 3.5

A beginning student of geometry wonders whether the following two assertions are true:

a. If a pair of angles are supplementary, then they must be adjacent.

b. If the exterior sides of a pair of adjacent angles form a straight line, then the angles are supplementary.

Comment on whether you think the statements are true or false. If you suspect that one or both are false, produce a diagram to help support your belief.

SOLUTION

a. A pair of supplementary angles, as the figures below illustrate, may be either adjacent or nonadjacent. The assertion is, therefore, false.

b. The assertion is true since, as illustrated below, a straight line is formed, implying that the sum of the measures of the adjacent angles is 180. Consequently, the angles are supplementary.

The statement presented in Example 3.5*b* may be formally stated as a theorem.

THEOREM 3.1

If the exterior sides of a pair of adjacent angles form a straight line, then the angles are supplementary.

If two lines intersect, four angles are formed, as shown. Angles 1 and 3 are called vertical angles; angles 1 and 4 are *not* vertical angles since they are adjacent. Notice that vertical angles are "opposite" one another. Angles 2 and 4 are also vertical angles.

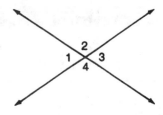

DEFINITION OF VERTICAL ANGLES

Vertical angles are pairs of nonadjacent (opposite) angles formed by two intersecting lines.

 EXAMPLE 3.6 Name all pairs of vertical angles in the accompanying diagram.

SOLUTION
Angle pairs 1 and 4, 2 and 5, 3 and 6.

Theorems Relating to Complementary, Supplementary, and Vertical Angles

In the diagrams below, angles A and B are each complementary to angle C.

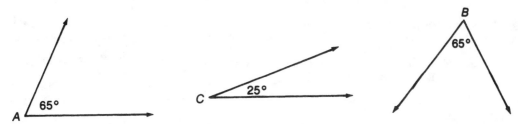

We may conclude that angle A must be equal in measure (or congruent to) angle B. By the same reasoning, if angles A and B were each *supplementary* to angle C, they would necessarily be congruent to each other.

Now, suppose angle A is complementary to angle C and angle B is complementary to angle D; furthermore, angles C and D are congruent:

If m∡C = 20, then m∡A = 70. Since angles C and D are congruent, m∡D = 20 and m∡B = 70. Hence, angles A and B are congruent. If the original relationship specified that the angles were *supplementary* to a pair of congruent angles, the identical conclusion would result.

THEOREM 3.2

If two angles are complementary (or supplementary) to the same angle or to congruent angles, then they are congruent.

 Present a formal two-column proof.

GIVEN: $\overline{LM}$ bisects ∡KMJ,
∡1 is complementary to ∡2,
∡4 is complementary to ∡3.
PROVE: ∡1 ≅ ∡4.

SOLUTION

PROOF:

Statements	Reasons
1. $\overline{LM}$ bisects ∡KMJ.	1. Given.
2. ∡2 ≅ ∡3.	2. A bisector divides an angle into two congruent angles.
3. ∡1 is complementary to ∡2. ∡4 is complementary to ∡3.	3. Given.
4. ∡1 ≅ ∡4.	4. If two angles are complementary to congruent angles, then they are congruent.

 Present a formal two-column proof.

GIVEN: Lines ℓ and m intersect at point P.
PROVE: ∡1 ≅ ∡3.

SOLUTION

PROOF:

Statements	Reasons
1. Lines ℓ and m intersect at point P.	1. Given.
2. ∡1 is supplementary to ∡2, ∡3 is supplementary to ∡2.	2. If the exterior sides of a pair of adjacent angles form a straight line, then the angles are supplementary.
3. ∡1 ≅ ∡3.	3. If two angles are supplementary to the same angle, then they are congruent.

Notice, in the formal proofs presented in Examples 3.7 and 3.8, that factual statements are placed in their logical sequence and numbered in the "Statements" column. The reason used to support each statement receives a corresponding number and is written in the "Reasons" column. In addition to the Given, the "Reasons" column may include only previously stated definitions, postulates, and theorems. Keep in mind that, once a theorem is proved, it may be included in the repertoire of statements that may be used in the "Reasons" column of subsequent proofs.

In approaching a proof, you need to plan thoughtfully and then organize the necessary steps in the "Statements" column. To maintain a logical train of thought, it is sometimes helpful to concentrate first on completing the *entire* "Statement" column; then, to complete the proof, the corresponding reasons may be written.

In the phrasing of theorems written in support of statements in the "Reasons" column, we will sometimes use the expressions "congruent" and "equal in measure" interchangeably. This will simplify our work, avoiding the need to change from one expression to the other and then back again. For example, Theorem 3.2 may, if convenient, be used in the following form:

If two angles are complementary (or supplementary) to the same angle or to congruent angles, then they are *equal in measure*.

Returning to Example 3.8, we observe that angles 1 and 3 are vertical angles and are congruent. Using the same approach, we could establish that vertical angles 2 and 4 are congruent, leading to Theorem 3.3.

THEOREM 3.3

Vertical angles are congruent.

 a. Find the value of x.

b. Find the measures of angles AEC, DEB, DEA, and BEC.

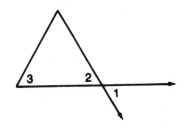

SOLUTION

a. By Theorem 3.3:
$$3x - 18 = 2x + 5$$
$$3x = 2x + 23$$
$$x = 23$$

b. $m\angle AEC = m\angle DEB = 3x - 18$
$$= 3(23) - 18$$
$$= 51$$

Since angles AEC and DEA are supplementary,
$$m\angle DEA = 180 - 51 = 129$$
$$m\angle DEA = m\angle BEC = 129$$

 Present a formal two-column proof.

GIVEN: $\angle 2 \cong \angle 3$.

PROVE: $\angle 1 \cong \angle 3$.

SOLUTION

PROOF: Statements	Reasons
1. $\angle 1 \cong \angle 2$.	1. Vertical angles are congruent.
2. $\angle 2 \cong \angle 3$.	2. Given.
3. $\angle 1 \cong \angle 3$.	3. Transitive property of congruence.

SUMMARY

Two angles are congruent if they are:
- Vertical angles formed by two intersecting lines.
- Complements of the same or of congruent angles.
- Supplements of the same or of congruent angles.

Definitions and Theorems Relating to Right Angles and Perpendiculars

Recall that a *right angle* is an angle of measure 90. The following theorems are useful in proving other theorems:

THEOREM 3.4

All right angles are congruent.

INFORMAL PROOF: Let angles 1 and 2 be right angles. We must show that $\angle 1 \cong \angle 2$. Since angles 1 and 2 are right angles, $m\angle 1 = 90$ and $m\angle 2 = 90$. Therefore, $m\angle 1 = m\angle 2$, so $\angle 1 \cong \angle 2$.

THEOREM 3.5

If two angles are congruent and supplementary, then each is a right angle.

INFORMAL PROOF: Let angles 1 and 2 be congruent and supplementary. We must show that angles 1 and 2 are right angles.

- Since they are supplementary, $m\angle 1 + m\angle 2 = 180$
- By substitution, $\quad m\angle 1 + m\angle 1 = 2\,m\angle 1 = 180$
- Therefore, $\quad\quad\quad\quad m\angle 1 = 90 = m\angle 2$
- Hence, angles 1 and 2 are right angles.

Two lines, two segments, or a line and a segment that intersect to form a right angle are said to be *perpendicular*. If line ℓ is perpendicular to line m, we may write $\ell \perp m$, where the symbol $\perp$ is read as "is perpendicular to." See Figure 3.2.

A *perpendicular bisector* of a line segment is a line or segment that is perpendicular to the given segment at its midpoint. See Figure 3.3.

(a) **(b)** **(c)**

FIGURE 3.2 Perpendiculars.

$\overline{AX} \cong \overline{XB}$ and $\ell \perp \overline{AB}$ at point X.

FIGURE 3.3 A perpendicular bisector.

DEFINITION OF PERPENDICULAR LINES

Perpendicular lines are lines that intersect to form right angles. If a line is perpendicular to a segment and intersects the segment at its midpoint, then the line is called the *perpendicular bisector* of the segment.

EXAMPLE 3.11

GIVEN: $\overline{AC} \perp \overline{BC}$, angles 1 and 2 are adjacent.
PROVE: $m\angle 1 + m\angle 2 = 90$.

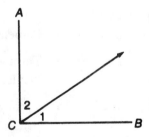

SOLUTION

PROOF:

Statements	Reasons
1. $\overline{AC} \perp \overline{BC}$.	1. Given.
2. $\angle ACB$ is a right angle.	2. Perpendicular segments intersect to form a right angle.
3. $m\angle ACB = 90$.	3. A right angle has measure 90.
4. $m\angle ACB = m\angle 1 + m\angle 2$.	4. Angle Addition Postulate.
5. $m\angle 1 + m\angle 2 = 90$.	5. Substitution.

Since the sum of the measures of angles 1 and 2 is 90, angles 1 and 2 are complementary. Example 3.11 proves the following theorem:

THEOREM 3.6

If the exterior sides of a pair of adjacent angles are perpendicular, then the angles are complementary.

EXISTENCE OF PERPENDICULARS

How many perpendiculars can be drawn to a given line? A line has an infinite number of perpendiculars; however, it can be proven that through a particular point *on* the line, there exists exactly one perpendicular to the given line. At a point *not* on a line, we postulate that there exists exactly one line (or segment) that passes through the point and is perpendicular to the line. These situations are represented in Figure 3.4*a*, *b*, and *c*, respectively. Observe that Figure 3.4*c* also illustrates that when a perpendicular intersects a line, *four* right angles are formed.

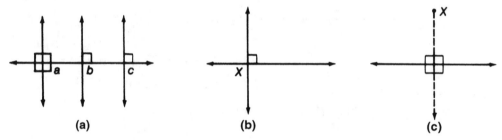

FIGURE 3.4 Existence of perpendiculars.

These results are summarized below.

Facts about Perpendiculars

Theorem 3.7 Perpendicular lines intersect to form four right angles.

Theorem 3.8 Through a given point on a line, there exists exactly one perpendicular to the given line.

Postulate 3.1 Through a given point *not* on a line, there exists exactly one perpendicular to the given line.

DISTANCE

The term *distance* in geometry is always interpreted as the *shortest* path between two points. In Figure 3.5 the distance between points P and Q is the length of segment $\overline{PQ}$. A point that is exactly the same distance from two other points is said to be *equidistant* from the two points. The midpoint of a segment, for example, is equidistant from the end points of the segment. See Figure 3.6.

FIGURE 3.5 The shortest path between two points is represented by the segments joining the points, rather than any zigzag or circular route.

FIGURE 3.6 In each figure, point *M* is equidistant from point *P* and *Q*.

The shortest distance from a point not on a line to the line is measured by the length of the segment drawn from the point perpendicular to the line.

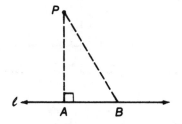

FIGURE 3.7 The distance from point *P* to line ℓ is the perpendicular segment $\overline{PA}$ rather than any other segment such as $\overline{PB}$.

DEFINITIONS OF DISTANCE

- The *distance between two points* is the length of the segment joining the points.
- The *distance between a line and a point not on the line* is the length of the perpendicular segment drawn from the point to the line.

METHODS FOR PROVING LINES ARE PERPENDICULAR

By using the definition of perpendicularity, we may conclude that two lines are perpendicular if they intersect to form a right angle. It will sometimes be convenient in subsequent work with perpendicular lines to use an alternative method. In Figure 3.8 it seems intuitively clear that, if we continue to rotate line ℓ in a clockwise fashion, eventually lines ℓ and *m* will be perpendicular, and this will be true when angles 1 and 2 are congruent.

FIGURE 3.8

This leads to the following theorem:

THEOREM 3.9

If two lines intersect to form congruent adjacent angles, then the lines are perpendicular.

Prove Theorem 3.8.

SOLUTION

In approaching the proof of a theorem in which a diagram is not provided, we must:

1. Identify the Given from the information contained in the "if clause" of the statement of the theorem
2. Identify what is to be proved from the relationship proposed in the "then clause" of the theorem
3. Use the information obtained in Steps 1 and 2 to draw an appropriate diagram
4. Organize our thoughts by planning the sequence of statements that must be followed in logically progressing from the Given to the final conclusion
5. Write the formal proof

GIVEN: Lines ℓ and m intersect at k, $\angle 1 \cong \angle 2$.
PROVE: $\ell \perp m$.

PROOF:

Statements	Reasons
1. Lines ℓ and m intersect at k so that $\angle 1 \cong \angle 2$.	1. Given.
2. $\angle 1$ is supplementary to $\angle 2$.	2. If the exterior sides of a pair of adjacent angles form a straight line, then the angles are supplementary.
3. $\angle 1$ is a right angle.	3. If two angles are congruent and supplementary, then each is a right angle.
4. $\ell \perp m$	4. If two lines intersect to form a right angle, then the lines are perpendicular. (This follows from the reverse of the definition of perpendicularity.)

To Prove That the Two Lines Are Perpendicular:
- Show that the lines intersect to form a right angle.
- Show that the lines intersect to form a pair of congruent adjacent angles (Theorem 3.9).

Other Types of Proofs in Mathematics

Although the formal, two-column deductive proof is the dominant mode of proof found in introductory courses in geometry, it is not the only type of proof used in mathematics. In general, proofs in mathematics may proceed either directly or indirectly. The two-column proof uses a direct approach, progressing from the Given to the Prove in step-by-step fashion. In an indirect method of proof, all possible outcomes are identified and proved false or contradictory *except one*, implying that the remaining possibility must be true by the process of elimination.

The format of the proof may also vary. The two-column deductive proof is usually referred to as a "formal" proof since each statement is listed with a corresponding justification, following a very structured format. An "informal" proof usually takes paragraph form with only the essential ingredients of the proof explained. Informal proofs were given for Theorems 3.4 and 3.5.

 Using the postulate "Exactly one line may be drawn between two given points," use an indirect method of proof to establish the following theorem:

THEOREM 3.10

If two lines intersect, then they intersect in exactly one point.

SOLUTION

An Informal Proof Using the Indirect Method:

Let lines ℓ and m represent the intersecting lines. There are two possibilities: (1) the lines intersect in exactly one point, and (2) they intersect in more than one point. Assume the latter is true and the lines intersect in *two* points, say P and Q. This implies that both lines ℓ and m pass through points P and Q, contradicting Postulate 1.1 which asserts that one and only one line may be drawn between two points. Therefore, the remaining conclusion—that there is exactly one point of intersection—must be true.

Another type of proof is "proof by counterexample." We may be able to prove a statement is false simply by showing that it fails in one specific instance. For example, consider the assertion "All pairs of supplementary angles are right angles." We can prove that this statement is false merely by producing an angle pair that satisfies the hypothesis (the angles are supplementary) but contradicts the conclusion (the angles are right angles). The angle pair 100 and 80 is one such pair. The assertion is therefore false.

REVIEW EXERCISES FOR CHAPTER 3

1. For the accompanying diagram, list all pairs of vertical and adjacent angles.

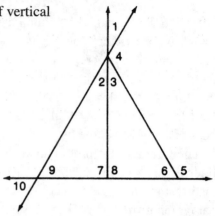

2. The measure of an angle is 5 times as great as the measure of its complement. Find the measure of the angle. $150°$

3. The measure of an angle exceeds 3 times its supplement by 4. Find the measure of the angle.

4. The measure of the supplement of an angle is 3 times as great as the measure of the angle's complement. Find the measure of the angle.

5. The difference between the measures of an angle and its complement is 14. Find the measure of the angle and its complement.

6. Find the measure of an angle if it is 12 less than twice the measure of its complement.

7. Find the value of x:

(a)
$(4x + 5)°$
$(x + 32)°$

$4x+5 = x+32$
$4x = 27$
$3x = 27$ $x=9$

(b)
$(x + 2)°$
$(3x + 18)°$

$(x+2) + (3x+18) = 180$
$4x + 20 = 180$
$4x = 160$
$x = 40$

(c)
$(2x - 5)°$
$(3x + 5)°$

$2x-5 + 3x+5 = 90$
$5x = 90$ $x=18$

8. $\overleftrightarrow{XY}$ and $\overleftrightarrow{AB}$ intersect at point C. If $m\angle XCB = 4x - 9$ and $m\angle ACY = 3x + 29$, find $m\angle XCB$.

$3x+29=4x-9 \quad x=20 \; \boxed{-71}$ A———C———B

In Exercises 9 to 13, state whether each statement is true or false. Prove a statement is false by providing a counterexample.

9. Complements of vertical angles are congruent.

true

10. If two lines intersect, then the bisectors of a pair of adjacent angles are perpendicular to each other.

11. If two lines intersect to form congruent angles, then the lines are perpendicular.

12. If an angle is congruent to its supplement, then it is a right angle.

13. If a point C is equidistant from points A and B, then point C is the midpoint of the segment that joins A and B.

14. GIVEN: $\angle 1 \cong \angle 4$.
 PROVE: $\angle 2 \cong \angle 3$.

15. GIVEN: $\overline{BD}$ bisects $\angle ABC$.
 PROVE: $\angle 1 \cong \angle 2$.

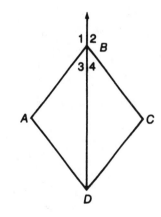

16. GIVEN: $\angle 3$ is complementary to $\angle 1$,
 $\angle 4$ is complementary to $\angle 2$.
 PROVE: $\angle 3 \cong \angle 4$.

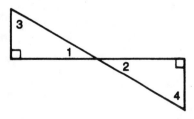

17. GIVEN: $\overline{AB} \perp \overline{BD}$, $\overline{CD} \perp \overline{BD}$,
 $\measuredangle 2 \cong \measuredangle 4$.
 PROVE: $\measuredangle 1 \cong \measuredangle 3$.

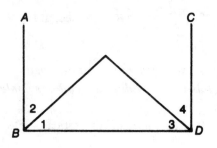

18. GIVEN: $\overline{KL} \perp \overline{JM}$,
 $\overline{KL}$ bisects $\measuredangle PLQ$.
 PROVE: $\measuredangle 1 \cong \measuredangle 4$.

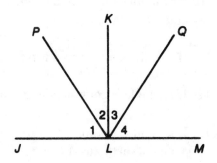

19. GIVEN: $\overline{NW} \perp \overline{WT}$,
 $\overline{WB} \perp \overline{NT}$,
 $\measuredangle 4 \cong \measuredangle 6$.
 PROVE: $\measuredangle 2 \cong \measuredangle 5$.

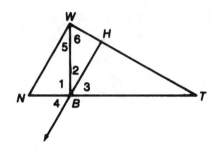

20. GIVEN: $\overline{MT}$ bisects $\measuredangle ETI$,
 $\overline{KI} \perp \overline{TI}$, $\overline{KE} \perp \overline{TE}$,
 $\measuredangle 3 \cong \measuredangle 1$, $\measuredangle 5 \cong \measuredangle 2$.
 PROVE: $\measuredangle 4 \cong \measuredangle 6$.

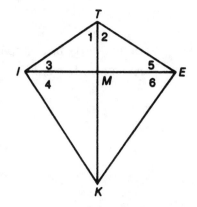

CHAPTER 4

PARALLEL LINES

WHAT YOU WILL LEARN

Lines in a plane either intersect or are parallel. When two parallel lines are cut by a third line, called a *transversal*, eight angles are formed. In this chapter you will learn:

- that special relationships exist between pairs of angles formed when parallel lines are cut by a transversal;
- how you can prove two lines are parallel;
- that the controversial Parallel Postulate guarantees that, through a given point, a line can be drawn parallel to another line.

SECTIONS IN THIS CHAPTER
Planes and Lines
Properties of Parallel Lines
Converses and Methods of Proving Lines Parallel
The Parallel Postulate

Planes and Lines

Any flat surface such as a chalkboard or floor may be thought of as a *plane*. A plane has these properties:

1. It has length and width, but not depth or thickness.
2. Its length and width may be extended indefinitely in each direction.

Figure 4.1 illustrates that two lines may lie in same plane or in different planes. Lines, segments, rays, or points that lie in the same plane are said to be *coplanar*. We assume that lines are coplanar unless otherwise stated.

Coplanar Lines

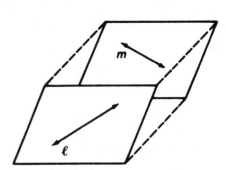

Noncoplanar Lines (Called *Skew* Lines)

FIGURE 4.1

When a pair of lines are drawn, the plane is divided into distinct regions. The region bounded by both lines is referred to as the *interior* region; the remaining outside areas are *exterior* regions. See Figure 4.2.

Exterior

Interior

Exterior

FIGURE 4.2

A line that intersects two or more lines in different points is called a *transversal*. In Figure 4.3 line *t* represents a transversal since it intersects lines ℓ and *m* at two distinct points. Notice that at each point of intersection four angles are formed. The angles that lie in the interior region—angles 1, 2, 5, and 6—are referred to as *interior angles*. Angles 3, 4, 7, and 8 lie in the exterior region and are called *exterior angles*.

Table 4.1 further classifies angle pairs according to their position relative to the transversal.

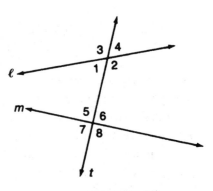

FIGURE 4.3

TABLE 4.1

Typing of Angle Pair	Distinguishing Features	Examples in Figure 4.3
Alternate interior angles	Angles are interior angles. Angles are on opposite sides of the transversal. Angles do not have the same vertex.	Angles 1 and 6; Angles 2 and 5.
Corresponding angles	One angle is an interior angle; the other angle is an exterior angle. Angles are on the same side of the transversal. Angles do not have the same vertex.	Angles 3 and 5; Angles 4 and 6; Angles 1 and 7; Angles 2 and 8.
Alternate exterior angles	Angles are exterior angles. Angles are on opposite sides of the transversal. Angles do not have the same vertex.	Angles 3 and 8; Angles 4 and 7.

Referring to the diagram, name all pairs of:

a. Alternate interior angles
b. Corresponding angles
c. Alternate exterior angles

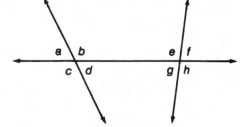

SOLUTION

a. Alternate interior angle
 pairs: *b* and *g; d* and *e.*
b. Corresponding angle
 pairs: *a* and *e; b* and *f; c* and *g; d* and *h.*
c. Alternate exterior angle
 pairs: *a* and *h; c* and *f.*

In analyzing diagrams, alternate interior angle pairs can be recognized by their Z shape; corresponding angles form an F shape. The Z and F shapes, however, may be turned so that the letter may appear reversed or upside down. See Figure 4.4.

Properties of Parallel Lines

Wires strung on telephone poles, bars on a prison cell, and lines drawn on a sheet of note paper are examples of *parallel* lines. We postulate that a pair of coplanar lines, if extended indefinitely, will eventually either intersect or never meet. Lines that never intersect are defined to be parallel.

In Figure 4.5, we may concisely express the fact that line ℓ *is parallel to* line *m* by using the notation $\ell \parallel m$; in Figure 4.6, the notation $\ell \nparallel m$ expresses the fact that line ℓ is *not* parallel to line *m*.

FIGURE 4.4

FIGURE 4.5 **FIGURE 4.6**

DEFINITION OF PARALLEL LINES

Parallel lines are lines in the same plane that never intersect.

NOTATION: $\ell \| m$ is read as "Line ℓ is parallel to line m."

It follows from the preceding definition that:

• Portions (segment or rays) of parallel lines are parallel.

If line $\ell \parallel m$, then $\overline{AB} \parallel \overline{CD}$.

• Extensions of parallel segments or rays are parallel.

If $\overline{AB} \parallel \overline{CD}$, then $\overleftrightarrow{AB} \parallel \overleftrightarrow{CD}$.

Suppose, in Figure 4.7, that $\overline{AD} \parallel \overline{BC}$ and $\overline{AB} \parallel \overline{CD}$. To indicate this information in the diagram, arrowheads that point in the same direction are used. See Figure 4.8.

FIGURE 4.7

FIGURE 4.8

| EXAMPLE 4.2 | Mark the diagram with the information supplied |

GIVEN: $\overline{SR} \parallel \overline{TW}$,
$\overline{RL} \parallel \overline{TM}$,
$\overline{ST} \cong \overline{RW}$.

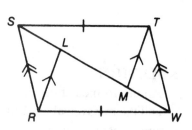

SOLUTION

In Figure 4.9a, lines ℓ and m are not parallel and the alternate interior angles do *not* measure the same. On the other hand, in Figure 4.9b, where line ℓ has been drawn parallel to line m, all pairs of acute angles look congruent and all pairs of obtuse angles appear congruent. This suggests that there may be a relationship between certain pairs of angles formed by parallel lines. Since this observation cannot be proved with our existing knowledge, it is postulated.

(a)

(b)

FIGURE 4.9

POSTULATE 4.1

If two lines are parallel, then their alternate interior angles are congruent.

| EXAMPLE 4.3 | Given that the indicated lines are parallel, determine the value of *x*. |

a.

b.

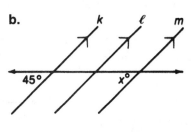

SOLUTION

a. From Postulate 4.1,

$$3x - 40 = 2x - 10$$
$$3x = 2x + 30$$
$$x = 30$$

b. Using vertical angles and Postulate 4.1:

EXAMPLE 4.4 If lines ℓ and m are parallel, and $m\angle a = 60$, find the measure of each of the numbered angles.

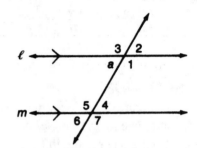

SOLUTION

Angles formed at the intersection of line ℓ and the transversal are either vertical or supplementary angle pairs.

By Postulate 4.1, alternate interior angles are equal in measure:

$$m\angle a = m\angle 4 = 60$$
$$m\angle 1 = m\angle 5 = 120$$

And since vertical angles are equal in measure,

$$m\angle 6 = m\angle 4 = 60$$
$$m\angle 7 = m\angle 5 = 120$$

The completed diagram is as follows:

Look closely at the relationships between the special angle pairs derived in Example 4.4. The following theorems are suggested:

THEOREM 4.1

If two lines are cut by a transversal, then any pair of angles are either congruent or supplementary.

THEOREM 4.2

If two lines are parallel, then their corresponding angles are congruent.

For example, if $\ell \parallel m$, then $\angle a \cong \angle b$.

THEOREM 4.3

If two lines are parallel, then interior angles on the same side of the transversal are supplementary.

For example, if $\ell \parallel m$, then $\angle e$ and $\angle f$ are supplementary.

 EXAMPLE 4.5 In the accompanying diagram, parallel lines HE and AD are cut by transversal, $\overleftrightarrow{BF}$ at points G and C, respectively. If m$\angle HGF = 5n$ and m$\angle BCD = 2n + 66$, find the value of n.

SOLUTION

Since vertical angles are equal in measure,

$$m\angle EGC = m\angle FGH = 5n.$$

Angles *EGC* and *BCD* are corresponding angles and are equal in measure since $\overleftrightarrow{HE}$ and $\overleftrightarrow{AD}$ are parallel. Therefore:

$$
\begin{aligned}
m\angle EGC &= m\angle BCD \\
5n &= 2n + 66 \\
5n - 2n &= 66 \\
\frac{3n}{3} &= \frac{66}{3} \\
n &= 22
\end{aligned}
$$

Prove Theorem 4.2.

GIVEN: $\ell \parallel m$.
PROVE: $\angle a \cong \angle b$.

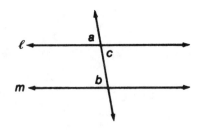

SOLUTION

PLAN: Introduce $\angle c$ in the diagram so that angles *b* and *c* are alternate interior angles. Then use Postulate 4.1 and vertical angles to establish the conclusion.

PROOF:

Statements	Reasons
1. $\ell \parallel m$.	1. Given.
2. $\angle a \cong \angle c$.	2. Vertical angles are congruent.
3. $\angle c \cong \angle b$.	3. If two lines are parallel, then their alternate interior angles are congruent.
4. $\angle a \cong \angle b$.	4. Transitive property.

Note that we have introduced the notion of including a formal statement of a plan of attack. An indication of a "plan" and the corresponding annotation of the diagram is suggested before you begin to write the statements and reasons in the two-column proof.

GIVEN: $\overline{RS} \parallel \overline{TW}, \overline{TS} \parallel \overline{WX}$.
PROVE: $\angle S \cong \angle W$.

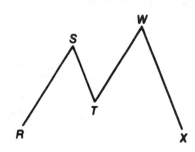

SOLUTION

PLAN: By alternate interior angles,
$\angle S \cong \angle T$ and $\angle T \cong \angle W$.
Therefore, $\angle S \cong \angle W$.

PROOF:

Statements	Reasons
1. $\overline{RS} \parallel \overline{TW}$.	1. Given.
2. $\angle S \cong \angle T$.	2. If two lines are parallel, then their alternate interior angles are congruent.
3. $\overline{TS} \parallel \overline{WX}$.	3. Given.
4. $\angle T \cong \angle W$.	4. Same as reason 2.
5. $\angle S \cong \angle W$.	5. Transitive property of congruence.

A few words of warning: Alternate interior and corresponding exterior angle pairs are formed whenever two or more lines are intersected by a transversal, regardless of whether the lines in the original set are parallel. These angle pairs, therefore, may or may *not* be congruent. We are entitled to conclude that alternate interior, corresponding and alternate exterior angle pairs are congruent only if it is first determined that the lines intersected by the transversal are parallel.

SUMMARY

If two lines are parallel, then any pair of angles formed are either congruent or supplementary:
- All acute angles are congruent, and all obtuse angles are congruent.
- Any pair of acute and obtuse angles are supplementary.

In particular:
- Alternate interior angles are congruent.

- Corresponding angles are congruent.

- Alternate exterior angles are congruent.

- Interior angles on the same side of the transversal are supplementary.

$a + b = 180°$

Converses and Methods of Proving Lines Parallel

Geometry depends on preciseness of language as well as on logical reasoning. Often the meaning of a sentence can be changed dramatically simply by switching the positions of words or phrases in the sentence. As an example, consider the statement:

If I am 12 years old, then I am not eligible to vote.

A new statement can be formed by interchanging the single-underlined words in the "if clause" with the double-underlined words in the "then clause:"

If I am not eligible to vote, then I am 12 years old.

The new statement formed is called the *converse* of the original statement. The relationship between the original statement and its converse is illustrated below.

Original Statement: If condition, then conclusion

Converse: If _____ , then _____

Our example illustrates that the converse of a true statement may not be assumed to be true. If I am not eligible to vote, I can be 13 years old, 14 years old, or perhaps not a citizen.

As another example, form the converse of the statement "If yesterday was Saturday, then today is Sunday." The converse is "If today is Sunday, then yesterday was Saturday." In this instance the original statement and its converse are true. The only generalization that can be made regarding the truth or falsity of the converse is that it may be either true or false.

Why then do we mention converses? Converses are important in mathematics since they provide *clues*—they suggest avenues to explore that may lead to the discovery of true relationships. In the discussion on properties of parallel lines we studied how to prove that certain types of angle relationships are true, *given* that lines are parallel. In this section we examine the converse situation. How can we prove that lines are parallel, *given* that certain types of angle relationships are true?

A logical starting point is to investigate the validity of the statement obtained by taking the converse of Postulate 4.1:

POSTULATE 4.1	CONVERSE
If two lines are parallel, then their alternate interior angles are congruent.	If a pair of alternate interior angles are congruent, then the lines are parallel.

In forming the converse, the phrase "a pair of" was inserted; if one pair of alternate interior angles are congruent, so must the other pair be congruent.

The truth of the converse is consistent with experience—if it is known that a pair of alternate interior angles are congruent, then the lines must be parallel. See Figure 4.10.

FIGURE 4.10

On the basis of our current knowledge, the converse cannot be proved. We therefore postulate it.

POSTULATE 4.2 CONGRUENT ALTERNATE INTERIOR ANGLES IMPLY PARALLEL LINES POSTULATE

If a pair of alternate interior angles are congruent, then the lines are parallel.

GIVEN: $\overline{BC} \parallel \overline{AD}$ ∠2 ≅ ∠3.

PROVE: $\overline{AB} \parallel \overline{CD}$.

SOLUTION

PLAN: To show $\overline{AB} \parallel \overline{CD}$, first prove ∠1 ≅ ∠2 and then use the Given to conclude that ∠1 ≅ ∠3. By Postulate 4.2 the line segments must be parallel.

PROOF:

Statements	Reasons
1. $\overline{BC} \parallel \overline{AD}$.	1. Given.
2. ∠1 ≅ ∠2.	2. If two lines are parallel, then their corresponding angles are congruent.
3. ∠2 ≅ ∠3.	3. Given.
4. ∠1 ≅ ∠3.	4. Transitive property of congruence.
5. $\overline{AB} \parallel \overline{CD}$.	5. If a pair of alternate interior angles are congruent, then the lines are parallel (Postulate 4.2).

Additional methods of proving lines parallel may be derived from the converses of Theorems 4.2 and 4.3. The following theorems can be proved using Postulate 4.2:

THEOREM 4.4

If a pair of corresponding angles are congruent, then the lines are parallel.

THEOREM 4.5

If a pair of alternate exterior angles are congruent, then the lines are parallel.

THEOREM 4.6

If a pair of interior angles on the same side of the transversal are supplementary, then the lines are parallel.

GIVEN: $\overleftrightarrow{AB} \perp t$, $\overleftrightarrow{CD} \perp t$.

PROVE: $\ell \parallel m$.

SOLUTION

PLAN: Perpendicular lines intersect to form right angles; all right angles are congruent. Therefore, the lines are parallel by either Theorem 4.4 or Postulate 4.2. (Theorem 4.6 may also be used.)

PROOF:

Statements	Reasons
1. $\overleftrightarrow{AB} \perp t$ and $\overleftrightarrow{CD} \perp t$.	1. Given.
2. Angles 1 and 2 are right angles.	2. Perpendicular lines intersect to form right angles.
3. $\angle 1 \cong \angle 2$.	3. All right angles are congruent.

The Parallel Postulate

One of the most controversial postulates is concerned with how many parallel lines can be drawn to a line through a point not on the line. See Figure 4.11. Again we must appeal to intuition and experience. We cannot prove that there is exactly one such line, so we postulate that this is the case. Since this postulate affirms the existence of parallel lines, it is commonly referred to as the Parallel Postulate.

FIGURE 4.11

POSTULATE 4.3 THE PARALLEL POSTULATE

Through a given point not on a line, exactly one line may be drawn parallel to the line.

The Parallel Postulate is of great historic interest. Euclid, a Greek mathematician who lived in approximately 300 B.C., is credited with organizing the various aspects of geometry into a coherent discipline and then writing the first major treatise on the sub-

ject, called the *Elements*. In the *Elements* Euclid presents 10 postulates on which the development of geometry is based. The Parallel Postulate corresponds to the fifth postulate listed in the *Elements*; as a result it is also referred to as Euclid's Fifth Postulate. Actually, the statement of Postulate 4.3 most closely resembles the version used by the mathematician John Playfair (1748–1819).

The Parallel Postulate implies the existence of parallel lines, and from the time of Euclid mathematicians have been somewhat skeptical that this assumption represents a self-evident truth. For 2000 years mathematicians have attempted to provide a satisfactory proof of Euclid's Fifth Postulate; all efforts have proved unsuccessful.

In the 1800s two mathematicians, Bolyai and Lobachevsky, used a slightly different approach. They independently investigated how Euclidean geometry would change if the Parallel Postulate was relaxed and it was assumed that more than one line can be drawn through a point not on a line and parallel to the line. In the mid 1800s, Riemann took yet another approach. He postulated that, through a given point not on a line, *no* line could be drawn parallel to the line.

Geometries that tamper with Euclid's Fifth Postulate are collectively referred to as non-Euclidean geometries. Although interesting, their investigation goes well beyond the scope of this book.

EXAMPLE 4.10

GIVEN: $\ell \parallel m$.

FIND: $m \angle ABC$.

SOLUTION

Using the Parallel Postulate, through point *B* draw a line parallel to line ℓ (and, therefore, line *m*).

$\therefore m \angle ABC = 100$

REVIEW EXERCISES FOR CHAPTER 4

In Exercises 1 to 5 classify each pair of angles labeled a *and* b *as alternate interior, corresponding, or alternate exterior. (Suggestion: First identify the transversal and then look for a Z or F shape; remember that the angle pair may be neither alternate interior nor corresponding.)*

1. alternate interior

2. corresponding

3. corresponding

4. Find the values of x and y.

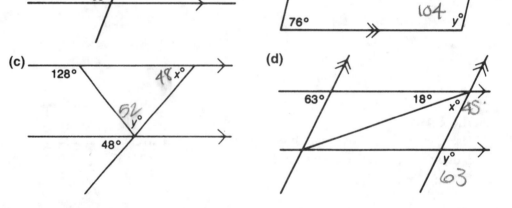

(a) 112° $x°$ $y°$ 68°

68°

(b) $x°$ 76°

104

76° 104 $y°$

(c) 128° 48 $x°$

52 $y°$

48°

(d) 63° 18° $x°$ 45

$y°$ 63

5. Given $\ell \parallel m$, find the value of x.

(a) ℓ $(4x - 30)°$

$x = 8$

m $(2x + 14)°$

$2X = 16$ 4 X-30=2x+14
$2x = -16$
-8

(b) ℓ

$(3x - 40)°$

$x = 55$

$x°$

m

(c) ℓ $x°$

40° 100° m

4X-40 60°

6. Two parallel lines are cut by a transversal. Find the measures of the angles if a pair of interior angles on the same side of the transversal:
 (a) Are represented by $(5x - 32)°$ and $(x + 8)°$.
 (b) Have measures such that one angle is 4 times the measure of the other.

7. Find the value of x.
 (a)

$35°$

55

$x°$

$\ell \parallel m$

 (b)

$(3x + 15)°$

$(5x - 5)°$

$X = 10$

$3x + 15 = 5x - 5$

$3x + 20 = 5x$

$\overleftrightarrow{AB} \parallel \overleftrightarrow{CD}$

8. In the accompanying diagram, $\overleftrightarrow{AB} \parallel \overleftrightarrow{CD}$ and $\overrightarrow{FG}$ bisects $\angle EFD$. If $m\angle EFG = x$ and $m\angle FEG = 4x$, find x.

$x = 45°$

41

9. In the accompanying diagram, $\overleftrightarrow{AB} \parallel \overleftrightarrow{CD}$ and $\overline{EF}$ bisects $\angle AFG$.
 (a) If $m\angle 1 = 100$, find the measure of each of the numbered angles.
 (b) If $m\angle 3 = 4x - 9$ and $m\angle 5 = x + 19$, find the measure of each of the numbered angles.

$3 = 127$

$5 = 53$

50 130 100 80

13 12 1 2

130 10 115 80 4 3 100

50 6

50 7 5 80

80 9 8 100

$5x + 10 = 180$

10. Form the converse of each of the following statements. State whether the converse is true or false.
 (a) If I live in New York, then I live in the United States.
 (b) If two angles are congruent, then the angles are equal in measure.
 (c) If two angles are vertical angles, then they are congruent.
 (d) If two angles are complementary, then the sum of their measures is 90.
 (e) If two angles are adjacent, then they have the same vertex.
 (f) If two lines are perpendicular to the same line, then they are parallel.

11. GIVEN: $\overline{BA} \parallel \overline{CF}, \overline{BC} \parallel \overline{ED}$.
 PROVE: $\angle 1 \cong \angle 2$.

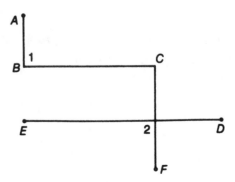

12. GIVEN: $\overline{LJ} \parallel \overline{WK} \parallel \overline{AP}$, $\overline{PL} \parallel \overline{AG}$.
 PROVE: $\angle 1 \cong \angle 2$.

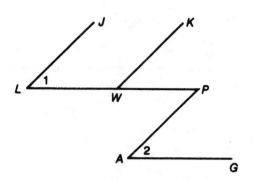

For Exercise 13, choose the correct numbered response to complete the sentence.

13. Two parallel lines are cut by a transversal. The bisectors of a pair of interior angles on the same side of the transversal intersect to form an angle that is
 (1) always acute
 (2) always right
 (3) always obtuse
 (4) either acute or obtuse, but never right

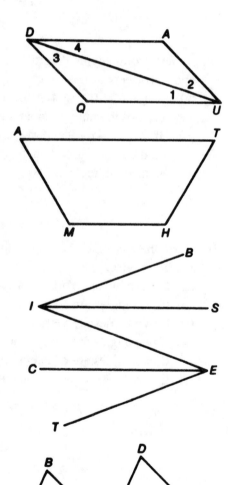

14. GIVEN: $\overline{QD} \parallel \overline{UA}$,
 $\overline{QU} \parallel \overline{DA}$.
 PROVE: $\angle QUA \cong \angle ADQ$.

15. GIVEN: $\overline{AT} \parallel \overline{MH}$,
 $\angle M \cong \angle H$.
 PROVE: $\angle A \cong \angle T$.

16. GIVEN: $\overline{IB} \parallel \overline{ET}$,
 $\overline{IS}$ bisects $\angle EIB$,
 $\overline{EC}$ bisects $\angle TEI$.
 PROVE: $\angle BIS \cong \angle TEC$.

17. GIVEN: $\angle B \cong \angle D$, $\overline{BA} \parallel \overline{DC}$.
 PROVE: $\overline{BC} \parallel \overline{DE}$.

18. GIVEN: $k \parallel \ell$, $\angle 5 \cong \angle 8$.
 PROVE: $j \parallel \ell$.

19. GIVEN: $\angle K \cong \angle P$,
 $\angle J + m\angle P = 180$.
 PROVE: $\overline{KL} \parallel \overline{JP}$.

20. GIVEN: $\overline{AB} \perp \overline{BC}$,
 $\angle ACB$ is complementary
 to $\angle ABE$.
 PROVE: $\overleftrightarrow{AC} \parallel \overleftrightarrow{EBD}$.

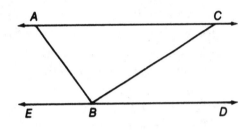

21. GIVEN: $\overline{AG} \parallel \overline{BC}$,
 $KH \parallel BC$,
 $\angle 1 \cong \angle 2$.
 PROVE: $\overline{HK} \perp \overline{AB}$.

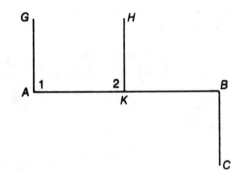

22. Prove that, if two lines are each parallel to a third line, then the lines are parallel to each other.

23. Prove that, if the rays that bisect a pair of alternate interior angles are parallel, then the lines are parallel.

24. Prove that, if two lines are parallel, then the rays that bisect a pair of corresponding angles are parallel.

ANGLES OF A POLYGON

WHAT YOU WILL LEARN

A triangle is the simplest type of polygon. This chapter looks at angle-sum relationships in a triangle. These relationships are then generalized to other polygons, which may have any number of sides. In particular, you will learn:

- the way to name and classify polygons;
- the way to use the properties of parallel lines to prove that the measures of the three angles of a triangle must add up to 180;
- the relationship between the measures of the exterior and interior angles of a triangle;
- the way to find the sum of the measures of the interior and exterior angles of a polygon of any number of sides;
- the special angle relationships in equilateral and equiangular polygons.

SECTIONS IN THIS CHAPTER

The Anatomy of a Polygon

Angles of a Triangle

Exterior Angles of a Triangle

Angles of a Polygon

The Anatomy of a Polygon

Until now we have focused on points, segments, lines, rays, and angles. In this chapter we begin our study of geometric figures. A "closed" geometric figure whose sides are line segments is called a *polygon*, provided that each side intersects another side at its end point.

A polygon is named by choosing any vertex and then writing the letters corresponding to each vertex consecutively, proceeding in either a clockwise or counterclockwise fashion.

In Figure 5.1*a* and 5.1*b*, *ABCDE* is a polygon. However, in Figure 5.1*a*, the measure of angle *AED* is greater than 180; in Figure 5.1*b*, angle *AED* measures less than 180. Figure 5.1*a* is a nonconvex polygon, while 5.1*b* is a *convex* polygon. In this chapter, whenever the term *polygon* is used, it refers to a convex polygon.

▌ DEFINITION OF CONVEX POLYGON
▌ A *convex polygon* is a polygon in which the measure of each angle is less than 180.

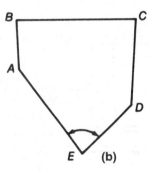

FIGURE 5.1 Convex versus nonconvex polygons.

PARTS OF A POLYGON

There are four important terms related to a polygon: *interior angle, side, vertex,* and *diagonal*. These terms are illustrated in Figures 5.2 and 5.3.

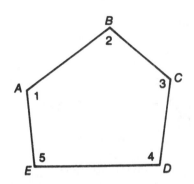

VERTICES: *A, B, C, D,* and *E*
SIDES: $\overline{AB}, \overline{BC}, \overline{CD}, \overline{DE}$ and $\overline{AD}$
INTERIOR ANGLES: 1, 2, 3, 4, and 5

FIGURE 5.2 Some parts of a polygon.

A segment that joins a pair of *nonadjacent* vertices of a polygon is called a *diagonal*. See Figure 5.3. Clearly, there exists a relationship between the number of sides of a polygon and the number of diagonals that can be drawn.

DIAGONALS: $\overline{AC}, \overline{BD}, \overline{EC}, \overline{EB}$ and $\overline{AD}$

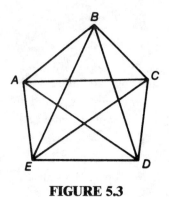

FIGURE 5.3

THEOREM 5.1

If a polygon has N sides, then $\frac{1}{2}N(N-3)$ diagonals can be drawn.

INFORMAL
PROOF

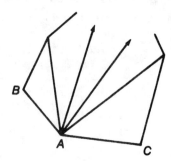

Consider a polygon having N sides. Draw segments from any vertex, say A, to each of the other $N-1$ vertices, resulting in $N-1$ segments. However, this number also includes the segments that coincide with sides $\overline{AB}$ and $\overline{AC}$. Hence, subtracting the two segments, $N-3$ segments remain that correspond to diagonals. This analysis may be repeated with each of the N vertices, yielding a total of $N(N-3)$ diagonals. But, each diagonal has been counted twice, since its end points are a vertex of the polygon. Therefore, the total number of distinct diagonals is $\frac{1}{2}N(N-3)$.

How many diagonals can be drawn in a polygon having seven sides?

SOLUTION

Number of diagonals $= \frac{1}{2}N(N-3)$
For $N = 7$,
Number of diagonals $= \frac{1}{2}(7)(7-3) = \frac{1}{2}(7)(4) = 14$

CLASSIFYING POLYGONS

A polygon may be classified according to the number of sides it has. The most commonly referred to polygons are as follows:

Polygon Name	Number of Sides
Triangle	3
Quadrilateral	4
Pentagon	5
Hexagon	6
Octagon	8
Decagon	10
Duodecagon	12

A polygon may also be categorized according to whether all of its angles are equal in measure and/or all of its sides have the same length.

DEFINITIONS OF POLYGONS HAVING PARTS EQUAL IN MEASURE

- An *equiangular* polygon is a polygon in which each angle has the same measure.
- An *equilateral* polygon is a polygon in which each side has the same length.
- A *regular* polygon is a polygon that is both equiangular and equilateral.

 EXAMPLE 5.2 Give an example of a regular quadrilateral.

SOLUTION

A square is both equiangular and equilateral. A square is, therefore, a regular quadrilateral.

 EXAMPLE 5.3 Give an example of a quadrilateral that is equiangular but *not* necessarily equilateral.

SOLUTION

A rectangle.

Angles of a Triangle

One of the most familiar geometric relationships is that the sum of the measures of the angles of a triangle is 180. See Figure 5.4. An experiment that informally proves this relationship can be performed by drawing any triangle, tearing off angles 1 and 3 (see Figure 5.5), and then aligning the edges (sides) of the angle about angle 2 so that a straight line is formed. Since a straight angle has measure 180, the sum of the measures of the angles of the triangle must be 180.

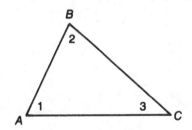

FIGURE 5.4 $m\angle 1 + m\angle 2 + m\angle 3 = 180$.

FIGURE 5.5 "Proving" that the sum of the measures of the angles of a triangle is 180. It is shown that angles 1, 2, and 3 can be aligned to form a straight angle.

THE TRIANGLE ANGLE-SUM THEOREM

THEOREM 5.2 SUM OF THE ANGLES OF A TRIANGLE THEOREM

The sum of the measures of the angles of a triangle is 180.

Since in reality we cannot "tear up" a triangle, the proof of Theorem 5.2 introduces the concept of drawing an extra (auxiliary) line in the original diagram.

GIVEN: $\triangle ABC$ (read as "triangle ABC").
PROVE: $m\angle 1 + m\angle 2 + m\angle 3 = 180$.
PLAN: Draw an auxiliary line ℓ through B and parallel to $\overline{AC}$. The sum of the measures of angles 4, 2, and 5 is 180. Use alternate interior angles and substitution to obtain the desired relationship.

PROOF: Statements	Reasons
1. Through point B, draw line ℓ parallel to $\overline{AC}$.	1. Through a point not on a line exactly one line may be drawn parallel to the line.
2. $m\angle 4 + m\angle 2 + m\angle 5 = 180$.	2. A straight angle has measure 180.
3. $m\angle 1 = m\angle 4, m\angle 3 = m\angle 5$.	3. If two lines are parallel, then their alternate interior angles are equal in measure
4. $m\angle 1 + m\angle 2 + m\angle 3 = 180$.	4. Substitution.

 EXAMPLE 5.4 The measures of the angles of a triangle are in the ratio 2:3:4. Find the measure of each angle.

SOLUTION

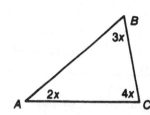

$$2x + 3x + 4x = 180$$
$$9x = 180$$
$$x = 20$$
$$m\angle A = 2x = 40$$
$$m\angle B = 3x = 60$$
$$m\angle C = 4x = 80$$

EXAMPLE 5.5

Find the value of x.

SOLUTION

$m\angle BCA = 30,$
$m\angle EDF = 45,$
$m\angle DGC = 105,$
so $x = 105.$

EXAMPLE 5.6

Given $\ell \parallel m$ and $\overline{CD} \perp \overline{AB}$, find the value of x.

SOLUTION

$m\angle A = 35, m\angle ADC = 90$
Hence $35 + 90 + x = 180$
$x = 180 - 125$
$= 55$

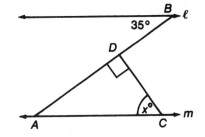

COROLLARIES OF THE TRIANGLE ANGLE-SUM THEOREM

A *corollary* is a theorem that is closely related to a previously proved theorem.

COROLLARY 5.2.1

The measure of each angle of an equiangular triangle is 60.

INFORMAL
PROOF

$3x = 180$
$x = \dfrac{180}{3} = 60$

COROLLARY 5.2.2

A triangle may have at most one right or one obtuse angle.

INFORMAL
PROOF

Use an indirect method of proof—suppose a triangle had two such angles. Then the sum of their measures would be greater than or equal to 180. What about the measure of the third angle of the triangle?

COROLLARY 5.2.3

The acute angles of a right triangle are complementary.

INFORMAL
PROOF

$$90 + x + y = 180$$
$$x + y = 90$$

COROLLARY 5.2.4

If two angles of a triangle are congruent to two angles of another triangle, then the remaining pair of angles are also congruent.

INFORMAL
PROOF

$$\begin{aligned}
m\angle A &+ m\angle B + m\angle C = 180 \\
-m\angle X &+ m\angle Y + m\angle Z = 180 \\
m\angle A - m\angle X &+ 0 \quad + 0 \quad\ = 0
\end{aligned}$$

Hence, $m\angle A = m\angle X$.

EXAMPLE 5.7 Use Corollary 5.2.4 to help construct a formal proof for the following problem:

GIVEN: $\overline{BD}$ bisects $\angle ABC$,
$\overline{BD} \perp \overline{AC}$.

PROVE: $\angle A \cong \angle C$.

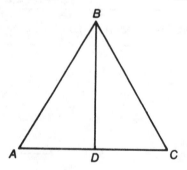

SOLUTION

PLAN: 1. Mark the diagram with the given information. Number the angles to make it easier to refer to them.
2. Since two angles of $\triangle ADB$ are congruent to two angles of $\triangle CDB$, the third pair of angles must be congruent. Hence $\angle A \cong \angle C$.

PROOF:	Statements	Reasons
	1. $\overline{BD}$ bisects $\angle ABC$.	1. Given.
	2. $\angle 1 \cong \angle 2$.	2. A bisector divides an angle into two congruent angles.
	3. $\overline{BD} \perp \overline{AC}$.	3. Given.
	4. Angles 3 and 4 are right angles.	4. Perpendicular lines intersect to form right angles.
	5. $\angle A \cong \angle C$.	5. If two angles of a triangle are congruent to two angles of another triangle, then the remaining pair of angles are congruent.

Exterior Angles of a Triangle

At each vertex of the triangle drawn in Figure 5.6, the sides have been extended, forming additional angles. Angles that are adjacent and supplementary to the interior angle at each vertex are called *exterior* angles of the triangle. Notice that at each vertex there are two such exterior angles:

Vertex	Interior Angle	Exterior Angles*
A	1	4 and 6
B	2	7 and 9
C	3	10 and 12

* Angles 5, 8, and 11 form a vertical pair with the interior angle and are therefore *not* exterior angles.

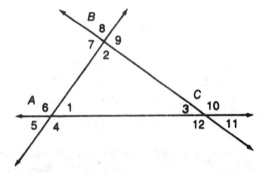

FIGURE 5.6 Exterior angles of a triangle.

Although it is possible to draw two exterior angles at each vertex, when we refer to the exterior angles of a polygon, we will normally mean only one at each vertex—that is, three in the case of a triangle. The fact that this may be done in several different ways is unimportant. Notice in Figure 5.7 that the exterior angle and the interior angle at a vertex form an adjacent pair of angles such that their exterior sides form a straight line.

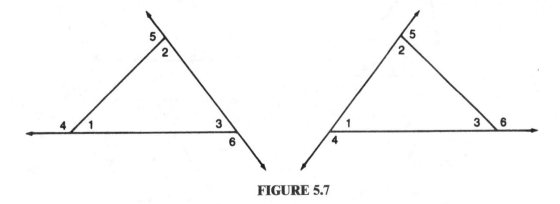

FIGURE 5.7

DEFINITION OF EXTERIOR ANGLE OF A TRIANGLE

An *exterior angle* of a triangle (or polygon) is an angle adjacent to an interior angle formed in such a way that their exterior sides form a straight line.

EXAMPLE 5.8
The measures of two angles of a triangle are 80 and 60. Find the sum of the measures of the exterior angles (one exterior angle at a vertex) of the triangle.

SOLUTION

First we note that the third angle of the triangle has measure 40 (180 – 140). Using the fact that, at each vertex, interior and exterior angles are supplementary, we deduce that the measures of the exterior angles are as shown in the diagram:

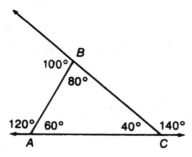

Then we sum: 120 + 100 + 140 = 360.

In the next section we will state and prove that the sum of the measures of the exterior angles of any polygon, regardless of the number of sides, is 360.

In Example 5.8 observe that the measure of each exterior angle is equal to the sum of the measures of the two nonadjacent (remote) interior angles.

THEOREM 5.3 EXTERIOR ANGLE OF A TRIANGLE THEOREM

The measure of an exterior angle of a triangle is equal to the sum of the measures of the two nonadjacent interior angles.

EXAMPLE 5.9
Find the value of x.

a.

b.

c.

d.

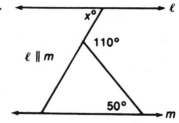

SOLUTION

a. $x = 48 + 52 = 100$

b. $110 = 2x + 30 + 60$
$110 = 2x + 90$
$20 = 2x$
$x = 10$

c. $3x - 10 = x + 15 + 45$
$\quad\quad\;\; = x + 60$
$\quad 3x = x + 70$
$\quad 2x = 70$
$\quad\; x = 35$

d. $x + 50 = 110$
$\quad\quad\; x = 60$

We have already established that m∡1 = m∡b + m∡c, as shown in the figure at the right. Since m∡b and m∡c represent positive numbers, if we delete one of these quantities from the equation, we no longer have an equality. The following inequalities result:

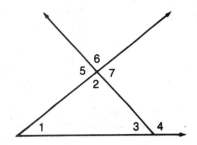

m∡1 = m∡b + m∡c, so m∡1 > m∡b
m∡1 = m∡b + m∡c, so m∡1 > m∡c

The symbol > is read as "is greater than" (the symbol < is read as "is less than"). We now state the corresponding theorem, which will prove useful in future work involving inequality relationships in triangles.

THEOREM 5.4 EXTERIOR ANGLE INEQUALITY THEOREM

The measure of an exterior angle of a triangle is greater than the measure of either nonadjacent interior angle.

EXAMPLE 5.10

Inside the box □ insert the correct symbol (either >, <, or =) so that the resulting statement is true. If it is not possible to draw a conclusion based on the given diagram, then write "cannot be determined."

a. m∡4 □ m∡1
b. m∡2 □ m∡6
c. m∡3 □ m∡2
d. m∡7 □ m∡1
e. m∡5 □ m∡2
f. m∡4 □ m∡1 + m∡2

SOLUTION

a. >
b. =
c. Cannot be determined

d. >
e. Cannot be determined
f. =

Angles of a Polygon

The sum of the measures of the angles of a polygon with three sides is 180. What is the sum of the measures of the angles of a polygon having four sides? Five sides? One hundred sides? To answer these questions, we need to derive a formula that gives the relationship between the sum of the angles of a polygon and the number of its sides. Our strategy is to break down any polygon into an equivalent set of triangles, as shown in Figure 5.8.

Figure 5.8 suggests that a polygon of n sides can be separated into $n - 2$ triangles. Since there are 180 degrees in the sum of the angle measures of each such triangle, it follows that the sum of the angles in the polygon can be found by multiplying $(n - 2)$ by 180.

THEOREM 5.5 SUM OF THE INTERIOR ANGLES OF A POLYGON THEOREM

The sum of the measures of the interior angles of a polygon having n sides is $180(n - 2)$.

 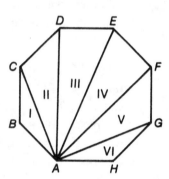

FIGURE 5.8 Separating a polygon into triangles.

EXAMPLE 5.11 Find the sum of the measures of the interior angles of an octagon.

SOLUTION
For $n = 8$, sum $= 180(n - 2) = 180(8 - 2) = 180(6) = 1,080$.

EXAMPLE 5.12 If the sum of the measures of the angles of a polygon is 900, determine the number of sides.

SOLUTION

For sum = 900,

$$900 = 180(n-2)$$
$$\frac{900}{180} = n-2$$
$$5 = n-2$$
$$n = 7 \text{ sides}$$

We know that the sum of the measures of the *exterior* angles of a triangle is 360. Let's prove that this sum remains the same for any polygon.

THEOREM 5.6 SUM OF THE EXTERIOR ANGLES OF A POLYGON THEOREM

The sum of the measures of the exterior angles of any polygon (one exterior angle at a vertex) is 360.

INFORMAL An *n*-sided polygon has *n* vertices.
PROOF At each vertex the sum of the interior and exterior angles is 180:

At vertex 1: Interior angle 1 + exterior angle 1 = 180
At vertex 2: Interior angle 2 + exterior angle 2 = 180

 • • • •
 • • • •
 • • • •

At vertex N: Interior angle *n* + exterior angle *n* = 180

Adding: Sum of interior angles + sum of exterior angles = 180*n*
Substituting: 180(*n* − 2) + sum of exterior angles = 180*n*
Simplifying: 180*n* − 360 + sum of exterior angles = 180*n*
 −360 + sum of exterior angles = 0

Sum of exterior angles = 360

If a polygon is equiangular, then each interior angle will have the same measure, implying that each exterior angle has the same measure. Thus, in a *regular* polygon the measure of each exterior angle is equal to the sum of the measures of the exterior angles, 360, divided by the number of sides.

For Regular Polygons

- Exterior angle = $\dfrac{360}{n}$
- Interior angle = 180 − Exterior angle

NG
78 ANGLES OF A POLYGON

EXAMPLE 5.13 Find the measure of each interior angle and each exterior angle of a regular decagon.

SOLUTION

Method 1: Sum $= 180(n-2)$
$= 180(10-2)$
$= 180(8)$
$= 1,440$

Since there are 10 interior angles, each of which is identical in measure,

Interior angle $= \dfrac{1,440}{10} = 144$

Since interior and exterior angles are supplementary,

Exterior angle $= 180 - 144 = 36$

Method 2: First determine the measure of an exterior angle:

Exterior angle $= \dfrac{360}{10} = 36$

Interior angle $= 180 - 36 = 144$

EXAMPLE 5.14 The measure of each interior angle of a regular polygon is 150. Find the number of sides.

SOLUTION

We use a method similar to the approach illustrated in Method 2 of Example 5.13. Since the measure of each interior angle is 150, the measure of an exterior angle is $180 - 150$, or 30. Therefore,

$$30 = \frac{360}{n}, \quad \text{so } n = 12$$

SUMMARY OF GENERAL PRINCIPLES

- An *n*-sided polygon has *n* vertices and *n* interior angles. At each vertex an exterior angle may be drawn by extending one of the sides.
- If the polygon is regular, then the measures of the interior angles are equal and the measures of the exterior angles are equal.
- The sum of the measures of the *interior* angles of *any* polygon is given by this formula:

 Sum $= 180(n-2)$.

 The sum of the measures of the *exterior* angles of any polygon, regardless of the number of sides, is 360.
- To find the measures of the interior and exterior angles of a *regular* polygon, given the number of sides (or vice versa), we use the following relationships:

 Exterior angle $= \dfrac{360}{n}$

 and

 Interior angle $= 180 -$ exterior angle

REVIEW EXERCISES FOR CHAPTER 5

1. Find the value of x:

(a)

(b)

(c)

(d)

(e)

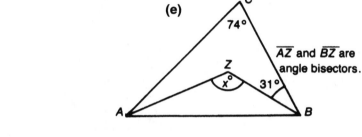

$\overline{AZ}$ and $\overline{BZ}$ are angle bisectors.

2. Given that $\ell \parallel m$, $m\angle 2 = 110$ and $m\angle 6 = 70$, find the measure of each of the remaining numbered angles.

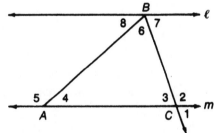

3. A triangle may be classified according to the measure of its largest angle. A triangle that contains an obtuse angle is called an *obtuse triangle*; if it contains a right angle it is referred to as a *right triangle*; an *acute triangle* is a triangle in which each angle is an acute angle.

 For each of the following, the measures of the angles of $\triangle ABC$ are represented in terms of x. Find the value of x and classify the triangle as acute, right, or obtuse.

 (a) $m\angle A = 3x + 8$
 $m\angle B = x + 10$
 $m\angle C = 5x$

 (b) $m\angle A = x + 24$
 $m\angle B = 4x + 17$
 $m\angle C = 2x - 15$

 (c) $m\angle A = 3x - 5$
 $m\angle B = x + 14$
 $m\angle C = 2x - 9$

For Exercises 4 to 6 find the value of x.

4.

5.

$\ell \parallel m$

6.

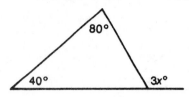

$\overleftrightarrow{EA} \parallel \overleftrightarrow{FB}$,

$\overline{AC}$ bisects ∡EAB,

$\overline{BC}$ bisects ∡FBA.

7. Find the measure of angle *RWT*.

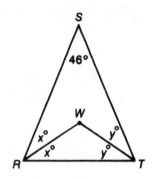

8. Find the sum of the measures of the interior angles of a polygon having:
 (a) 4 sides (b) 6 sides (c) 9 sides (d) 13 sides

9. Find the number of sides of a polygon if the sum of the measures of the interior angles is:
 (a) 1,800 (b) 2,700 (c) 540 (d) 2,160

10. Find the measure of the remaining angle of each of the following figures, given the measures of the other interior angles.
 (a) Quadrilateral: 42, 75, and 118
 (b) Pentagon: 116, 138, 94, 88
 (c) Hexagon: 95, 154, 80, 145, 76

11. Find the measure of each interior angle of a regular polygon having:
 (a) 5 sides (b) 24 sides (c) 8 sides (d) 15 sides

12. Find the number of sides of a regular polygon if the measure of an interior angle is:
 (a) 162 (b) 144 (c) 140 (d) 168

13. Which of the following *cannot* represent the measure of an exterior angle of a regular polygon?

 (1) 72 (2) 15 (3) 27 (4) 45

14. Find the number of sides in a polygon if the sum of the measures of the interior angles is 4 times as great as the sum of the measures of the exterior angles.

15. Find the number of sides in a regular polygon if:
 (a) The measure of an interior angle is 3 times the measure of an exterior angle.
 (b) The measure of an interior angle equals the measure of an exterior angle.
 (c) The measure of an interior angle exceeds 6 times the measure of an exterior angle by 12

16. GIVEN: $\overline{AB} \perp \overline{BD}$,
 $\overline{ED} \perp \overline{BD}$.
 PROVE: $\angle A \cong \angle E$.

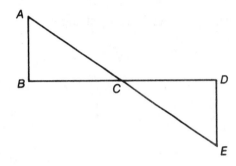

17. GIVEN: $\overline{AC} \perp \overline{CB}$,
 $\overline{DE} \perp \overline{AB}$.
 PROVE: $\angle 1 \cong \angle 2$.

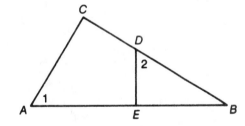

18. GIVEN: $\overline{JP} \perp \overline{PR}$,
 $\overline{LK} \perp \overline{KM}$,
 $\overline{KM} \parallel \overline{PR}$.
 PROVE: $\overline{KL} \parallel \overline{PJ}$.

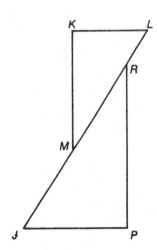

19. GIVEN: $\overline{AB} \perp \overline{BC}$,
 $\overline{DC} \perp \overline{BC}$,
 $\overline{DE} \perp \overline{AC}$.

 PROVE: $\angle 1 \cong \angle 2$.

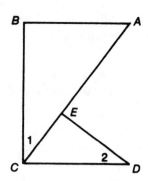

20. GIVEN: $\overleftrightarrow{AX} \parallel \overleftrightarrow{CY}$,
 $\overline{BA}$ bisects $\angle CAX$,
 $\overline{BC}$ bisects $\angle ACY$.

 PROVE: $\angle ABC$ is a right angle.

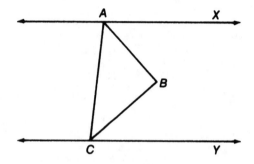

CHAPTER 6

PROVING TRIANGLES ARE CONGRUENT

WHAT YOU WILL LEARN

Triangles that can be made to coincide by placing one on top of the other are said to be *congruent*. If the triangles cannot be "moved," how can we tell whether two triangles are congruent? Do we need to know that all of the angles and all of the sides of one triangle are congruent to the corresponding parts of a second triangle before we can conclude that the two triangles are congruent? In this chapter you will learn:

* different shortcut methods for proving triangles congruent that require only that a particular combination of angles and sides of one triangle be proved congruent to the corresponding combination of parts of the second triangle;
* the fact that two *right triangles* are congruent if the hypotenuse and leg of one right triangle are congruent to the corresponding parts of the second right triangle.

SECTIONS IN THIS CHAPTER
Correspondences and Congruent Triangles
Proving Triangles Congruent: SSS, SAS and ASA Postulates
Proving Overlapping Triangles Congruent
Proving Triangles Congruent: AAS and Hy-Leg Methods

83

Correspondences and Congruent Triangles

On a triple blind date, Steve, Bob, and Charles accompanied Jane, Lisa, and Kris. To indicate which young man from the set of gentlemen is being paired with which young lady, the following notation may be used:

Steve ↔ Jane Bob ↔ Lisa Charles ↔ Kris

Such a pairing of the members of one group with the members of another group is called a *correspondence*. A correspondence may also be established between the vertices of two triangles, as shown in Figure 6.1.

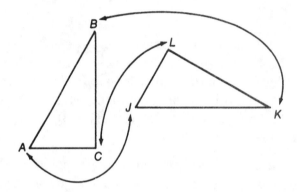

FIGURE 6.1 Correspondence: $A \leftrightarrow J, B \leftrightarrow K, C \leftrightarrow L$.

This correspondence can be expressed concisely by using the notation: $\triangle ABC \leftrightarrow$ $\triangle JKL$. The order in which the vertices are written matters since it defines the pairing of vertices:

$$\triangle A\ B\ C \leftrightarrow \triangle J\ K\ L$$

The correspondence may be written in more than one way. For example, $\triangle CAB \leftrightarrow$ $\triangle LJK$ defines the same correspondence as $\triangle ABC \leftrightarrow \triangle JKL$ since, in each case, the same pair of vertices are matched. On the other hand, $\triangle ABC \leftrightarrow \triangle JKL$ and $\triangle ACB \leftrightarrow$ $\triangle LKJ$ define two different correspondences.

A specified correspondence also serves to define a set of corresponding angles and a set of corresponding sides. If $\triangle RST \leftrightarrow \triangle XYZ$, angle pairs R and X, S and Y, and T and Z are corresponding angles. Segments whose end points are corresponding vertices determine the corresponding sides:

$$\triangle R\ S\ T \leftrightarrow \triangle X\ Y\ Z \qquad \triangle R\ S\ T \leftrightarrow \triangle X\ Y\ Z \qquad \triangle R\ S\ T \leftrightarrow \triangle X\ Y\ Z$$
$$\overline{RS} \leftrightarrow \overline{XY} \qquad\qquad \overline{ST} \leftrightarrow \overline{YZ} \qquad\qquad \overline{RT} \leftrightarrow \overline{XZ}$$

Corresponding sides are $\overline{RS}$ and $\overline{XY}$, $\overline{ST}$ and $\overline{YZ}$, and $\overline{RT}$ and $\overline{XZ}$. Drawing the actual triangles in Figure 6.2, we see that *corresponding sides lie opposite corresponding angles.*

FIGURE 6.2 Corresponding sides lie opposite corresponding angles.

> **DEFINITIONS OF CORRESPONDING PARTS OF TRIANGLES**
> * Corresponding *angles* are pairs of angles whose vertices are paired in a given correspondence between two triangles.
> * Corresponding *sides* are pairs of segments whose end points are vertices that are paired in a given correspondence between two triangles. Corresponding sides of triangles lie opposite corresponding angles.

 Given the correspondence $\triangle BAW \leftrightarrow \triangle TFK$, name all pairs of corresponding angles and sides.

SOLUTION

$\measuredangle B \leftrightarrow \measuredangle T$	$\overline{BA} \leftrightarrow \overline{TF}$
$\measuredangle A \leftrightarrow \measuredangle F$	$\overline{AW} \leftrightarrow \overline{FK}$
$\measuredangle W \leftrightarrow \measuredangle K$	$\overline{BW} \leftrightarrow \overline{TK}$

 Rewrite the correspondence given in Example 6.1 so that the same pairing of elements is maintained.

SOLUTION

There is more than one way of writing the same correspondence including, $\triangle WAB \leftrightarrow \triangle KFT$.

CONGRUENT TRIANGLES

Congruence means "same size" and "same shape." Plane congruent figures can be made to coincide. Two segments, for example, can be made to coincide (assuming, of course, that we could move the segments) if they have the same length. Two angles can be made to coincide if they have the same shape, that is, if they have the same measure. In order for geometric *figures* to be congruent, they must have the same size *and* the same shape.

> **DEFINITION OF CONGRUENT TRIANGLES**
> Two triangles are *congruent* if their vertices can be paired so that two conditions are met:
> 1. All pairs of corresponding angles are congruent.
> 2. All pairs of corresponding sides are congruent.

The definition states that, in order for two triangles to be congruent, six pairs of parts must be congruent: three pairs of angles and three pairs of sides. A similar definition may be used for congruent polygons of any number of sides.

EXAMPLE 6.3 Are the following two triangles congruent? If they are, write the correspondence between the triangles that establishes the congruence.

SOLUTION

Yes. $\triangle PEG \cong \triangle TRY$.

INCLUDED ANGLES AND SIDES

With respect to angles A and C in Figure 6.3, side $\overline{AC}$ is called an *included* side; or side $\overline{AC}$ is said to be *included* between angle A and angle C. With respect to angles B and C, $\overline{BC}$ is the included side. What is the included side between angles A and B? The answer is $\overline{AB}$.

Since sides $\overline{AB}$ and $\overline{CB}$ intersect at B, angle B is called an *included angle*; or angle B is said to be *included* between side

FIGURE 6.3

$\overline{AB}$ and side $\overline{CB}$. With respect to sides $\overline{BA}$ and $\overline{CA}$, angle A is the included angle. What angle is included between sides $\overline{BC}$ and AC? If you answer "angle C," you have the right idea.

For the accompanying figure:

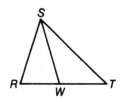

a. Use three letters to name the angle included between sides:
 (**i**) $\overline{RS}$ and $\overline{WS}$

 (**ii**) $\overline{SW}$ and $\overline{TW}$

b. Name the segment that is included between:
 (**i**) Angles *WST* and *WTS*
 (**ii**) Angles *SWR* and *SRW*

SOLUTION

 a. (**i**) $\measuredangle RSW$ (**ii**) $\measuredangle TWS$ **b.** (**i**) $\overline{TS}$ (**ii**) $\overline{RW}$

Proving Triangles Congruent: SSS, SAS, and ASA Postulates

Much of our work in geometry will be devoted to trying to prove that two triangles are congruent, given certain facts about the triangles. Fortunately, there are several short-cut methods for proving triangles congruent. Rather than proving triangles congruent by demonstrating that they agree in *six* pairs of parts, it is possible to conclude that a pair of triangles are congruent if they agree in *three* pairs of parts, *provided that these parts are a particular set of three pairs of congruent angles and sides.*

TRIANGLES THAT AGREE IN THREE SIDES

Using a ruler, draw two triangles so that the three sides of the first triangle have the same length as the corresponding sides of the second triangle. For example, see Figure 6.4.

FIGURE 6.4

After the triangles are drawn, use a pair of scissors to cut out △II. You should be able to demonstrate that the two triangles can be made to coincide. In other words, if three sides of one triangle are congruent to three sides of another triangle, the corresponding angles of the triangles are "forced" to be congruent; therefore, the triangles are congruent. This is postulated as follows:

SIDE-SIDE-SIDE (SSS) POSTULATE

If the vertices of two triangles can be paired so that three sides of one triangle are congruent to the corresponding sides of the second triangle, then the two triangles are congruent.

GIVEN: $\overline{AB} \cong \overline{BC}$,
M is the midpoint of $\overline{AC}$.

PROVE: $\triangle AMB \cong \triangle CMB$.

SOLUTION

PLAN: 1. Mark the diagram with the Given. Note that an "x" is used to indicate that side $\overline{BM}$ is congruent to itself
2. After the three pairs of congruent sides are identified, write the formal proof.

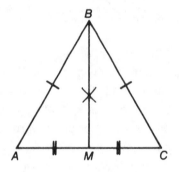

PROOF:

Statements	Reasons
1. $\overline{AB} \cong \overline{BC}$. (Side)	1. Given.
2. M is the midpoint of $\overline{AC}$.	2. Given.
3. $\overline{AM} \cong \overline{CM}$. (Side)	3. A midpoint divides a segment into two congruent segments.
4. $\overline{BM} \cong \overline{BM}$. (Side)	4. Reflexive property of congruence.
5. $\triangle AMB \cong \triangle CMB$.	5. SSS Postulate.

THE SIDE-ANGLE-SIDE (SAS) AND THE ANGLE-SIDE-ANGLE (ASA) POSTULATES

In addition to the SSS Postulate, several other shortcut methods can be used to prove triangles congruent. The Side-Angle-Side Postulate is illustrated in Figure 6.5. If it is known that two sides and the included angle of one triangle are congruent to the corresponding parts of another triangle, then we may conclude that the triangles are congruent. Because the congruent angle must be sandwiched in between the congruent pairs of sides, this method is commonly referred to as the SAS Postulate.

FIGURE 6.5

$$\overline{AB} \cong \overline{XY} \quad \boxed{S}$$
$$\angle B \cong \angle Y \quad \boxed{A}$$
$$\overline{BC} \cong \overline{YZ} \quad \boxed{S}$$

SIDE-ANGLE-SIDE (SAS) POSTULATE

If the vertices of two triangles can be paired so that two sides and the included angle of one triangle are congruent to the corresponding parts of the second triangle, then the two triangles are congruent.

Similarly, if two angles and the included side (ASA) of one triangle are congruent to the corresponding parts of another triangle, then the triangles are congruent. See Figure 6.6.

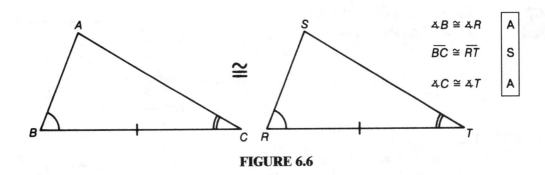

$$\angle B \cong \angle R \quad \boxed{A}$$
$$\overline{BC} \cong \overline{RT} \quad \boxed{S}$$
$$\angle C \cong \angle T \quad \boxed{A}$$

FIGURE 6.6

ANGLE-SIDE-ANGLE (ASA) POSTULATE

If the vertices of two triangles can be paired so that two angles and the included side of one triangle are congruent to the corresponding parts of the second triangle, then the two triangles are congruent.

EXAMPLE
6.6

GIVEN: C is the midpoint of $\overline{BE}$,
 $\angle B \cong \angle E$.
PROVE: $\triangle ABC \cong \triangle DEC$.

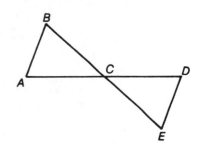

SOLUTION

PLAN: 1. Mark the diagram with the Given *and* any additional parts, such as the vertical angle pair, that can be deduced to be congruent based on the diagram.
2. Examine the diagram marked in Step 1 to decide which method of proving triangles congruent to use. In this case, use ASA.
3. Write the formal proof.

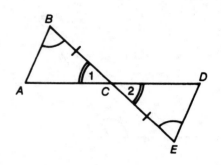

PROOF:

Statements	Reasons
1. $\angle B \cong \angle E$. (Angle)	1. Given.
2. C is the midpoint of $\overline{BE}$.	2. Given.
3. $\overline{BC} \cong \overline{EC}$. (Side)	3. A midpoint divides a segment into two congruent segments.
4. $\angle 1 \cong \angle 2$. (Angle)	4. Vertical angles are congruent.
5. $\triangle ABC \cong \triangle DEC$.	5. ASA Postulate.

 EXAMPLE 6.7

GIVEN: $\overline{AB} \parallel \overline{CD}, \overline{AB} \cong \overline{CD}$.
PROVE: $\triangle DAB \cong \triangle BCD$.

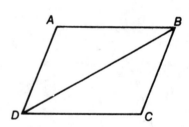

SOLUTION

PLAN: 1. Mark the diagram.
2. Decide on the method to use. In this example, SAS is appropriate.
3. Write the formal proof.

PROOF:

Statements	Reasons
1. $\overline{AB} \cong \overline{CD}$. (Side)	1. Given.
2. $\overline{AB} \parallel \overline{CD}$.	2. Given.
3. $\angle 1 \cong \angle 2$. (Angle)	3. If two lines are parallel, then their alternate interior angles are congruent.
4. $\overline{BD} \cong \overline{BD}$. (Side)	4. Reflexive property of congruence.
5. $\triangle DAB \cong \triangle BCD$.	5. SAS Postulate.

<div style="border: 1px solid black;">

To Prove Two Triangles Congruent

1. Mark the diagram with the Given.
2. Mark any additional parts that are congruent such as vertical angles, or sides (or angles) shared by both triangles.
3. If helpful, label angles with numbers. This will make the job of writing the proof easier.
4. Decide which method of proving triangles congruent to use.
5. Write the formal proof. Next to each statement in which a required side or angle is established as being congruent, write "(Side)" or "(Angle)." When you look back at your proof, this notation will help you verify that you have satisfied the necessary conditions of the congruence postulate being used.

</div>

Proving Overlapping Triangles Congruent

Sometimes proofs appear to be more complicated than they actually are simply because the triangles are drawn so that one overlaps the other. For example, consider the following problem:

GIVEN: $\overline{AB} \cong \overline{DC}$,
$\overline{AB} \perp \overline{BC}, \overline{DC} \perp \overline{BC}$.
PROVE: $\triangle ABC \cong \triangle DCB$.

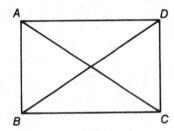

Often the analysis of the problem can be made easier by using pencils of different colors to outline the triangles to be proved congruent. Alternatively, it is sometimes helpful to redraw the diagram, "sliding" the triangles apart and then proceeding to mark the diagram:

PLAN: Show $\triangle ABC \cong \triangle DCB$ by using SAS. Notice that $\overline{BC}$ is a side of both triangles. Overlapping triangles often share the same side or angle.

EXAMPLE
6.8

GIVEN: $\overline{RA} \perp \overline{PQ}, \overline{QB} \perp \overline{PR},$
$\overline{PA} \cong \overline{PB}.$

PROVE: $\triangle PAR \cong \triangle PBQ.$

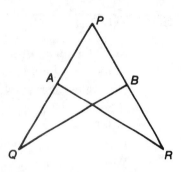

SOLUTION

PLAN: 1. "Separate" the triangles, and mark corresponding congruent parts:

2. Note that angle P is common to both triangles.
3. Prove triangles congruent by ASA.

PROOF:

Statements	Reasons
1. $\angle P \cong \angle P.$ (Angle)	1. Reflexive property of congruence.
2. $\overline{PA} \cong \overline{PB}.$ (Side)	2. Given
3. $\overline{RA} \perp \overline{PQ}$ and $\overline{QB} \perp \overline{PR}.$	3. Given.
4. Angles 1 and 2 are right angles.	4. Perpendicular lines intersect to form right angles.
5. $\angle 1 \cong \angle 2.$ (Angle)	5. All right angles are congruent.
6. $\triangle PAR \cong \triangle PBQ.$	6. ASA Postulate.

Sometimes the addition or subtraction property must be applied to obtain a pair of congruent corresponding parts that will be needed in order to establish the congruence of a pair of triangles. Recall that we may only add or subtract the *measures* of segments or angles. To reduce the number of steps in a proof, we will freely convert from congruence to measure (in order to perform the arithmetic operation) and then convert back to congruence. This procedure is illustrated in Steps 3 and 6 of following example.

EXAMPLE
6.9

GIVEN: $\overline{AB} \cong \overline{DE}, \overline{AD} \cong \overline{FC},$
$\overline{AB} \parallel \overline{DE}.$

PROVE: $\triangle ABC \cong \triangle DEF.$

SOLUTION

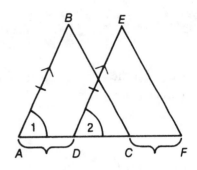

PLAN: Mark the diagram with the Given. NOTE: We may use SAS, *but* we must first establish within the proof that $\overline{AC} \cong \overline{DF}$ by adding the measure of $\overline{DC}$ to AD and CF.

PROOF:

Statements	Reasons
1. $\overline{AB} \cong \overline{DE}$. (Side)	1. Given.
2. $\angle 1 \cong \angle 2$. (Angle)	2. If two lines are parallel, then their corresponding angles are congruent.
3. $AD = FC$.	3. Given.
4. $DC = DC$.	4. Reflexive property of equality.
5. $AD + DC = FC + DC$.	5. Addition property of equality.
6. $\overline{AC} \cong \overline{FD}$. (Side)	6. Substitution.
7. $\triangle ABC \cong \triangle DEF$.	7. SAS Postulate.

EXAMPLE 6.10

GIVEN: $\angle ABE \cong \angle CBD$,
$\angle BDE \cong \angle BED$,
$\overline{BD} \cong \overline{BE}$.

PROVE: $\triangle DAB \cong \triangle ECB$.

SOLUTION

PLAN: Mark the diagram. NOTE: We may use ASA, but we must establish within the proof that $\angle 3 \cong \angle 4$ by subtracting the measure of angle DBE from the measures of angles ABE and CBD.

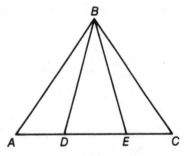

PROOF:

Statements	Reasons
1. $\angle BDE \cong \angle BED$.	1. Given.
2. $\angle 1 \cong \angle 2$. (Angle)	2. Supplements of congruent angles are congruent.
3. $\overline{BD} \cong \overline{BE}$. (Side)	3. Given.
4. $m\angle ABE = m\angle CBD$.	4. Given.
5. $m\angle DBE = m\angle DBE$.	5. Reflexive property of equality.
6. $m\angle ABE - m\angle DBE = m\angle CBD - m\angle DBE$.	6. Subtraction property of equality.
7. $\angle 3 \cong \angle 4$. (Angle)	7. Substitution.
8. $\triangle DAB \cong \triangle ECB$.	8. ASA Postulate.

Proving Triangles Congruent: AAS and Hy-Leg Methods

Consider the following pair of triangles with the corresponding pairs of congruent parts already marked:

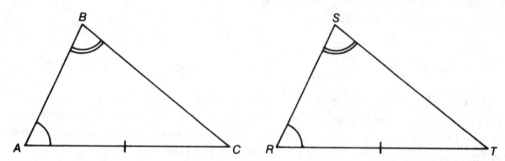

Are the triangles congruent? The triangles agree in two angles and the side *opposite* one of them (AAS), so SSS, SAS, or ASA cannot be directly applied in establishing the congruence of the triangles. However, by Corollary 5.2.4, the third pair of angles of the triangles must be congruent. It follows that the triangles are congruent by application of the ASA Postulate. Thus, angle-angle-side ≅ angle-angle-side means that the two triangles are congruent.

ANGLE-ANGLE-SIDE (AAS) THEOREM

If the vertices of two triangles can be paired so that two angles and the side opposite one of them in one triangle are congruent to the corresponding parts of the second triangle, then the two triangles are congruent.

GIVEN: $\overline{AB} \perp \overline{BD}, \overline{AC} \perp \overline{CD}$,
$\angle 1 \cong \angle 2$.

PROVE: $\triangle ABD \cong \triangle ACD$.

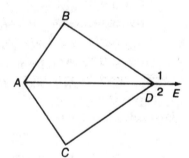

SOLUTION

OUTLINE OF Angles 3 and 4 are congruent since
PROOF: they are supplements of congruent
angles. Angles 5 and 6 are congruent
since all right angles are congruent.
The triangles are, therefore, congru-
ent by the AAS Theorem.

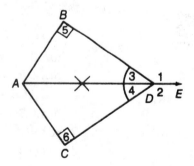

Two definitions related to a right triangle are necessary before we consider the next (and final) method for proving triangles congruent.

In a right triangle, the side opposite the right angle is called the *hypotenuse* and each of the two remaining sides is called a leg. Notice in Figure 6.7 that the legs are perpendicular to each other.

FIGURE 6.7

Next, consider two right triangles in which the hypotenuses and one of the legs (Figure 6.8) have the same lengths. This information is sufficient to conclude that the two *right* triangles are congruent. This way of proving triangles congruent is called the Hypotenuse-Leg (Hy-Leg) method.

FIGURE 6.8

HYPOTENUSE-LEG (HY-LEG) POSTULATE

If the vertices of two *right* triangles can be paired so that the hypotenuse and leg of one triangle are congruent to the corresponding parts of the second *right* triangle, then the two *right* triangles are congruent.

This method is applicable only to pairs of *right* triangles. Before the Hy-Leg Postulate can be applied, you need to know that the triangles involved are right triangles.

EXAMPLE
6.12

GIVEN: $\overline{AB} \perp \overline{BC}, \overline{AD} \perp \overline{DC},$
 $\overline{AB} \cong \overline{AD}.$

PROVE: $\triangle ABC \cong \triangle ADC.$

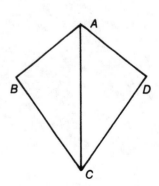

SOLUTION

PLAN: 1. Mark the diagram.
2. Use the Hy-Leg Postulate.

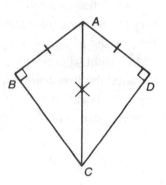

PROOF:

Statements	Reasons
1. $\overline{AC} \cong \overline{AC}$. (Hy)	1. Reflexive property of congruence.
2. $\overline{AB} \cong \overline{AD}$. (Leg)	2. Given.
NOTE: Before we can use the Hy-Leg Postulate we must establish that triangles ABC and ADC are right triangles.	
3. $\overline{AB} \perp \overline{BC}, \overline{AD} \perp \overline{DC}$.	3. Given.
4. Angles B and D are right angles.	4. Perpendicular lines intersect to form right angles.
5. Triangles ABC and ADC are right triangles.	5. A triangle that contains a right angle is a right triangle.
6. $\triangle ABC \cong \triangle ADC$.	6. Hy-Leg Postulate.

SUMMARY

- You may conclude that two triangles are congruent if it can be shown that:

1. Three sides of one triangle are congruent to the corresponding parts of the other triangle.

2. Two sides and the included angle of one triangle are congruent to the corresponding parts of the other triangle.

METHOD: SSS Postulate

METHOD: SAS Postulate

Summary (*Continued*)

3. Two angles and the included side of one triangle are congruent to the corresponding parts of the other triangle.

4. Two angles and the side opposite one of them are congruent to the corresponding parts of the other triangle.

METHOD: ASA Postulate

METHOD: AAS Theorem

- You may conclude that two *right* triangles are congruent if the hypotenuse and either leg of one triangle are congruent to the corresponding parts of the other right triangle.

METHOD: Hy-Leg Postulate

or

METHOD: Hy-Leg Postulate

- You may *not* conclude that two triangles are congruent when:
 1. Two sides and an angle that is *not* included of one triangle are congruent to the corresponding parts of the other triangle.

SSA ≇ SSA

2. Three angles of one triangle are congruent to the corresponding parts of the other triangle.

AAA ≇ AAA

REVIEW EXERCISES FOR CHAPTER 6

1. GIVEN: $\overline{BM} \perp \overline{AC}$, M is the midpoint of $\overline{AC}$.
 PROVE: $\triangle ABM \cong \triangle CBM$.

2. GIVEN: $\overline{RT}$ bisects angles STW and SRW.
 PROVE: $\triangle RST \cong \triangle RWT$.

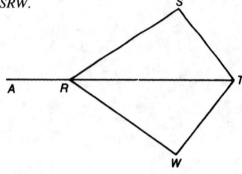

3. GIVEN: $\overline{EF} \parallel \overline{AB}$, $\overline{ED} \parallel \overline{BC}$,
 $\overline{AD} \cong \overline{FC}$.
 PROVE: $\triangle ABC \cong \triangle FED$.

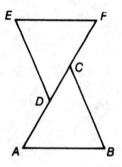

Use the accompanying diagram to solve Exercises 4 and 5.

4. GIVEN: $\angle R \cong \angle T$,
 $\overline{SR} \cong \overline{ST}$.
 PROVE: $\triangle SRH \cong \triangle STE$.

5. GIVEN: $\overline{TE} \perp \overline{RS}$, $\overline{RH} \perp \overline{ST}$,
 $\overline{EW} \cong \overline{HW}$.
 PROVE: $\triangle EWR \cong \triangle HWT$.

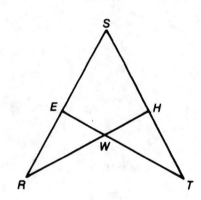

6. GIVEN: $\overline{QU} \parallel \overline{DA}, \overline{QU} \cong \overline{DA}$.
 PROVE: $\triangle UXQ \cong \triangle DXA$.

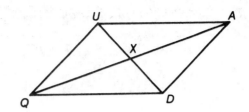

Use the accompanying diagram to solve Exercises 7 and 8.

7. GIVEN: $\overline{JK} \perp \overline{KT}, \overline{ET} \perp \overline{KT}$,
 $\overline{KV} \cong \overline{TL}, \overline{JL} \cong \overline{EV}$.
 PROVE: $\triangle JKL \cong \triangle ETV$.

8. GIVEN: $\overline{JK} \perp \overline{KT}, \overline{ET} \perp \overline{KT}$,
 $\angle 1 \cong \angle 2, \overline{KL} \cong \overline{TV}$.
 PROVE: $\triangle JKL \cong \triangle ETV$.

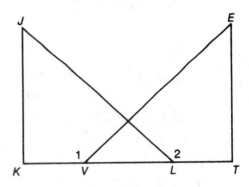

9. GIVEN: $\overline{AR}$ bisects $\angle FRI$,
 $\angle 1 \cong \angle 2$,
 $\angle RFI \cong \angle RIF$.
 PROVE: $\triangle AFR \cong \triangle AIR$.

Use the accompanying diagram to solve Exercises 10 and 11.

10. GIVEN: S is the midpoint of $\overline{RT}$,
 $\overline{SW} \cong \overline{SP}, \angle RSW \cong \angle TSP$.
 PROVE: $\triangle TSW \cong \triangle SRP$.

11. GIVEN: $\overline{TW} \cong \overline{SP}, \overline{RP} \parallel \overline{SW}$,
 $\overline{SP} \parallel \overline{TW}$.
 PROVE: $\triangle TSW \cong \triangle SRP$.

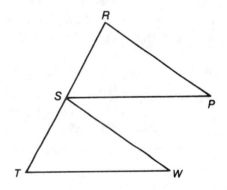

Use the accompanying diagram to solve Exercises 12 and 13.

12. GIVEN: $\overline{AB} \cong \overline{DE}$, M is the midpoint of $\overline{BE}$,
 $\overline{AM} \cong \overline{KM}, \overline{DM} \cong \overline{KM}$.
 PROVE: $\triangle ABM \cong \triangle DEM$.

13. GIVEN: $\overline{KM}$ is the $\perp$ bisector of $\overline{BE}$,
 $\overline{KM}$ bisects $\angle AMD$,
 $\overline{AB} \parallel \overline{KM} \parallel \overline{DE}$.
 PROVE: $\triangle ABM \cong \triangle DEM$.

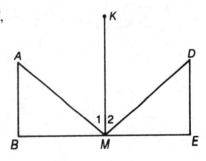

Use the accompanying diagram to solve Exercises 14 and 15.

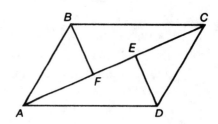

14. GIVEN: $\overline{BD} \perp \overline{AC}, \overline{AF} \perp \overline{BC},$
$\overline{FC} \cong \overline{DC}.$
PROVE: $\triangle AFC \cong \triangle BDC.$

15. GIVEN: $\overline{AD} \cong \overline{BF},$
$\angle BAD \cong \angle ABF.$
PROVE: $\triangle BAD \cong \triangle ABF.$

16. GIVEN: $\overline{BF} \perp \overline{AC}, \overline{DE} \perp \overline{AC},$
$\overline{AB} \cong \overline{DC}, \overline{AE} \cong \overline{CF}.$
PROVE: $\triangle AFB \cong \triangle CED.$

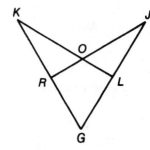

Use the accompanying diagram to solve Exercises 17 and 18.

17. GIVEN: $\overline{JR} \perp \overline{KG}, \overline{KL} \perp \overline{JG},$
$\overline{KL} \cong \overline{JR}.$
PROVE: $\triangle KLG \cong \triangle JRG.$

18. GIVEN: $\overline{KG} \cong \overline{JG}, \overline{JR} \perp \overline{KG}, \overline{KL} \perp \overline{JG},$
R is the midpoint of $\overline{KG},$
L is the midpoint of $\overline{JG}.$
PROVE: $\triangle KOR \cong \triangle JOL.$

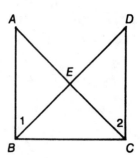

19. GIVEN: $\overline{AB} \perp \overline{BC}, \overline{DC} \perp \overline{BC},$
$\overline{DB}$ bisects $\angle ABC,$
$\overline{AC}$ bisects $\angle DCB,$
$\overline{EB} \cong \overline{EC}.$
PROVE: $\triangle BEA \cong \triangle CED.$

20. Prove that, if two triangles are congruent to the same triangle, then they are congruent to each other.

APPLYING CONGRUENT TRIANGLES

WHAT YOU WILL LEARN

You can prove that two segments or two angles are congruent if you can first show that these pairs of segments or angles are contained in congruent triangles. In this chapter you will learn:

- the way to prove pairs of segments or angles congruent using congruent triangles;
- the difference between drawing segments determined, overdetermined, and underdetermined;
- that every triangle has three altitudes and three medians;
- that the base angles of an isosceles triangle are congruent, and, conversely, the sides opposite congruent angles of a triangle are congruent;
- the way to use one pair of congruent triangles to prove a second pair of triangles congruent.

SECTIONS IN THIS CHAPTER

Using Congruent Triangles to Prove Segments and Angles Congruent

Using Congruent Triangles to Prove Special Properties of Lines

Classifying Triangles and Special Segments

The Isosceles Triangle

Double Congruence Proofs

Using Congruent Triangles to Prove Segments and Angles Congruent

Based on the information marked in Figure 7.1, what conclusion can be drawn about how the measures of angles *R* and *H* compare?

FIGURE 7.1

At first glance, there may seem to be insufficient information to draw any conclusion about these angles. However, by the SAS Postulate, $\triangle PLR \cong \triangle GMH$. We need to recall the *definition* of congruent triangles: If two triangles have all *six* pairs of corresponding parts congruent, then the triangles are congruent. In this case, since we know that the two triangles are congruent, we may use the reverse of the definition of congruent triangles to conclude that *all* pairs of corresponding sides and *all* pairs of corresponding angles must be congruent. Since angles *R* and *H* are corresponding angles, they must be congruent.

This reasoning represents an extremely useful method for proving segments or angles congruent, provided that they are parts of triangles that can be proved congruent. *After* a pair of triangles are proved congruent, then *any* pair of corresponding parts may be stated to be congruent, based on the principle that *corresponding parts of congruent triangles are congruent*, abbreviated as CPCTC.

EXAMPLE 7.1

GIVEN: $\overline{AB} \cong \overline{AD}$,
$\overline{BC} \cong \overline{DC}$.

PROVE: $\angle B \cong \angle D$.

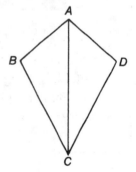

SOLUTION

PLAN: Angles *B* and *D* are corresponding parts of triangles *ABC* and *ADC*, respectively. After proving these triangles congruent, we may then conclude that the desired pair of angles are congruent. Marking the diagram suggests that the triangles can be proved congruent by the SSS Postulate.

PROOF:

Statements	Reasons
1. $\overline{AB} \cong \overline{AD}$ and $\overline{BC} \cong \overline{DC}$.	1. Given
2. $\overline{AC} \cong \overline{AC}$.	2. Reflexive property of congruence.
3. $\triangle ABC \cong \triangle ADC$.	3. SSS Postulate.
4. $\angle B \cong \angle D$.	4. CPCTC.

EXAMPLE
7.2

GIVEN: $\overline{AB} \cong \overline{DC}$,
$\overline{AB} \perp \overline{BC}, \overline{DC} \perp \overline{BC}$.

PROVE: $\overline{AC} \cong \overline{DB}$.

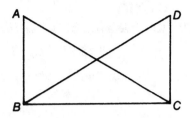

SOLUTION

PLAN: $\overline{AC}$ and $\overline{DB}$ are sides of triangles ABC and DCB, respectively. After proving these triangles congruent, we may then conclude that the desired pair of segments are congruent. Marking the diagram suggests that the SAS method be applied.

PROOF:

Statements	Reasons
1. $\overline{AB} \cong \overline{DC}$. (Side)	1. Given.
2. $\overline{AB} \perp \overline{BC}$ and $\overline{DC} \perp \overline{BC}$.	2. Given.
3. Angles ABC and DCB are right angles.	3. Perpendicular lines intersect to form right angles.
4. $\angle ABC \cong \angle DCB$. (Angle)	4. All right angles are congruent.
5. $\overline{BC} \cong \overline{BC}$. (Side)	5. Reflexive property of congruence.
6. $\triangle ABC \cong \triangle DCB$.	6. SAS Postulate.
7. $\overline{AC} \cong \overline{DB}$.	7. CPCTC.

As the diagrams and problems become more complicated, we apply the following four-step procedure.

To Prove Segments and/or Angles Congruent Using Congruent Triangles

1. IDENTIFY. Identify the pair of triangles that contain the parts that need to be proved congruent.
2. PLAN. Plan how to prove the selected pair of triangles congruent by first marking the diagram with the Given. Then mark any additional pairs of parts that may be congruent as a result of vertical angles, perpendicular lines, parallel lines, supplements (complements) of the same (or congruent) angles, or a common angle or side.
3. SELECT. Select the method of congruence to be used.
4. WRITE. Write the two-column formal proof.

We illustrate this procedure in the next example.

EXAMPLE
7.3

GIVEN: $\overline{MP} \cong \overline{ST}$,
$\overline{MP} \parallel \overline{ST}$,
$\overline{PL} \cong \overline{RT}$.

PROVE: $\overline{RS} \parallel \overline{LM}$.

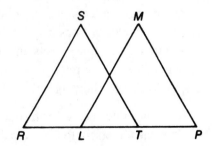

SOLUTION

To prove a pair of line segments parallel, we must usually prove that an appropriate pair of angles are congruent.

1. Looking at the diagram, we see that, if $\overline{RS}$ is to be parallel to $\overline{LM}$, we must prove that $\angle SRT \cong \angle MLP$. This implies that the triangles that contain these angles must be proved congruent. Therefore, we must prove $\triangle RST \cong \triangle LMP$.
2. Mark the diagram with the Given.
3. Use the SAS Postulate.
4. Write the formal proof.

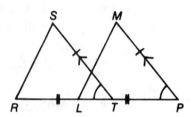

PROOF:

Statements	Reasons
1. $\overline{MP} \cong \overline{ST}$. (Side)	1. Given.
2. $\overline{MP} \parallel \overline{ST}$.	2. Given.
3. $\angle MPL \cong \angle STR$. (Angle)	3. If two lines are parallel, then their corresponding angles are congruent.
4. $\overline{PL} \cong \overline{RT}$. (Side)	4. Given.
5. $\triangle RST \cong \triangle LMP$.	5. SAS Postulate.
6. $\angle SRT \cong \angle MLP$.	6. CPCTC.
7. $\overline{RS} \parallel \overline{LM}$.	7. If two lines have their corresponding angles congruent, then they are parallel.

Notice that our approach is based on developing a plan that works backward, beginning with the Prove. Once the solution path is clear, we write the proof, working forward.

Using Congruent Triangles to Prove Special Properties of Lines

In the previous section we illustrated how congruent triangles can be used to prove a pair of lines are parallel. Similarly, by first showing that an appropriate pair of segments or angles are congruent, a line may be demonstrated to be an angle or segment bisector or a pair of lines may be proved to be perpendicular.

To prove that a line bisects an angle or segment, we must show that the line divides the angle or segment into two congruent parts. A pair of lines may be proved to be perpendicular by showing either of the following:

- The lines intersect to form right angles.
- The lines intersect to form a pair of congruent adjacent angles.

For example, from the accompanying diagram, we may conclude that:

- $\overline{BX}$ bisects $\overline{AC}$ by first proving that $\overline{AX} \cong \overline{XC}$.
- $\overline{BX}$ bisects $\angle ABC$ by first proving that $\angle 1 \cong \angle 2$.
- $\overline{BX} \perp \overline{AC}$ by first proving that $\angle 3 \cong \angle 4$ (that is, by showing a pair of adjacent angles are congruent).

EXAMPLE 7.4

GIVEN: $\angle 1 \cong \angle 2$,
$\overline{RM} \cong \overline{TM}$.
PROVE: $\overline{SM}$ bisects $\angle RST$.

SOLUTION

PLAN: Our goal is to prove $\angle RSM \cong \angle TSM$ by proving $\triangle RSM \cong \triangle TSM$. Marking the diagram suggests that the SAS Postulate be applied.

PROOF:

Statements	Reasons
1. $\overline{RM} \cong \overline{TM}$. (Side)	1. Given.
2. $\angle 1 \cong \angle 2$. (Angle)	2. Given.
3. $\overline{SM} \cong \overline{SM}$. (Side)	3. Reflexive property of congruence.
4. $\triangle RSM \cong \triangle TSM$.	4. SAS Postulate.
5. $\angle RSM \cong \angle TSM$.	5. CPCTC.
6. $\overline{SM}$ bisects $\angle RST$.	6. A segment that divides an angle into two congruent angles is an angle bisector. (NOTE. This is the *reverse* of the definition of angle bisector.)

EXAMPLE 7.5

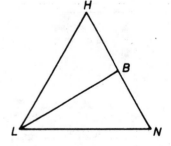

GIVEN: $\overline{LH} \cong \overline{LN}$, $\overline{LB}$ bisects $\overline{HN}$.
PROVE: $\overline{LB} \perp \overline{HN}$.

SOLUTION

PLAN: Show $\angle LBH \cong \angle LBN$ by showing $\triangle LBH \cong \triangle LBN$ using SSS.

PROOF:

Statements	Reasons
1. $\overline{LH} \cong \overline{LN}$. (Side)	1. Given.
2. $\overline{LB}$ bisects $\overline{HN}$.	2. Given.
3. $\overline{BH} \cong \overline{BN}$. (Side)	3. A bisector divides a segment into two congruent segments.
4. $\overline{LB} \cong \overline{LB}$. (Side)	4. Reflexive property of congruence.
5. $\triangle LBH \cong \triangle LBN$.	5. SSS Postulate.
6. $\angle LBH \cong \angle LBN$.	6. CPCTC.
7. $\overline{LB} \perp \overline{HN}$.	7. If two lines intersect to form congruent adjacent angles, then the lines are perpendicular.

SUMMARY

- To prove a line bisects a segment (or an angle) show that it divides the segment (or angle) into two congruent segments (or angles).
- To prove a line is perpendicular to another line show that the lines meet to form right angles or, equivalently, that a pair of adjacent angles are congruent.

Classifying Triangles and Special Segments

In addition to classifying a triangle as acute, right, or obtuse by the measures of its angles, we may classify a triangle according to the number of its sides that are congruent. See Figure 7.2.

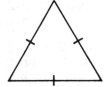

A *scalene* triangle has no congruent sides.

An *isosceles* triangle has at least two congruent sides.

An *equilateral* triangle has three congruent sides.

FIGURE 7.2

Notice that an equilateral triangle is also isosceles. Some further definitions regarding the parts of an isosceles triangle are given in Figure 7.3.

FIGURE 7.3 Legs are the congruent sides; vertex angle is the angle included between the legs; base is the side opposite the vertex angle; base angles are the angles that include the base and lie opposite the legs.

DRAWING AUXILIARY LINES (SEGMENTS)

To be able to complete a proof, it may be necessary to draw another line. This "extra" line is sometimes referred to as an *auxiliary line*. For example, an auxiliary line is needed to prove that the sum of the measures of the angles of a triangle is 180. Recall that this proof requires that a line be drawn that satisfies two conditions: it passes through a vertex of the given triangle, and at the same time it is parallel to the side opposite this vertex.

Sometimes it may be necessary to draw other types of auxiliary lines. Given one or more conditions, is it always possible to draw an auxiliary line that satisfies these conditions? If so, how many different lines can be drawn that satisfy the given set of conditions?

A particular set of geometric conditions may make the existence of a desired auxiliary line (or segment) *determined, underdetermined*, or *overdetermined*. An auxiliary line is said to be:

- *Determined* if exactly one line can be drawn that satisfies the given set of conditions. For example, an auxiliary line drawn so that it bisects a given angle is determined since every angle has exactly one bisector. Figure 7.4 illustrates that, through a point not on a line segment, a line can be drawn parallel to the segment, perpendicular to the segment, or to the midpoint of the segment.

- *Underdetermined* if too few conditions are given, so that more than one line can be drawn to satisfy the conditions. For example, in Figure 7.5 more than one line can be drawn through vertex B of triangle ABC so that the line intersects side $\overline{AC}$.

FIGURE 7.4 Each of these lines (or segments) is *determined* since, based on the given conditions, *exactly one* such line (segment) can be drawn.

- *Overdetermined* if too many conditions are given, so that it is not necessarily possible to draw one line that simultaneously meets all of these conditions. For example, in Figure 7.6 we cannot be sure that the line we draw through vertex *B* of triangle *ABC* will be a perpendicular bisector of side $\overline{AC}$.

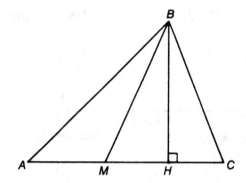

FIGURE 7.5 The line is *underdetermined* since more than one such line may be drawn.

FIGURE 7.6 The line is *overdetermined* since we cannot be assured that the perpendicular dropped from point *B* will also intersect $\overline{AC}$ at its midpoint.

For the accompanying figure, classify each line you draw as underdetermined, determined, or overdetermined.

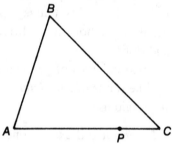

 a. Draw $\overline{BP}$.
 b. Through point *B* draw a line that bisects $\overline{AC}$.
 c. Through point *B* draw a line that intersects $\overline{AC}$.
 d. Draw a line through point *A* and parallel to $\overline{BC}$.
 e. Draw $\overline{BP}$ so that $\overline{BP}$ bisects angle *ABC*.
 f. Through point *B* draw a line that is perpendicular to $\overline{AC}$.

SOLUTION

(a) Determined	**(b)** Determined	**(c)** Underdetermined
(d) Determined	**(e)** Overdetermined	**(f)** Determined

ALTITUDES AND MEDIANS

In every triangle a median and an altitude can be drawn from any vertex to the side opposite that vertex.

- A *median* of a triangle is a segment drawn from a vertex of the triangle to the midpoint of the opposite side.

- An *altitude* of a triangle is a segment drawn from a vertex of the triangle perpendicular to the opposite side or, as in the accompanying diagram, to the opposite side extended.

Altitudes and medians of triangles are determined segments. In any triangle, three medians and three altitudes can be drawn.

EXAMPLE 7.7 Prove that the median drawn to the base of an isosceles triangle bisects the vertex angle.

SOLUTION

When confronted with a verbal statement of a problem requiring a formal proof, our first concern is to draw a suitable diagram. The following general approach is suggested:

1. Rewrite, if necessary, the problem statement in "if . . . then" form. For example,

 PROVE: If a median is drawn to the base of an isosceles triangle, then the vertex angle is bisected.

2. Identify the hypothesis (the Given), which is contained in the "If clause."
3. Identify the conclusion (the Prove), which is contained in the "then clause."
4. Draw and label an appropriate diagram:

 GIVEN: $\overline{AB} \cong \overline{AC}$,
 $\overline{AM}$ is a median to side $\overline{BC}$.
 PROVE: $\overline{AM}$ bisects $\angle BAC$.

5. Proceed as usual by arriving at a plan before writing the formal two-column proof.

 PLAN: $\overline{BM} \cong \overline{CM}$ since a median bisects the segment to which it is drawn. $\triangle AMB \cong \triangle AMC$ by the SSS Postulate. Angles 1 and 2 are congruent by CPCTC. $\overline{AM}$ bisects angle BAC (a segment that divides an angle into two congruent angles is an angle bisector). The formal two-column proof is left for you.

The Isosceles Triangle

THE BASE ANGLES THEOREM

After drawing several isosceles triangles (Figure 7.7), you may suspect that there is a relationship between the measures of the angles that lie opposite the congruent sides

THEOREM 7.1 BASE ANGLES THEOREM

If two sides of a triangle are congruent, then the angles opposite those sides are congruent.

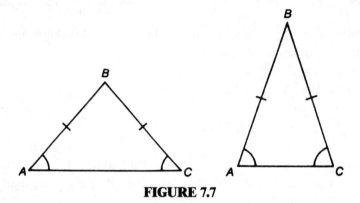

FIGURE 7.7

The proof is easy after an auxiliary line is drawn.

GIVEN: $\overline{AB} \cong \overline{CB}$.

PROVE: $\angle A \cong \angle C$.

PLAN: From past experience we look to prove angle A congruent to angle C by proving that these angles are corresponding angles of congruent triangles. To form two triangles that can be proved congruent, we may draw the bisector of angle ABC, which will intersect the base at some point, say R. The resulting triangles may be proved congruent by the SAS Postulate. (NOTE: It is also possible to prove this theorem by drawing an altitude or median to side $\overline{AC}$.)

PROOF:

Statements	Reasons
1. $\overline{AB} \cong \overline{CB}$. (Side)	1. Given.
2. Draw the bisector of angle ABC, naming the point at which it intersects $\overline{AC}$, point R.	2. An angle has exactly one bisector.
3. $\angle ABR \cong \angle CBR$. (Angle)	3. An angle bisector divides an angle into two congruent angles.
4. $\overline{BR} \cong \overline{BR}$. (Side)	4. Reflexive property of congruence.
5. $\triangle ABR \cong \triangle CBR$.	5. SAS Postulate.
6. $\angle A \cong \angle C$.	6. CPCTC.

EXAMPLE 7.8 The measure of the vertex angle of an isosceles triangle is three times as great as the measure of a base angle. Find the measure of a base angle of the triangle.

SOLUTION

$3x + x + x = 180$
$5x = 180$
$x = 36$

EXAMPLE 7.9

GIVEN: $\overline{SR} \cong \overline{ST}$,
$\overline{MP} \perp \overline{RS}, \overline{MQ} \perp \overline{ST}$,
M is the midpoint of $\overline{RT}$.
PROVE: $\overline{MP} \cong \overline{MQ}$.

SOLUTION

PLAN: By application of Theorem 7.1,
$\angle R \cong \angle T$. Marking the diagram suggests that triangles MPR and MQT may be proved congruent by using the AAS Theorem.

PROOF:

Statements	Reasons
1. $\overline{SR} \cong \overline{ST}$.	1. Given.
2. $\angle R \cong \angle T$. (Angle)	2. If two sides of a triangle are congruent, then the angles opposite those sides are congruent.
3. $\overline{MP} \perp \overline{RS}, \overline{MQ} \perp \overline{ST}$.	3. Given.
4. Angles MPR and MQT are right angles.	4. Perpendicular lines intersect to form right angles.
5. $\angle MPR \cong \angle MQT$. (Angle)	5. All right angles are congruent.
6. M is the midpoint of $\overline{RT}$.	6. Given.
7. $\overline{RM} \cong \overline{TM}$. (Side)	7. A midpoint divides a segment into two congruent segments.
8. $\triangle MPR \cong \triangle MQT$.	8. AAS Theorem.
9. $\overline{MP} \cong \overline{MQ}$.	9. CPCTC.

PROVING A TRIANGLE IS ISOSCELES

The converse of the Base Angles Theorem is also a useful theorem.

THEOREM 7.2 CONVERSE OF THE BASE ANGLES THEOREM

If two angles of a triangle are congruent, then the sides opposite are congruent.

GIVEN: ∡A ≅ ∡C.
PROVE: $\overline{AB}$ ≅ $\overline{CB}$.

PLAN: Draw the bisector of angle B, intersecting side $\overline{AC}$ at R. The resulting pair of triangles may be proven congruent by the AAS Theorem. It follows that $\overline{AB}$ ≅ $\overline{CB}$ by CPCTC. (Other auxiliary lines, such as a median or altitude, may also be drawn.)

 Theorem 7.2 is particularly useful in proving that a triangle is isosceles. To prove that a triangle is isosceles, show either of the following:
- A pair of sides are congruent.
- A pair of angles are congruent (since by Theorem 7.2 the sides opposite must be congruent).

EXAMPLE 7.10

GIVEN: $\overline{WR}$ ∥ $\overline{ST}$,
 $\overline{WR}$ bisects ∡SRE.
PROVE: △SRT is isosceles.

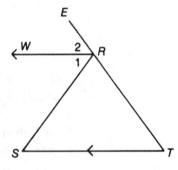

SOLUTION

PLAN: Show the base angles are congruent to each other by showing that each is congruent to one of the pairs of angles formed by the angle bisector. Since the base angles are congruent to congruent angles, they are congruent to each other and the triangle is isosceles.

PROOF:

Statements	Reasons
1. $\overline{WR}$ bisects ∡ SRE.	1. Given.
2. ∡1 ≅ ∡2.	2. A bisector divides an angle into two congruent angles.
3. $\overline{WR}$ ∥ $\overline{ST}$.	3. Given.
4. ∡1 ≅ ∡S.	4. If two lines are parallel, then their alternate interior angles are congruent.
5. ∡2 ≅ ∡T.	5. If two lines are parallel, then their corresponding angles are congruent.
6. ∡S ≅ ∡T.	6. Transitive property of congruence.
7. △SRT is isosceles.	7. A triangle that has a pair of congruent angles is isosceles.

 EXAMPLE 7.11 Prove that, if two altitudes of a triangle are congruent, then the triangle is isosceles.

SOLUTION

GIVEN: $\overline{CD}$ is the altitude to $\overline{AB}$,
$\overline{AE}$ is the altitude to $\overline{BC}$,
$\overline{CD} \cong \overline{AE}$.

PROVE: $\triangle ABC$ is isosceles.

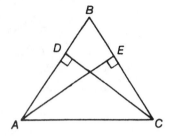

PLAN: Our goal is to show $\angle BAC = \angle BCA$ by proving $\triangle ADC \cong \triangle CEA$. Marking the diagram suggests the Hy-Leg Postulate be used:

$$\overline{AC} \cong \overline{AC} \text{ (Hy)} \quad \text{and} \quad \overline{CD} \cong \overline{AE} \text{ (Leg)}$$

PROOF:

Statements	Reasons
1. $\overline{CD}$ is the altitude to $\overline{AB}$, $\overline{AE}$ is the altitude to $\overline{BC}$.	1. Given.
2. Triangles ADC and CEA are right triangles.	2. A triangle that contains a right angle is a right triangle. (NOTE: This step consolidates several obvious steps.)
3. $\overline{CD} \cong \overline{AE}$. (Leg)	3. Given.
4. $\overline{AC} \cong \overline{AC}$. (Hy)	4. Reflexive property of congruence.
5. $\triangle ADC \cong \triangle CEA$.	5. Hy-Leg Postulate.
6. $\angle BAC \cong \angle BCA$.	6. CPCTC.
7. $\triangle ABC$ is isosceles.	7. A triangle that has a pair of congruent angles is isosceles.

Double Congruence Proofs

In some problems it may appear that not enough information is provided in the Given to prove a pair of triangles congruent. Upon closer examination, however, it may be possible to prove another pair of triangles congruent in order to obtain congruent corresponding parts, which can then be used to prove the original pair of triangles congruent. These problems tend to be difficult and will require a certain amount of trial-and-error work on your part.

 EXAMPLE 7.12

GIVEN: $\overline{AB} \cong \overline{CB}$,
E is the midpoint of $\overline{AC}$.

PROVE: $\triangle AED \cong \triangle CED$.

SOLUTION

PLAN: 1. Based on the Given, △AED could be proved congruent to △CED if it was known that ∡AED was congruent to ∡CED.

2. Triangles AEB and CEB contain these angles as parts and can be proved congruent by the SSS Postulate.

3. By CPCTC, ∡AED ≅ ∡CED.

4. △AED ≅ △CED by SAS.

PROOF:

Statements	Reasons
Part I. To Prove △AEB ≅ △CEB:	
1. $\overline{AB} \cong \overline{CB}$. (Side)	1. Given.
2. E is the midpoint of $\overline{AC}$.	2. Given.
3. $\overline{AE} \cong \overline{CE}$. (Side)	3. A midpoint divides a segment into two congruent segments.
4. $\overline{BE} \cong \overline{BE}$.	4. Reflexive property of congruence.
5. △AEB ≅ △CEB.	5. SSS Postulate.
Part II. To Prove △AED ≅ △CED:	
6. ∡AED ≅ ∡CED. (Angle)	6. CPCTC.
7. $\overline{DE} \cong \overline{DE}$. (Side)	7. Reflexive property of congruence.
8. △AED ≅ △CED.	8. SAS Postulate.

EXAMPLE 7.13

GIVEN: $\overline{BC} \cong \overline{AD}$,
$\overline{BC} \parallel \overline{AD}$,
$\overline{AR} \cong \overline{CS}$.

PROVE: $\overline{BR} \cong \overline{DS}$.

SOLUTION

PLAN: 1. The desired pair of segments can be proven congruent if it can be proven that △BRS ≅ △DSR. (*NOTE:* Although the desired pair of segments are also contained in triangles ARB and CSD, efforts to prove these triangles congruent would prove fruitless.)

2. By first proving △ARD ≅ △CSB we may obtain the congruent parts necessary to prove the desired pair of triangles congruent. Triangles ARD and CSB are congruent by SAS:

3. By CPCTC, $\overline{RD} \cong \overline{SB}$ and ∡1 ≅ ∡2. Since supplements of congruent angles are congruent, angles 3 and 4 are congruent.

4. $\overline{RS} \cong \overline{RS}$ so △BRS ≅ △DSR by SAS, and $\overline{BR} \cong \overline{DS}$ by CPCTC.

The formal proof is left for you.

EXAMPLE 7.14

GIVEN: $\overline{AB} \cong \overline{AC}$, $\overline{BD} \cong \overline{CE}$,
$\overline{BF}$ and $\overline{CG}$ are drawn perpendicular to $\overline{AD}$ and $\overline{AE}$, respectively.

PROVE: $\overline{DF} \cong \overline{EG}$.

SOLUTION

First prove $\triangle ABD \cong \triangle ACE$ in order to obtain an additional pair of congruent parts needed to prove that $\triangle DFB \cong \triangle EGC$. Here is a detailed plan:

PLAN: 1. $\triangle ABD \cong \triangle ACE$ by SAS since $\overline{AB} \cong \overline{AC}$, $\angle ABD \cong \angle ACE$ (supplements of the congruent base angles are congruent), and $\overline{BD} \cong \overline{CE}$. Therefore, $\angle D \cong \angle E$.

2. $\triangle DFB \cong \triangle EGC$ by AAS since $\angle D \cong \angle E$, $\angle DFB \cong \angle EGC$ (right angles are congruent), and $\overline{BD} \cong \overline{CE}$. Therefore, $\overline{DF} \cong \overline{EG}$ by CPCTC.

REVIEW EXERCISES FOR CHAPTER 7

1. Find the value of x:

2. In the accompanying diagram, $\overline{BDE}$, $\overleftrightarrow{AB} \parallel \overleftrightarrow{CD}$, and $\overleftrightarrow{DC}$ bisects $\angle ADE$. Triangle ABD must be

 (1) scalene (3) equiangular

 (2) isosceles (4) right

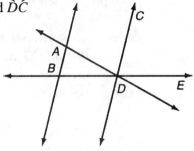

3. GIVEN: $\overline{FH}$ bisects $\angle GHJ$,
 $\overline{GH} \cong \overline{JH}$.

 PROVE: $\angle 1 \cong \angle 2$.

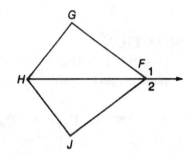

4. GIVEN: $\overline{AC} \cong \overline{BD}$,
 $\overline{AB} \perp \overline{BC}$, $\overline{DC} \perp \overline{BC}$.

 PROVE: $\angle 1 \cong \angle 2$.

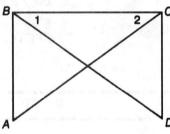

Use the accompanying diagram to solve Exercises 5 and 6.

5. GIVEN: $\angle J \cong \angle K$,
 $\overline{PJ} \cong \overline{PK}$.

 PROVE: $\overline{JY} \cong \overline{KX}$.

6. GIVEN: $\overline{PX} \cong \overline{PY}$,
 $\overline{XJ} \cong \overline{YK}$.

 PROVE: $\overline{KX} \cong \overline{JY}$.

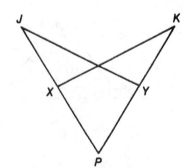

7. GIVEN: $\overline{UT} \parallel \overline{DW}$, $\overline{UT} \cong \overline{DW}$,
 $\overline{QW} \cong \overline{AT}$.

 PROVE: $\overline{UQ} \parallel \overline{AD}$.

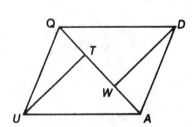

8. GIVEN: $\angle S \cong \angle H$,
 $\overline{SR} \perp \overline{RW}$, $\overline{HW} \perp \overline{RW}$,
 $\overline{ST} \cong \overline{HT}$.

 PROVE: T is the midpoint of $\overline{RW}$.

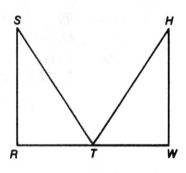

9. GIVEN: $\overline{AB} \cong \overline{CB}$, $\overline{AD} \cong \overline{CD}$.
 PROVE: $\overline{DB}$ bisects $\angle ADC$.

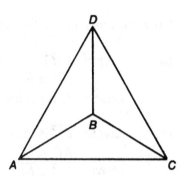

10. GIVEN: $\overline{BA} \perp \overline{MA}$, $\overline{CD} \perp \overline{MD}$,
 M is the midpoint of $\overline{BC}$.
 PROVE: $\overline{BC}$ bisects $\overline{AD}$.

11. GIVEN: $\angle 1 \cong \angle 2$,
 $\overrightarrow{HK}$ bisects $\angle RHN$,
 $\overline{HR} \cong \overline{HN}$.
 PROVE: $\overline{HK} \perp \overline{RN}$.

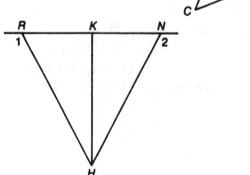

12. GIVEN: $\angle 1$ is supplementary to $\angle 2$,
 $\overline{RC} \cong \overline{AT}$,
 $\overline{SR} \cong \overline{AB}$,
 $\overline{ST} \cong \overline{BC}$.
 PROVE: 1. $\overline{ST} \perp \overline{TR}$.
 2. $\overline{BC} \perp \overline{AC}$.

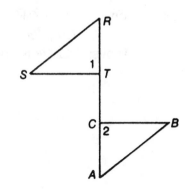

13. GIVEN: $\overline{RS} \perp \overline{SL}$, $\overline{RT} \perp \overline{LT}$,
 $\overline{RS} \cong \overline{RT}$.
 PROVE: 1. $\triangle RLS \cong \triangle RLT$.
 2. $\overline{WL}$ bisects $\measuredangle SWT$.

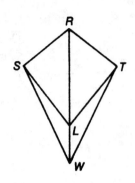

Use the accompanying diagram to solve Exercises 14 and 15.

14. GIVEN: $\overline{PM}$ is the altitude to $\overline{KL}$,
 $\overline{PK} \cong \overline{PL}$.
 PROVE: M is the midpoint of $\overline{KL}$.

15. GIVEN: $\overline{PM}$ is the median to $\overline{KL}$,
 $\overline{KP} \cong \overline{LP}$.
 PROVE: $\overline{PM} \perp \overline{KL}$.

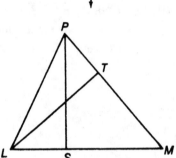

16. GIVEN: $\overline{PS}$ and $\overline{LT}$ are altitudes
 to sides $\overline{LM}$ and $\overline{PM}$, respectively.
 $\measuredangle LPT \cong \measuredangle SLP$.
 PROVE: $\overline{PS} \cong \overline{LT}$.

17. GIVEN: $\overline{XL} \cong \overline{XP}$,
 $\overline{XL} \perp \overline{TR}$, $\overline{XP} \perp \overline{TS}$,
 X is the midpoint of $\overline{RS}$.
 PROVE: $\triangle RTS$ is isosceles.

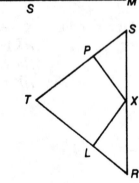

18. GIVEN: $\overline{QL} \cong \overline{QM}$,
 $\overline{LM} \parallel \overline{PR}$.
 PROVE: $\triangle PQR$ is isosceles.

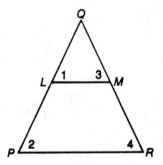

19. GIVEN: $\overline{LW} \perp \overline{TW}$, $\overline{FX} \perp \overline{XP}$,

$\overline{TF} \cong \overline{PL}$, $\overline{WL} \cong \overline{XF}$.

PROVE: $\triangle FML$ is isosceles.

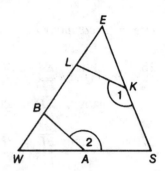

20. GIVEN: $\overline{SE} \cong \overline{SW}$,

$\angle 1 \cong \angle 2$,

$\overline{EL} \cong \overline{WB}$.

PROVE: $\overline{KL} \cong \overline{AB}$.

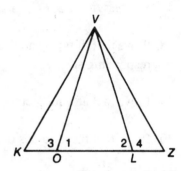

21. GIVEN: $\overline{OV} \cong \overline{LV}$,

$\overline{KO} \cong \overline{ZL}$.

PROVE: $\triangle KVZ$ is isosceles.

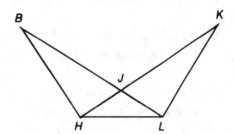

22. GIVEN: $\angle JHL \cong \angle JLH$,

$\overline{BH} \perp \overline{HJ}$, $\overline{KL} \perp \overline{LJ}$.

PROVE: $\overline{JB} \cong \overline{JK}$.

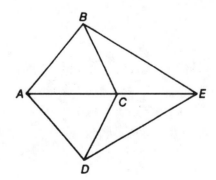

Use the accompanying diagram to solve Exercises 23 and 24.

23. GIVEN: $\overline{AB} \cong \overline{AD}$,

$\overline{EA}$ bisects $\angle DAB$.

PROVE: $\triangle BCE \cong \triangle DCE$.

24. GIVEN: $\overline{BE} \cong \overline{DE}$,

$\overline{BC} \cong \overline{DC}$.

PROVE: $\overline{CA}$ bisects $\angle DAB$.

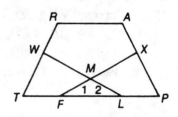

25. GIVEN: $\overline{AB} \parallel \overline{CD}$,
$\overline{AB} \cong \overline{CD}$,
$\overline{AL} \cong \overline{CM}$.

PROVE: $\angle CBL = \angle ADM$.

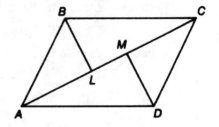

Use the accompanying diagram to solve Exercises 26 and 27.

26. GIVEN: $\angle FAC \cong \angle FCA$,
$\overline{FD} \perp \overline{AB}, \overline{FE} \perp \overline{BC}$.

PROVE: $\overline{BF}$ bisects $\angle DBE$.

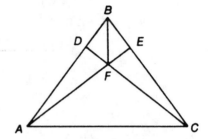

27. GIVEN: $\overline{BD} \cong \overline{BE}$,
$\overline{FD} \cong \overline{FE}$.

PROVE: $\triangle AFC$ is isosceles.

28. Prove that, if two triangles are congruent, then the altitudes drawn to a pair of corresponding sides are congruent.

29. Prove that an equilateral triangle is equiangular.

30. Prove that the altitudes drawn to the legs of an isosceles triangle are congruent.

31. Prove that the medians drawn to the legs of an isosceles triangle are congruent.

32. Prove that any point on the perpendicular bisector of a line segment is equidistant from the endpoints of the segment.

33. Prove that, if two points are each equidistant from the endpoints of a line segment, then the two points determine the perpendicular bisector of the line segment.

CUMULATIVE REVIEW EXERCISES: CHAPTERS 1–7

1. In an isosceles triangle, the measure of the vertex angle is 8 times the measure of one of the base angles. Find the number of degrees in the measure of a base angle of the triangle.

2. In the accompanying diagram, $\overleftrightarrow{AB}$, $\overleftrightarrow{CD}$, and $\overleftrightarrow{FH}$ intersect at E, and $\overleftrightarrow{AB}$ bisects $\angle FEC$. If m$\angle DEB = 25$, what is m$\angle DEF$?

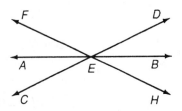

3. The measures of two complementary angles are represented by $2x$ and $3x - 10$. What is the value of x?

4. In the accompanying diagram, $\overleftrightarrow{AKB} \parallel \overleftrightarrow{CD}$, $\overline{AE} \perp \overline{CK}$, m$\angle KCD = 2x$, and m$\angle KAE = 3x$. What is m$\angle CKB$?

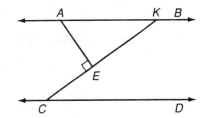

5. What is the measure of an exterior angle of a regular polygon having nine sides?

6. $\overleftrightarrow{AB} \parallel \overleftrightarrow{CD}$ and each line is intersected by $\overleftrightarrow{MN}$ at G and H, respectively. If m$\angle BGH = 2x + 50$ and m$\angle CHG = 5x - 70$, find x.

Use the accompanying diagram to solve Exercises 7 and 8.

7. GIVEN: $\angle 1 \cong \angle 2$,
 $\overline{AB} \cong \overline{BC}$,
 F is the midpoint of $\overline{AB}$,
 G is the midpoint of $\overline{BC}$.
 PROVE: $\overline{FD} \cong \overline{GE}$.

8. GIVEN: $\angle 1 \cong \angle 2$,
 $\overline{FD} \perp \overline{AC}$, $\overline{GE} \perp \overline{AC}$,
 $\overline{AE} \cong \overline{CD}$.
 PROVE: $\triangle ABC$ is isosceles.

9. GIVEN: $\overline{RS}$ intersects $\overleftrightarrow{ARB}$ and $\overleftrightarrow{CTS}$,
 $\overleftrightarrow{ARB} \parallel \overleftrightarrow{CTS}$,
 $\overline{RT}$ bisects $\angle BRS$,
 M is the midpoint of $\overline{RT}$,
 $\overline{SM}$ is drawn.
 PROVE: **a** $\overline{RS} \cong \overline{ST}$.
 b $\overline{SM}$ bisects $\angle RST$.

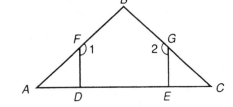

10. GIVEN: $\triangle ABC$, $\overline{CM}$ is the median to $\overline{AB}$, $\overline{CM}$ is extended to P so that $\overline{CM} \cong \overline{MP}$, and $\overline{AP}$ is drawn.

 PROVE: $AP \parallel CB$.

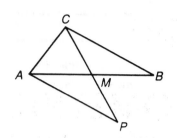

11. GIVEN: $\overline{ADFB}$, $\overline{AGC}$, $\overline{BEC}$, $\overline{AD} \cong \overline{BF}$, $\angle x \cong \angle y$, and $\angle A \cong \angle B$.

 PROVE: **a** $\triangle AFG \cong \triangle BDE$

 b $\overline{GC} \cong \overline{EC}$.

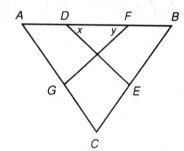

12. GIVEN: $\overline{KPQRL}$, $\overline{MQN}$, $\overline{KM}$, $\overline{NL}$, $\overline{MP}$, $\overline{NR}$, $\overline{KL}$ and $\overline{MN}$ bisect each other at Q, $\angle 1 \cong \angle 2$.

 PROVE: $\overline{PM} \cong \overline{NR}$.

13. GIVEN: $\triangle ABC$, $\overline{CEA} \cong \overline{CDB}$, $\overline{AD}$ and $\overline{BE}$ intersect at P, and $\angle PAB \cong \angle PBA$.

 PROVE: $\overline{PE} \cong \overline{PD}$.

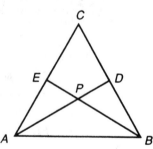

14. GIVEN: $\triangle ABC$, $\overline{AC} \cong \overline{BC}$, $\overline{CE} \cong \overline{CD}$, in $\triangle BCD$, $\overline{DF}$ is a median to $\overline{BC}$, in $\triangle ACE$, $\overline{EG}$ is a median to $\overline{AC}$.

 PROVE: $\overline{EG} \cong \overline{DF}$.

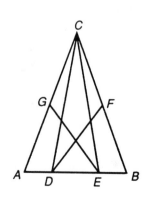

GEOMETRIC INEQUALITIES

WHAT YOU WILL LEARN

If angles or sides are not congruent, it is often helpful to compare the measures of these angles or sides using inequality statements. Geometric inequalities are often proved by an indirect method. In this chapter you will learn:

- the inequality relationship between the measure of an exterior angle of a triangle and the measure of either of the two remote interior angles;
- the relationship between unequal angle measures and unequal side lengths in a triangle;
- the way the length of any side of a triangle compares to the sum of the lengths of the other two sides;
- the application of the indirect method of proof.

SECTIONS IN THIS CHAPTER

Some Basic Properties of Inequalities

Inequality Relationships in a Triangle

The Indirect Method of Proof

Some Basic Properties of Inequalities

In everyday life we are constantly making comparisons. Which of two people earns more money? Which of two athletes can run faster? Which of two students has the higher grade average? Comparisons also play an important role in mathematics. In comparing two quantities, say a and b, there are exactly three possibilities, which are summarized in Table 8.1.

TABLE 8.1

Condition	Notation	Example
a is less than b.	$a < b$	$3 < 5$
a is equal to b.	$a = b$	$4 = 4$
a is greater than b.	$a > b$	$7 > 2$

The *direction* or *sense* of an inequality refers to whether the inequality symbol is pointing to the right ($>$) or to the left ($<$). For example:

- $a > b$ and $c > d$ are two inequality expressions that have the same direction or *sense* since they *both* indicate that one quantity is greater than the other.
- By changing $a > b$ to $a < b$ (or vice versa), we have reversed the direction or *sense* of the inequality.

We shall assume that a number of properties related to inequalities are true. See Table 8.2.

TABLE 8.2

Property	Formal Statement	Example
Addition	If $a < b$, then $a + c < b + c$.	$\begin{array}{r} 3 < 5 \\ +\,4 = 4 \\ \hline 7 < 9 \end{array}$
	If $a < b$ and $c < d$, then $a + c < b + d$.	$\begin{array}{r} 3 < 5 \\ +\,6 < 10 \\ \hline 9 < 15 \end{array}$
Subtraction	If $a < b$, then $a - c < b - c$.	$\begin{array}{r} 8 < 13 \\ -\,6 = -6 \\ \hline 2 < 7 \end{array}$
Multiplication	If $a < b$ and $c > 0$, then $ac < bc$.	$5 < 8$ $2(5) < 2(8)$ or $10 < 16$
Transitive	If $a < b$ and $b < c$, then $a < c$.	$4 < 7$ and $7 < 10$ then $7 < 10$.
Substitution	If $a + b < c$, and $b = d$, then $a + d < c$.	$x + y < 9$ $y = 5$ $x + 5 < 9$

Keep in mind that:

- Although in Table 8.2 each property is expressed in terms of the less than relation (<), the properties clearly hold for the greater than (and equal) relation.
- As a corollary to the multiplication property, "halves of unequals are unequal." For example, if $a < b$, then $a/2 < b/2$.
- $a < b$ and $b > a$ are equivalent expressions.

Here are two geometric inequalities that should be obvious:

1. If point X is *between* points A and B, then $AX < AB$ and $XB < AB$.

2. If ray $\overrightarrow{BX}$ is *between* rays $\overrightarrow{BA}$ and $\overrightarrow{BC}$ and $\overrightarrow{BX}$ lies in the interior of $\angle ABC$, then $m\angle ABX < m\angle ABC$ and $m\angle CBX < m\angle ABC$.

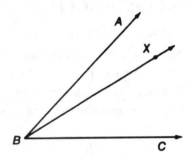

Inequality Relationships in a Triangle

In Figure 8.1, the shortest trip from city A to city B is the direct route represented by segment AB. This direct route must be less than any indirect route, such as that from city A to city C and then from city C to city B. In symbols:

$$AB < AC + CB$$

FIGURE 8.1

TRIANGLE INEQUALITY POSTULATE

The length of each side of a triangle must be less than the sum of the lengths of the other two sides.

This postulate provides us with a convenient method for determining whether a set of three numbers can represent the lengths of the sides of a triangle. All we need do is test that *each* number is less than the sum of the other two.

 EXAMPLE 8.1 Which of the following sets of numbers *cannot* represent the sides of a triangle?
 (1) 9, 40, 41 (2) 7, 7, 3 (3) 4, 5, 1 (4) 6, 6, 6

SOLUTION
 Choice (3): $4 < 5 + 1$ and $1 < 4 + 5$, but 5 is *not* less than $4 + 1$.

COMPARING ANGLES OF A TRIANGLE

The Base Angles Theorem (Theorem 7.1) tells us that, if two sides of a triangle are congruent, then the angles opposite these sides are congruent (see Figure 8.2a). What conclusion can be drawn if the two sides of the triangle are *not* congruent? In Figure 8.2b, clearly the angles cannot be congruent. Furthermore, it appears that the greater angle lies opposite the longer side.

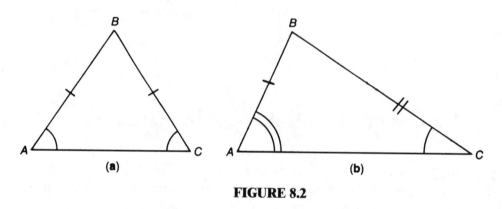

FIGURE 8.2

THEOREM 8.1 ≢ SIDES IMPLIES ≢ OPPOSITE ANGLES

If two sides of a triangle are not congruent, then the angles opposite these sides are not congruent, and the greater angle is opposite the longer side.

OUTLINE OF PROOF
 GIVEN: $BC > BA$.
 PROVE: $m\angle BAC > m\angle C$.

PLAN Since $BC > BA$, there exists a point D on $\overline{BC}$ such that $AB = DB$.
- Drawn $\overline{AD}$. (See diagram.)
- $m\angle 1 > m\angle C$ (Exterior Angle Inequality Theorem, Theorem 5.4.)
- $m\angle 1 = m\angle 2$ (Base Angles Theorem, Theorem 7.1.)
- $m\angle 2 > m\angle C$ (Substitution.)
- $m\angle BAC > m\angle 2$ (Definition of betweenness of rays; see diagram.)
- $m\angle BAC > m\angle C$ (Transitive property.)

The converse of Theorem 8.1 is also true.

THEOREM 8.2 ≠ ANGLES IMPLIES ≠ OPPOSITE SIDES

If two angles of a triangle are not congruent, then the sides opposite these angles are not congruent, and the longer side is opposite the greater angle.

EXAMPLE 8.2 In $\triangle ABC$, $AB = 3$, $BC = 5$, and $AC = 7$. What is the largest angle of the triangle? The smallest angle of the triangle?

SOLUTION
$\angle B$ is the largest angle.
$\angle C$ is the smallest angle.

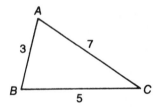

EXAMPLE 8.3 The measure of the vertex angle S of isosceles triangle RST is 80. What is the longest side of the triangle?

SOLUTION
$\overline{RT}$ is the longest side.

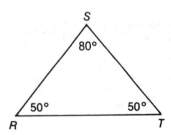

EXAMPLE 8.4 GIVEN: $\overline{BD}$ bisects $\angle ABC$.
PROVE: $AB > AD$.

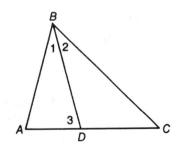

SOLUTION

PLAN: By Theorem 8.2, in order to prove $AB > AD$ we must first establish that $m\angle 3$ (the angle opposite $\overline{AB}$) is greater than $m\angle 1$ (the angle opposite $\overline{AD}$).

PROOF:

Statements	Reasons
1. $m\angle 3 > m\angle 2$.	1. The measure of an exterior angle of a triangle is greater than the measure of either nonadjacent interior angle.
2. $\overline{BD}$ bisects $\angle ABC$.	2. Given.
3. $m\angle 1 = m\angle 2$.	3. A bisector divides an angle into two angles having the same measure.
4. $m\angle 3 > m\angle 1$.	4. Substitution property of inequalities.
5. $AB > AD$.	5. If two angles of a triangle are not equal in measure, then the sides opposite are not equal and the longer side is opposite the greater angle.

The Indirect Method of Proof

If Sherlock Holmes had three suspects, he might arrive at the identity of the criminal *indirectly* by proving that two of the suspects could not possibly have been at the scene of the crime, thereby indirectly establishing the guilt of the only remaining suspect. Needless to say, the validity of Mr. Holmes' method is based on the assumption that he correctly surmised that there were exactly three suspects.

The indirect method of proof in mathematics is based on a similar strategy. To prove a statement *indirectly*, we account for all possibilities (suspects). We then consider each of the supposed possibilities (except the one to be proved) and show that each of these is actually impossible. The result is that the remaining possibility (the desired conclusion) must be true.

EXAMPLE 8.5 Use an indirect method of proof to show that a triangle has at most one obtuse angle.

SOLUTION

INDIRECT PROOF
• There are exactly two possibilities:
 1. The triangle has *at most* one obtuse angle.
 2. The triangle has *more than* one obtuse angle.

- Since our goal is to establish that the first possibility is true, we start by assuming the second possibility is true. We suppose that the triangle has two obtuse angles. Since the sum of these two obtuse angles exceeds 180, this situation is impossible.
- The second possibility is false. By elimination, the first possibility (which is the only one remaining) must be true.

To Prove a Statement Using the Indirect Method of Proof

1. Assume that the statement in the Prove is *not* true.
2. Show that this assumption contradicts a known fact.
3. Conclude that the statement in the Prove is true.

**EXAMPLE
8.6**

GIVEN: $\angle 1$ is not congruent to $\angle 2$.
PROVE: Line ℓ is not parallel to line m.

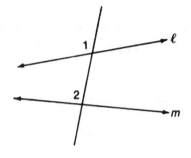

SOLUTION

Use an indirect method of proof:

1. There are two possibilities; the original statement and its negation: line ℓ is parallel to line m.
2. Assume the negation is true; that is, line ℓ *is* parallel to line m.
3. If the lines are parallel, then corresponding angles are congruent, so $\angle 1 \cong \angle 2$. But this contradicts the Given.
4. Since its negation is false, the statement "line ℓ is not parallel to line m" must be true since it is the only remaining possibility.

**EXAMPLE
8.7**

GIVEN: $\overline{AB} \cong \overline{DB}$.
PROVE: $\overline{AB}$ is not $\cong$ to $\overline{BC}$.

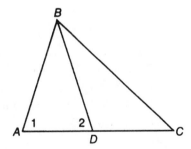

SOLUTION

INDIRECT Assume the negation of the desired conclusion is true; that is,
PROOF assume $\overline{AB}$ *is* $\cong$ to $\overline{BC}$. If $\overline{AB} \cong \overline{BC}$, then $\angle 1 \cong \angle C$. From the Given, $\angle 1 \cong \angle 2$. Hence, $\angle 2 \cong \angle C$. But this is impossible since the measure of an exterior angle of a triangle must be greater than the measure of either nonadjacent interior angle. Hence, the assumption is false and the only remaining possibility, that $\overline{AB}$ is not $\cong$ to $\overline{BC}$, must be true.

EXAMPLE
8.8

GIVEN: $\overline{TW} \perp \overline{RS}$, ∢1 is not ≅ to ∢2.

PROVE: $\overline{TW}$ is not the median to side $\overline{RS}$.

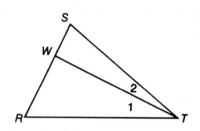

SOLUTION

INDIRECT
PROOF:
Assume the negation of the desired conclusion is true: assume $\overline{TW}$ is the median to side $\overline{RS}$. Then $\overline{WR} \cong \overline{WS}$ and $\triangle TWS$ is congruent to $\triangle TWR$ by SAS. By CPCTC, ∢1 ≅ ∢2. But this contradicts the Given. Hence, the only remaining possibility, that $\overline{TW}$ is *not* the median to side $\overline{RS}$, must be true.

Proofs that rely on the indirect method may also be organized in our familiar two-column format. The two-column format of the proof for Example 8.8 is as follows:

PROOF:

Statements	Reasons
1. ∢1 is *not* ≅ to ∢2.	1. Given.
2. Either $\overline{TW}$ is *not* the median to side $\overline{RS}$ or $\overline{TW}$ *is* the median to side $\overline{RS}$. *Assume $\overline{TW}$ is* the median to side $\overline{RS}$.	2. A statement is either true or false.
3. $\overline{WR} \cong \overline{WS}$. (Side)	3. A median divides a side into two congruent segments.
4. $\overline{TW} \perp \overline{RS}$.	4. Given.
5. Angles TWS and TWR are right angles.	5. Perpendicular lines meet to form right angles.
6. ∢TWS ≅ ∢TWR. (Angle)	6. All right angles are congruent.
7. $\overline{TW} \cong \overline{TW}$. (Side)	7. Reflexive property of congruence.
8. $\triangle TWS \cong \triangle TWR$.	8. SAS Postulate.
9. ∢1 ≅ ∢2.	9. CPCTC.
10. $\overline{TW}$ is *not* the median to side $\overline{RS}$.	10. Statement 9 contradicts statement 1. The assumption made in statement 2 must therefore be false. By elimination, statement 10 is established.

Another interesting application of the indirect method of proof provides an alternative means for developing the properties of parallel lines. Example 8.9 provides a proof for a statement that was originally postulated.

 EXAMPLE 8.9 If two lines are parallel, then their alternate interior angles are congruent.

GIVEN: $\ell \parallel m$.

PROVE: $\angle 1 \cong \angle 2$.

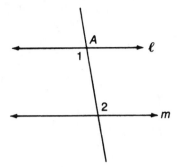

SOLUTION

Using an indirect method of proof, we assume that angle 1 is *not* congruent to angle 2. It is therefore possible to construct at point A a line k such that angle 3 is congruent to angle 2:

Since $\angle 3 \cong \angle 2$, line k is parallel to line m (if alternate interior angles are congruent, the lines are parallel). Through point A, line ℓ has been drawn parallel to line m (given) and line k has been drawn parallel to line m. This contradicts the Parallel Postulate, which says that exactly one such line can be drawn. Hence our assumption that angles 1 and 2 are not congruent leads to a contradiction, which implies that angles 1 and 2 must be congruent.

REVIEW EXERCISES FOR CHAPTER 8

1. In $\triangle ABC$, $m\angle A = 50$ and $m\angle B = 60$. Which is the longest side of the triangle?

2. In $\triangle ABC$, $m\angle A = 30$ and the measure of the exterior angle at B is 120. Which is the longest side of the triangle?

3. In right triangle ABC, altitude $\overline{CD}$ is drawn to hypotenuse $\overline{AB}$. Which is the longest side of $\triangle CDB$?

4. In $\triangle ABC$, $m\angle B = 120$, $m\angle A = 55$, and D is the point on $\overline{AC}$ such that $\overline{BD}$ bisects $\angle ABC$. Which is the longest side of $\triangle ABD$?

5. An exterior angle formed at vertex angle J of isosceles triangle JKL by extending leg $\overline{LJ}$ has a degree measure of 115. Which is the longest side of the triangle?

6. In $\triangle ABC$, $BC > AB$ and $AC < AB$. Which is the longest side of the triangle?

7. In $\triangle RST$ $m\angle R < m\angle T$ and $m\angle S > m\angle T$. Which is the largest angle of the triangle?

8. GIVEN: $\overline{AB} \cong \overline{BD}$,
 m∡5 > m∡6.

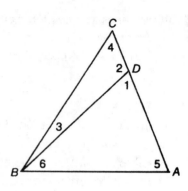

State whether each of the following inequality relationships is true or false.
 (a) m∡1 > m∡3. **(d)** m∡2 > m∡1.
 (b) m∡5 > m∡1. **(e)** BC > BA.
 (c) m∡3 < m∡ABC. **(f)** AB < AD.

9. In △ABC, AB > AC and BC > AC. Name the smallest angle of △ABC.

10. In △RST, ST > RT and RT > RS.
 (a) If one of the angles of the triangle is obtuse, which angle must it be?
 (b) If the measure of one of the angles of the triangle is 60, which angle must it be?

11. Determine whether each of the following sets of numbers can represent the lengths of the sides of a triangle.
 (a) 8, 17, 15 **(c)** 1, 1, 3
 (b) ½, ⅓, ⅙ **(d)** 6, 6, 7

12. GIVEN: $\overline{AB} \cong \overline{CB}$.
 PROVE: AB > BD.

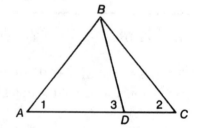

13. If the lengths of two sides of a triangle are 12 and 20 and the third side is represented by x, then
 (1) $x = 32$ (3) $x < 8$
 (2) $x > 32$ (4) $8 < x < 32$

14. GIVEN: m∡1 = m∡2.
 PROVE: AD > ED.

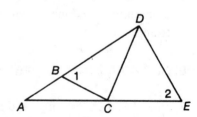

Use the accompanying diagram to solve Exercises 15 to 18.

15. GIVEN: Triangles *AEC* and *ABC*.
 PROVE: m∡4 > m∡ *AEC*.

16. GIVEN: *AC* > *BC*.
 PROVE: *AD* > *BD*.

17. GIVEN: *AD* > *BD*,
 AD bisects ∡*BAC*.
 PROVE: *AC* > *DC*.

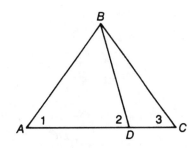

18. GIVEN: ∡1 ≅ ∡3.
 PROVE: m∡1 > m∡2.

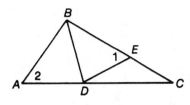

19. GIVEN: ∡1 ≅ ∡2.
 PROVE: *AB* ≇ *BC*.

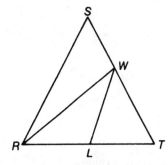

20. GIVEN: *RS* = *TS*.
 PROVE: *RW* ≠ *WL*.

21. GIVEN: △*ABC* is *not* isosceles,
 ∡*ADB* ≅ ∡*CDB*.
 PROVE: △*ADC* is *not* isosceles.

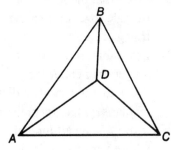

22. GIVEN: $AC > AB$,
 $\overline{DE} \cong \overline{CE}$.

 PROVE: $\overline{AB}$ is *not* parallel to $\overline{DE}$.

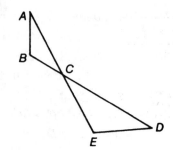

23. GIVEN: $\triangle ABC$ is scalene,
 $\overline{BD}$ bisects $\angle ABC$.

 PROVE: $\overline{BD}$ is *not* $\perp \overline{AC}$.

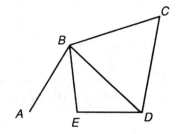

24. GIVEN: Quadrilateral $BCDE$, $\overline{BE}$ bisects $\angle ABD$,
 $\angle BDC \cong \angle C$, and $m\angle E > m\angle ABE$.

 PROVE: $BC > ED$.

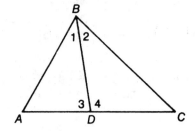

25. GIVEN: $\overline{AC} \cong \overline{BC}$, $\overline{AD} \cong \overline{BD}$, $\overline{AEC}$, and $\overline{BDE}$.

 PROVE: **a** $\angle CAD \cong \angle CBD$.
 b $AD > DE$.

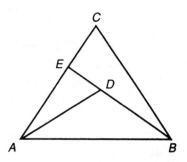

26. Prove that the length of the line segment drawn from any vertex of an equilateral triangle to a point on the opposite side is less than the length of any side of the triangle.

27. Prove that, if the vertex angle of an isosceles triangle is obtuse, then the base is longer than either leg.

28. Prove that the shortest distance from a point to a line is the length of the perpendicular segment from the point to the line.

29. Prove that an altitude of an acute scalene triangle *cannot* bisect the angle from whose vertex it is drawn.

SPECIAL QUADRILATERALS

WHAT YOU WILL LEARN

Quadrilaterals with one or two pairs of parallel sides have some interesting properties. In this chapter you will learn:

- the names and properties of special quadrilaterals;
- different ways of proving a quadrilateral is a parallelogram;
- different ways of proving a parallelogram is a rectangle, rhombus, or square;
- the relationship between a side of a triangle and the segment joining the midpoints of the two other sides;
- different ways of proving a trapezoid is isosceles.

Classifying Quadrilaterals

If you consider the relationships between the sides and the angle pairs of quadrilaterals to be genetic traits, then a family tree of quadrilaterals may be developed with a special name given to each type of quadrilateral that displays a special trait. This is illustrated in Figure 9.1.

The family tree of quadrilaterals shows that a quadrilateral has two major types of descendants. One, called a *trapezoid*, has exactly one pair of parallel sides. The other major type of quadrilateral has two pairs of parallel sides and is called a *parallelogram*.

DEFINITION OF A PARALLELOGRAM
A *parallelogram* is a quadrilateral having two pairs of parallel sides.
NOTATION: $\square ABCD$ is read as "parallelogram *ABCD*." The letters *A, B, C,* and *D* represent consecutive vertices of the quadrilateral, and the symbol that precedes these letters is a miniature parallelogram.

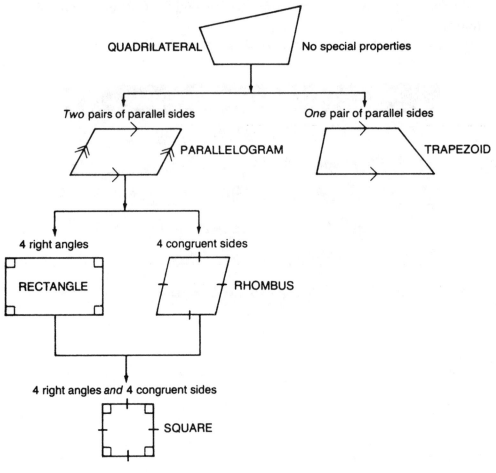

FIGURE 9.1

Of the special quadrilaterals illustrated in Figure 9.1, only the trapezoid is *not* a parallelogram. A rectangle, a rhombus, and a square are special types of parallelograms. The rest of this chapter is devoted to developing the properties of each of these figures.

Properties of a Parallelogram

ANGLES OF A PARALLELOGRAM

Consider parallelogram *ABCD* (Figure 9.2) in which the measure of angle *A* is 70. What are the measures of angles *B, C,* and *D*?

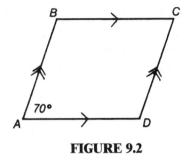

FIGURE 9.2

In parallelogram *ABCD*, $\overline{AB}$ may be considered to be a transversal, intersecting parallel line segments $\overline{AD}$ and $\overline{BC}$. Since interior angles on the same side of a transversal intersecting parallel lines are supplementary, m∡*B* = 110 (180 — 70 = 110). Similarly, angles *B* and *C* are supplementary, so m∡*C* = 70, and angles *A* and *D* are supplementary, so m∡*D* = 110.

Notice also that, since angles *A* and *C* are each supplementary to angle *B* (and angle *D*), they are congruent (recall that "If two angles are supplementary to the same angle, then they are congruent"). For the same reason, opposite angles *B* and *D* are congruent. The following two theorems summarize these results.

ANGLES OF A PARALLELOGRAM THEOREMS

THEOREM 9.1 Consecutive pairs of angles of a parallelogram are supplementary.

THEOREM 9.2 Opposite angles of a parallelogram are congruent.

 In parallelogram *ABCD* the measure of angle *B* is twice the measure of angle *A*. Find the measure of each angle of the parallelogram.

SOLUTION

Since angles *A* and *B* are consecutive angles of a parallelogram, by Theorem 9.1 they are supplementary. Hence, we may write

$$x + 2x = 180$$
$$3x = 180$$
$$x = 60$$
$$m\angle A = x = 60$$
$$m\angle B = 2x = 120$$
$$m\angle C = m\angle A = 60$$
$$m\angle D = m\angle B = 120$$

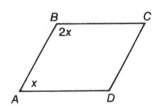

DIAGONALS AND SIDES OF A PARALLELOGRAM

If you draw either diagonal of a parallelogram, two triangles are formed. Is there a relationship between the triangles formed by a diagonal? Consider diagonal $\overline{BD}$, drawn in parallelogram $ABCD$, in Figure 9.3a. Notice that $\overline{BC}$, $\overline{BD}$, and $\overline{DA}$ form a Z. Since $\overline{BC}$ is parallel to $\overline{AD}$, $\angle 1 \cong \angle 2$. Similarly, $\overline{AB}$, $\overline{BD}$, and $\overline{DC}$ form a Z. Since, $\overline{AB} \parallel \overline{DC}$, $\angle 3 \cong \angle 4$. In addition, $\overline{BD} \cong \overline{BD}$, from which it follows that $\triangle BAD \cong \triangle DCB$ by ASA. Using the same approach it can be easily shown that in Figure 9.3b $\overline{AC}$ divides parallelogram $ABCD$ into two triangles such that $\triangle ABC \cong \triangle CDA$. These observations are summarized in Theorem 9.3.

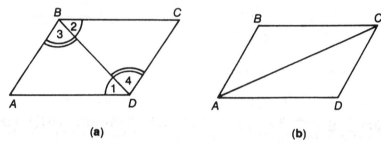

FIGURE 9.3

THEOREM 9.3

In a parallelogram either diagonal separates the parallelogram into two congruent triangles.

Theorem 9.3 allows us to draw some conclusions regarding the parts of the parallelogram. Applying the CPCTC principle, we may conclude that $\overline{AD} \cong \overline{BC}$ and $\overline{AB} \cong \overline{DC}$. This result is stated formally in Theorem 9.4.

THEOREM 9.4

Opposite sides of a parallelogram are congruent.

EXAMPLE 9.2

GIVEN: $\square ABCD$, diagonals $\overline{AC}$ and $\overline{BD}$ intersect at point E.

PROVE: (a) $\overline{AE} \cong \overline{EC}$
(b) $\overline{BE} \cong \overline{ED}$

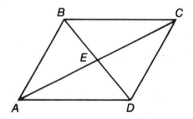

SOLUTION

PLAN: Prove $\triangle BEC \cong \triangle DEA$ by ASA.

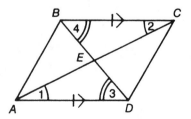

PROOF: Statements	Reasons
1. $\square ABCD$.	1. Given.
2. $\overline{BC} \parallel \overline{AD}$.	2. Opposite sides of a parallelogram are parallel.
3. $\angle 1 \cong \angle 2$. (Angle)	3. If two lines are parallel, then their alternate interior angles are congruent.
4. $\overline{AD} \cong \overline{BC}$. (Side)	4. Opposite sides of a parallelogram are congruent.
5. $\angle 3 \cong \angle 4$. (Angle)	5. Same as reason 3.
6. $\triangle BEC \cong \triangle DEA$.	6. ASA Postulate.
7. $\overline{AE} \cong \overline{EC}$ and $\overline{BE} \cong \overline{ED}$.	7. CPCTC.

Example 9.2 establishes Theorem 9.5.

THEOREM 9.5

The diagonals of a parallelogram bisect each other.

SUMMARY OF PROPERTIES OF A PARALLELOGRAM

DEFINITION

$\overline{AB} \parallel \overline{CD}$
and
$\overline{AD} \parallel \overline{BC}$

THEOREM 9.1

$a + b = 180$
$b + c = 180$
$c + d = 180$
$a + d = 180$

THEOREM 9.2

$\angle A \cong \angle C$
and
$\angle B \cong \angle D$

THEOREM 9.3

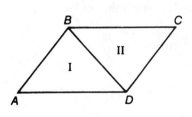

$\triangle I \cong \triangle II$

THEOREM 9.4

$\overline{AB} \cong \overline{CD}$
and
$\overline{AD} \cong \overline{BC}$

THEOREM 9.5

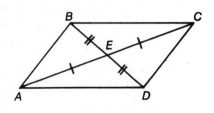

$\overline{AE} \cong \overline{EC}$
and
$\overline{BE} \cong \overline{ED}$

In a Parallelogram

1. Opposite sides are parallel. (Definition of parallelogram)
2. Consecutive angles are supplementary. (Theorem 9.1)
3. Opposite angles are congruent. (Theorem 9.2)
4. Opposite sides are congruent. (Theorem 9.4)
5. Diagonals bisect each other. (Theorem 9.5)

Properties of Special Parallelograms

An equiangular parallelogram is called a *rectangle*. An equilateral parallelogram is a *rhombus*. A *square* is a parallelogram that is both equiangular and equilateral. It will be convenient to use the following definitions in our work with these figures.

DEFINITIONS OF SPECIAL PARALLELOGRAMS
- A *rectangle* is a parallelogram having four right angles.
- A *rhombus* is a parallelogram having four congruent sides.
- A *square* is a rectangle having four congruent sides.

Rectangle Rhombus Square

EXAMPLE 9.3

GIVEN: Rectangle *ABCD*.
PROVE: $\overline{AC} \cong \overline{DB}$.

SOLUTION

OUTLINE OF PROOF: Prove $\triangle BAD \cong \triangle CDA$ by SAS.
- $\overline{AB} \cong \overline{CD}$ since opposite sides of a rectangle are congruent.
- Angles *BAD* and *CDA* are congruent since they are right angles. $\overline{AD} \cong \overline{AD}$.
- Since the triangles are congruent, $\overline{AC} \cong \overline{DB}$ by CPCTC.

Example 9.4 establishes Theorem 9.6.

THEOREM 9.6

The diagonals of a rectangle are congruent.

EXAMPLE 9.4

GIVEN: Rhombus *ABCD*.
PROVE: **a** ∡1 ≅ ∡2.
 b ∡3 ≅ ∡4.

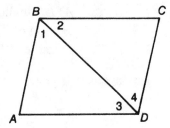

SOLUTION

OUTLINE OF
PROOF: Prove △*BAD* ≅ △*BCD* by SAS.
- $\overline{AB} \cong \overline{CB}$ since a rhombus is equilateral.
- Angles *A* and *C* are congruent since opposite angles of a rhombus are congruent.
- $\overline{AD} \cong \overline{CD}$ since a rhombus is equilateral.
- The desired angle pairs are congruent by the CPCTC principle.

Notice that diagonal $\overline{BD}$ bisects angles *B* and *D*. Using a similar approach, we can show that diagonal $\overline{AC}$ bisects angles *A* and *C*. Thus, the *diagonals of a rhombus bisect the four angles of the rhombus.*

EXAMPLE 9.5

Prove the diagonals of a rhombus are perpendicular to each other.

SOLUTION

GIVEN: Rhombus *ABCD*,
 diagonals $\overline{AC}$ and $\overline{BD}$ intersect at *E*.
PROVE: $\overline{AC} \perp \overline{BD}$.

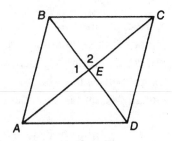

OUTLINE OF
PROOF: Prove △*AEB* ≅ △*CEB* (other triangle pairs may be selected) by SAS.
- $\overline{AB} \cong \overline{CB}$ since a rhombus is equilateral.
- Angles *ABE* and *CBE* are congruent by Theorem 9.7 since a diagonal of a rhombus bisects the angle formed at each vertex.
- $\overline{BE} \cong \overline{BE}$. By CPCTC, angles 1 and 2 are congruent.
- $\overline{AC} \perp \overline{BD}$ since the lines intersect to form a congruent pair of adjacent angles.

DIAGONALS OF A RHOMBUS THEOREMS

THEOREM 9.7 The diagonals of a rhombus bisect the four angles of the rhombus (see Example 9.4)

THEOREM 9.8 The diagonals of a rhombus are perpendicular to each other (see Example 9.5)

Since a rhombus includes all the properties of a parallelogram, the diagonals of a rhombus bisect each other. Each diagonal of a rhombus is the *perpendicular bisector of the other diagonal.*

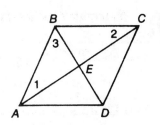

GIVEN: *ABCD* in the accompanying figure
is a rhombus and m ∡ 1 = 40. Find the
measure of each of the following angles:

(a) ∡2 (b) ∡3 (c) ∡*ADC*

SOLUTION

a. Triangle *ABC* is isosceles since $\overline{AB} \cong \overline{BC}$. Hence, the base angles of the triangle
must be congruent.

$$m\angle 1 = m\angle 2 = 40$$

b. In triangle *AEB*, angle *AEB* is a right angle since the diagonals of a rhombus are
perpendicular to each other. Since the sum of the measures of the angles of a tri-
angle is 180, the measure of angle 3 must be 50.

c. Since the diagonals of a rhombus bisect the angles of the rhombus, if m∡3 = 50,
then m∡*ABC* = 100. Since opposite angles of a rhombus are equal in measure,
m∡*ADC* must also equal 100.

Keep in mind that a rhombus is *not* necessarily a rectangle since it may or may not
contain four right angles. A rectangle is *not* necessarily a rhombus since it may or may
not contain four congruent sides. A square combines the properties of a rectangle with
the properties of a rhombus. The diagonals of a square are congruent, bisect its oppo-
site angles, and intersect at right angles.

GIVEN: Square *ABCD*,
$\overline{AE} \cong \overline{DF}$.
PROVE: $\overline{AF} \cong \overline{BE}$.

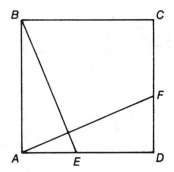

SOLUTION

PLAN: Prove △*BAE* ≅ △*ADF* by SAS.

PROOF:

Statements	Reasons
1. *ABCD* is a square.	1. Given.
2. $\overline{BA} \cong \overline{DA}$. (Side)	2. A square is equilateral.
3. Angles *A* and *D* are right angles.	3. A square contains four right angles.
4. ∡*A* ≅ ∡*D*. (Angle)	4. All right angles are congruent.
5. $\overline{AE} \cong \overline{DF}$. (Side)	5. Given.
6. △*BAE* ≅ △*ADF*.	6. SAS Postulate.
7. $\overline{AF} \cong \overline{BE}$.	7. CPCTC.

SUMMARY OF PROPERTIES OF A RECTANGLE, RHOMBUS, AND SQUARE

Property	Rectangle	Rhombus	Square
1. All the properties of a parallelogram?	Yes	Yes	Yes
2. Equiangular (4 right angles)?	Yes	No	Yes
3. Equilateral (4 congruent sides)?	No	Yes	Yes
4. Diagonals congruent?	Yes	No	Yes
5. Diagonals bisect opposite angles?	No	Yes	Yes
6. Diagonals perpendicular?	No	Yes	Yes

Proving a Quadrilateral Is a Parallelogram

The preceding sections developed the properties of quadrilaterals that were known to be parallelograms. We now consider the other side of the coin. How can we prove that a quadrilateral is a parallelogram? What is the minimum information required to justify the conclusion that a quadrilateral is a parallelogram?

Using the reverse of the definition of a parallelogram, we know that, if *both pairs of sides of a quadrilateral are parallel, then the quadrilateral is a parallelogram*. We may use this fact to establish alternative methods for proving that a quadrilateral is a parallelogram. For example, we may try drawing a quadrilateral in which the same pair of sides are both congruent and parallel. Imposing this condition forces the remaining pair of sides to be parallel.

THEOREM 9.9

If a quadrilateral has one pair of sides that are both parallel and congruent, then the quadrilateral is a parallelogram.

GIVEN: Quadrilateral $ABCD$,
$\overline{AD} \parallel \overline{BC}, \overline{AD} \cong \overline{BC}$.

PROVE: Quadrilateral $ABCD$ is a parallelogram.

PLAN: Show $\overline{AB} \parallel \overline{CD}$.

OUTLINE OF PROOF: Draw diagonal $\overline{AC}$. Prove $\triangle ABC \cong \triangle CDA$ by SAS.
- $\overline{BC} \cong \overline{AD}$ (Side)
- $\angle 1 \cong \angle 2$ (Angle)
- $\overline{AC} \cong \overline{AC}$ (Side)

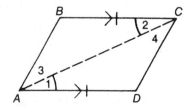

Angles 3 and 4 are congruent by CPCTC. This implies that $\overline{AB} \parallel \overline{CD}$ (since alternate interior angles are congruent, the lines are parallel). Since both pairs of sides of quadrilateral *ABCD* are parallel, *ABCD* is a parallelogram.

Are there any additional methods for proving a quadrilateral is a parallelogram? The converses of the theorems that state the properties of quadrilaterals that are parallelograms may offer some clues. Recall that these theorems take the general form

If a quadrilateral is a parallelogram, then a *certain property is true.*

The converse of this statement takes the form

If a quadrilateral has a *certain property*, then it is a parallelogram.

Keeping in mind that the converse of a theorem is *not* necessarily true, we need to investigate whether the converse holds for each special property of parallelograms. Table 9.1 summarizes the results of these investigations.

TABLE 9.1

If a Quadrilateral has ..	Then is it a Parallelogram (?)
• Congruent opposite sides	Yes, since $\triangle ABC \cong \triangle CDA$ by SSS. By CPCTC, angles 1 and 2 are congruent, which implies that $\overline{AB} \parallel \overline{CD}$. Similarly, angles 3 and 4 are congruent making $\overline{AD} \parallel \overline{BC}$. Since both pairs of sides are parallel, *ABCD* is a parallelogram.
• Congruent opposite angles	Yes. Since the sum of the interior angles of a quadrilateral is 360, $$\begin{aligned} x + y + x + y &= 360 \\ 2x + 2y &= 360 \\ 2(x + y) &= 360 \\ x + y &= 180 \end{aligned}$$ If interior angles on the same side of a transversal are supplementary, the lines are parallel. It follows that $\overline{AB} \parallel \overline{DC}$ and $\overline{AD} \parallel \overline{BC}$. *ABCD* is therefore a parallelogram.
• Diagonals that bisect each other 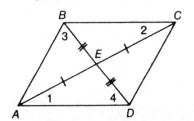	Yes. $\triangle AED \cong \triangle BEC$ by SAS. By CPCTC, $\overline{AD} \cong \overline{BC}$ and $\angle 1 \cong \angle 2$. It follows that $\overline{AD} \parallel \overline{BC}$. Since $\overline{AD}$ is both congruent and parallel to $\overline{BC}$, *ABCD* is a parallelogram (see Theorem 9.9).

Theorem 9.10 gives three additional methods for proving that a quadrilateral is a parallelogram.

THEOREM 9.10

MORE WAYS OF PROVING A QUADRILATERAL IS A PARALLELOGRAM

A quadrilateral is a parallelogram if any one of the following is true:
- Opposite sides are congruent.
- Opposite angles are congruent.
- Diagonals bisect each other.

Draw a diagram to help prove or disprove that a quadrilateral is a parallelogram if:
(a) One pair of opposite sides are congruent.
(b) One pair of sides are parallel.
(c) Two pairs of sides are congruent.

SOLUTION

a. Not necessarily a parallelogram:

b. Not necessarily a parallelogram:

c. Not necessarily a parallelogram:
NOTE: This type of figure is referred to as a *kite*

GIVEN: $\overline{BE} \perp \overline{AC}$, $\overline{DF} \perp \overline{AC}$,
$\overline{BE} \cong \overline{DF}$,
$\angle EBC \cong \angle FDA$.

PROVE: $ABCD$ is a parallelogram.

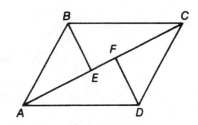

SOLUTION

PLAN: Prove $\overline{BC}$ is parallel and congruent to $\overline{AD}$ by first proving that
$\triangle BEC \cong \triangle DFA$ by ASA.

PROOF:

Statements	Reasons
1. $\angle EBC \cong \angle FDA$. (Angle)	1. Given.
2. $\overline{BE} \cong \overline{DF}$. (Side)	2. Given.
3. $\overline{BE} \perp \overline{AC}$ and $\overline{DF} \perp \overline{AC}$.	3. Given.

4. $\angle BEC \cong \angle DFA$. (Angle)	4. Perpendicular lines intersect to form right angles. All right angles are congruent. (*NOTE*: We have consolidated steps.)
5. $\triangle BEC \cong \triangle DFA$.	5. ASA Postulate.
6. $\overline{AD} \cong \overline{BC}$ and $\angle BCE \cong \angle DAF$.	6. CPCTC.
7. $\overline{AD} \parallel \overline{BC}$.	7. If alternate interior angles are congruent, then the lines are parallel.
8. Quadrilateral *ABCD* is a parallelogram.	8. If a quadrilateral has a pair of sides that are both parallel and congruent, then the quadrilateral is a parallelogram (Theorem 9.9).

EXAMPLE 9.10

GIVEN: $\square ABCD$,
 $\overline{AE}$ bisects $\angle BAD$,
 $\overline{CF}$ bisects $\angle BCD$.
PROVE: Quadrilateral *AECF* is a parallelogram.

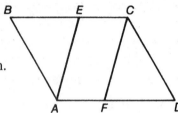

SOLUTION

PLAN: Prove $\triangle ABE \cong \triangle CDF$ by ASA. $AE = CF$ and $BE = DF$ by CPCTE. $EC = AF$ by subtraction. Hence, *AECF* is a parallelogram since opposite sides have the same length.

PROOF:

Statements	Reasons
1. *ABCD* is a parallelogram.	1. Given.
2. $m\angle BAD = m\angle DCB$.	2. Opposite angles of a parallelogram are equal in measure.
3. $\overline{AE}$ bisects $\angle BAD$. $\overline{CF}$ bisects $\angle BCD$.	3. Given.
4. $\angle BAE \cong \angle DCF$. (Angle)	4. Halves of equals are equal (and therefore congruent).
5. $\overline{AB} \cong \overline{DC}$. (Side)	5. Opposite sides of a parallelogram are congruent.
6. $\angle B \cong \angle D$. (Angle)	6. Same as reason 2.
7. $\triangle ABE \cong \triangle CDF$.	7. ASA Postulate.
8. $AE = FC$.	8. Corresponding sides of congruent triangles are equal in length.
To show EC = AF use substraction:	
9. $BE = DF$.	9. Same as reason 8.
10. $BC = AD$.	10. Opposite sides of a parallelogram are equal in length.

| 11. $EC = AF$. | 11. Subtraction property of equality. |
| 12. $AECF$ is a parallelogram. | 12. If the opposite sides of a quadrilateral are equal in length (that is, congruent), then the quadrilateral is a parallelogram. |

In the proof of Example 9.10, statements and corresponding reasons are expressed in terms of equality of measures rather than congruence. This is necessary since arithmetic operations (taking halves of equals and subtraction) were performed on these quantities.

SUMMARY

TO PROVE A QUADRILATERAL IS A PARALLELOGRAM
Show that any one of the following is true:
- Opposite sides are parallel.
- Opposite sides are congruent.
- Opposite angles are congruent.
- Diagonals bisect each other.
- A pair of sides are both parallel and congruent.

Sometimes we need to prove that a quadrilateral (or parallelogram) is a rectangle, rhombus, or square:

1. To prove a quadrilateral is a *rectangle*, show that it is a *parallelogram* having one of the following properties:
 - It contains a right angle.
 - The diagonals are congruent.
2. To prove a quadrilateral is a *rhombus*, show that it is a *parallelogram* having one of the following properties:
 - It contains a pair of congruent adjacent sides.
 - The diagonals intersect at right angles.
 - The diagonals bisect the vertex angles.
3. To prove that a quadrilateral is a *square*, show that it is either of the following:
 - A *rectangle* with an adjacent pair of congruent sides.
 - A *rhombus* with a right angle.

Applications of Parallelograms

In Figure 9.4a, *ABCD* is a rectangle. If $BD = 10$, what is the length of $\overline{AM}$? The diagonals of a rectangle are congruent, making $AC = 10$. Since the diagonals bisect each other, $AM = 5$ (and $MC = 5$). We can redraw the diagram as Figure 9.4b by deleting $\overline{BC}$, $\overline{MC}$, and $\overline{CD}$. Since *M* is the midpoint of $\overline{BD}$, $\overline{AM}$ is the median to hypotenuse $\overline{BD}$ of right triangle *BAD*. This suggests Theorem 9.11.

THEOREM 9.11

The length of the median drawn to the hypotenuse of a right triangle is one-half of the length of the hypotenuse.

FIGURE 9.4

The theorem that we will now establish states that if the midpoints of *any* two sides of a triangle are connected by a line segment, then this segment must be parallel to the remaining side of the triangle. Furthermore, its length must be exactly one-half the length of the remaining side of the triangle. This theorem may be stated formally as follows.

THEOREM 9.12 MIDPOINTS OF A TRIANGLE THEOREM

The line segment joining the midpoints of two sides of a triangle is parallel to the third side and is one-half its length.

GIVEN: D and E are midpoints of sides $\overline{AB}$ and $\overline{CB}$, respectively.

PROVE: **(a)** $\overline{DE} \parallel \overline{AC}$.

(b) $DE = \dfrac{1}{2} AC$.

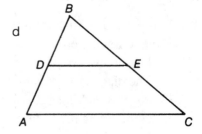

OUTLINE OF PROOF: Extend $\overline{DE}$ so that $\overline{DE} \cong \overline{EF}$, and then draw $\overline{CF}$.

Prove *ADFC* is a parallelogram by showing that $\overline{CF}$ is parallel and congruent to $\overline{AD}$ as follows:

- $\triangle DBE \cong \triangle FCE$ (by SAS).
- By CPCTC, angles 1 and 2 are congruent, implying $\overline{FC} \parallel \overline{AD}$. Also, $\overline{CF} \cong \overline{DB} \cong \overline{AD}$.
- Since *ADFC* is a parallelogram, $\overline{DE} \parallel \overline{AC}$. Also,

$$DE = \frac{1}{2} DF = \frac{1}{2} AC.$$

EXAMPLE 9.11 In triangle *RST*, *A* is the midpoint of $\overline{RS}$ and *B* is the midpoint of $\overline{RT}$.
 a. If *ST* = 18, find *AB*.
 b. If *AB* = 7, find *ST*.

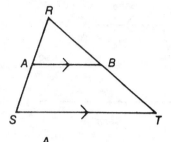

SOLUTION

 a. $AB = \frac{1}{2}\ ST = \frac{1}{2}(18) = 9.$

 b. $ST = 2(AB) = 2(7) = 14.$

EXAMPLE 9.12

GIVEN: Points *Q*, *R*, and *S* are midpoints.
PROVE: *PQRS* is a parallelogram.

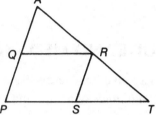

SOLUTION

PLAN: Show that $\overline{QR}$ is parallel and congruent to $\overline{PS}$.

PROOF:

Statements	Reasons
1. Points *Q*, *R*, and *S* are midpoints.	1. Given.
2. $\overline{QR} \parallel \overline{PT}$.	2. The line segment joining the midpoints of two sides of a triangle is parallel to the third side and is one-half its length.
3. $\overline{QR} \parallel \overline{PS}$.	3. Segments of parallel lines are parallel.
4. $QR = \frac{1}{2}\ PT$.	4. Same as reason 2.
5. $PS = \frac{1}{2}\ PT$.	5. A midpoint divides a segment into two congruent segments.
6. $\overline{QR} \cong \overline{PS}$.	6. Transitive property.
7. Quadrilateral *PQRS* is a parallelogram.	7. If a quadrilateral has a pair of sides that are both parallel and congruent, then the quadrilateral is a parallelogram.

Properties of a Trapezoid

Unlike a parallelogram, a trapezoid has exactly one pair of parallel sides. In Figure 9.5, quadrilaterals *ABCD* and *RSTW* are trapezoids. Quadrilateral *JKLM* is *not* a trapezoid since it does not have one pair of parallel sides.

DEFINITION OF A TRAPEZOID

A *trapezoid* is a quadrilateral that has exactly *one* pair of parallel sides. The parallel sides are called the *bases* of the trapezoid. The nonparallel sides are referred to as the *legs* of the trapezoid.

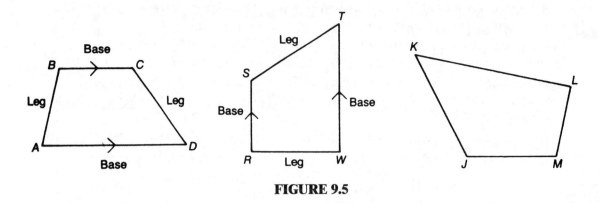

FIGURE 9.5

ALTITUDE AND MEDIAN OF A TRAPEZOID

Two additional terms that are used in connection with trapezoids are *altitude* and *median*. An *altitude* of a trapezoid (or for that matter, a parallelogram) is a segment drawn from any point on one of the parallel sides (base) perpendicular to the opposite side (other base). An infinite number of altitudes may be drawn in a trapezoid. See Figure 9.6.

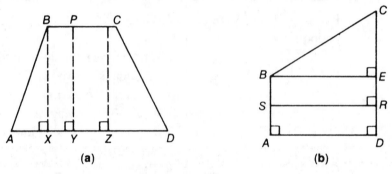

(a) **(b)**

FIGURE 9.6 *(a)* $\overline{BX}$, $\overline{PY}$, and $\overline{CZ}$ are examples of *altitudes*.
 (b) $\overline{AD}$, $\overline{SR}$, and $\overline{BE}$ are examples of *altitudes*.

A *median* of a trapezoid is the segment that joins the midpoints of the nonparallel sides (legs). A trapezoid has exactly one median. See Figure 9.7.

As Figure 9.7 seems to indicate, the median of a trapezoid appears to be parallel to the bases. In addition, there is a relationship between the length of the median and the lengths of the bases of the trapezoid. Our strategy in developing the properties of the median of a trapezoid is to apply Theorem 9.12. To accomplish this, we must draw an auxiliary segment so that the median of the trapezoid becomes the segment that joins the midpoints of a newly formed triangle.

FIGURE 9.7 **FIGURE 9.8**

In trapezoid *ABCD* of Figure 9.8, median $\overline{LM}$ has been drawn. By drawing line segment $\overline{BM}$ and extending it to meet the extension of line segment $\overline{AD}$, we form triangle *ABE*. Point *L* is the midpoint of side $\overline{AB}$ of triangle *ABE*. If we can show that point *M* is the midpoint of side $\overline{BE}$ of triangle *ABE*, then we can apply Theorem 9.12, which states that the line segment joining the midpoints of two sides of a triangle is parallel to the third side of the triangle. This would establish that median $\overline{LM}$ is parallel to $\overline{BC}$ and $\overline{AD}$.

To show that *M* is the midpoint of side $\overline{BE}$, we consider triangles *BMC* and *EMD*. They are congruent by ASA:

$$\angle BMC \cong \angle EMD \quad \text{(Angle)}$$
$$\overline{CM} \cong \overline{DM} \quad \text{(Side)}$$

and since $\overline{BC} \parallel \overline{AE}$,

$$\angle BCM \cong \angle EDM \quad \text{(Angle)}$$

By CPCTC, $\overline{BM} \cong \overline{EM}$. Hence, Theorem 9.12 applies.

Theorem 9.12 also informs us that the length of $\overline{LM}$ (the segment joining the midpoints of two sides of triangle *ABE*) must be one-half the length of the third side. Thus,

$$LM = \frac{1}{2}AE = \frac{1}{2}(AD + DE)$$

Since we have established that $\triangle BMC \cong \triangle EMD$, $\overline{DE} \cong \overline{BC}$. Applying the substitution principle, we have

$$LM = \frac{1}{2}(AD + BC) = \frac{1}{2}(\text{sum of lengths of bases})$$

The result of our investigation is stated formally in Theorem 9.13.

THEOREM 9.13 MEDIAN OF A TRAPEZOID THEOREM

The median of a trapezoid is parallel to the bases and has a length equal to one-half the sum of the lengths of the bases.

In trapezoid *ABCD* $\overline{BC} \parallel \overline{AD}$ and $\overline{RS}$ is the median.
 a. If *AD* = 13 and *BC* = 7, find *RS*.
 b. If *BC* = 6 and *RS* = 11, find *AD*.

SOLUTION

 a. $RS = \frac{1}{2}(13 + 7) = \frac{1}{2}(20) = 10$

 b. If median $= \frac{1}{2}(\text{sum of lengths of bases})$, then

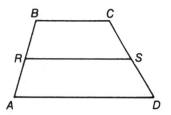

$$
\begin{aligned}
\text{Sum of lengths of bases} &= 2 \times \text{median} \\
AD + 6 &= 2 \times 11 \\
AD + 6 &= 22 \\
AD &= 22 - 6 \\
&= 16
\end{aligned}
$$

The length of the lower base of a trapezoid is 3 times the length of the upper base. If the median has a 24-inch length, find the lengths of the bases.

SOLUTION

$$\text{Let } a = \text{length of upper base.}$$
$$\text{Then } 3a = \text{length of lower base.}$$
$$\text{Since sum of lengths of bases} = 2 \times \text{median,}$$
$$a + 3a = 2 \times 24$$
$$4a = 48$$
$$a = \frac{48}{4}$$
$$= 12$$
$$3a = 36$$

SUMMARY

SPECIAL MEDIAN RELATIONSHIPS

* The length of the median drawn to the hypotenuse of a right triangle is one-half the length of the hypotenuse.
* The length of the median of a trapezoid may be found by taking the average of the lengths of the two bases:

 Median $= \frac{1}{2}$ (upper-base length + lower-base length)
* The median of a trapezoid is parallel to both bases.

THE ISOSCELES TRAPEZOID

If the legs of a trapezoid are congruent, then the trapezoid is called an *isosceles trapezoid*. See Figure 9.9. An *isosceles trapezoid* features some special properties not found in all trapezoids.

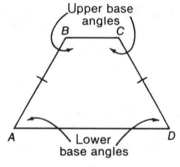

FIGURE 9.9

THEOREM 9.14

The lower (and upper) base angles of an isosceles trapezoid are congruent.

OUTLINE OF PROOF

GIVEN: Isosceles trapezoid *ABCD*.

PROVE: $\angle A \cong \angle D$.

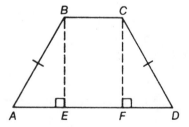

Draw altitudes $\overline{BE}$ and $\overline{CF}$, thereby forming triangles ABE and DCF. Prove these triangles congruent by applying the Hy-Leg method:

$$\overline{AB} \cong \overline{DC} \text{ (Hypotenuse)}$$

Since parallel lines are everywhere equidistant,

$$\overline{BE} \cong \overline{CF} \text{ (Leg)}$$

By CPCTC, $\angle A \cong \angle D$.

Since supplements of congruent angles are congruent, the upper base angles are congruent. We now turn our attention to the diagonals of an isosceles trapezoid.

THEOREM 9.15

The diagonals of an isosceles trapezoid are congruent.

OUTLINE OF PROOF

GIVEN: Isosceles trapezoid $ABCD$, diagonals $\overline{AC}$ and $\overline{DB}$.

PROVE: $\overline{AC} \cong \overline{DB}$.

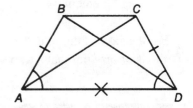

Diagonals $\overline{AC}$ and $\overline{DB}$ form a pair of overlapping triangles, $\triangle ACD$ and $\triangle DBA$. Prove these triangles congruent by using SAS. The included angles, angles BAD and CDA, are congruent as a result of Theorem 9.14. By CPCTC, $\overline{AC} \cong \overline{PB}$.

Properties of an Isosceles Trapezoid
1. It has all the properties of a trapezoid.
2. The legs are congruent. (Definition of isosceles trapezoid)
3. The base angles are congruent. (Theorem 9.14)
4. The diagonals are congruent. (Theorem 9.15)

EXAMPLE 9.15

Prove the opposite angles of an isosceles trapezoid are supplementary.

SOLUTION

GIVEN: Isosceles trapezoid $ABCD$.

PROVE: **a.** Angles A and C are supplementary.

b. Angles D and B are supplementary.

PROOF:

Statements	Reasons
1. *ABCD* is an isosceles trapezoid.	1. Given.
2. $\overline{AD} \parallel \overline{BC}$.	2. The bases of a trapezoid are parallel.
3. m∡D + m∡C = 180.	3. If two lines are parallel, then interior angles on the same side of the transversal are supplementary.
4. m∡ A = m∡D.	4. Base angles of an isosceles trapezoid are equal in measure.
5. m∡A + m∡C = 180.	5. Substitution.
6. Angles *A* and *C* are supplementary.	6. If the sum of the measures of two angles is 180, then the angles are supplementary.

In a similar fashion, it can be shown easily that angles *D* and *B* are supplementary.

For the sake of completeness we state without proof Theorem 9.16, which offers methods of proving that a trapezoid is an isosceles trapezoid.

THEOREM 9.16 WAYS OF PROVING A TRAPEZOID IS ISOSCELES

A trapezoid is an isosceles trapezoid if any one of the following is true:
1. The legs are congruent.
2. The base angles are congruent.
3. The diagonals are congruent.

The congruent-legs part of Theorem 9.16 follows directly from the reverse of the definition of an isosceles trapezoid. The congruent-base-angles part can be proved by dropping altitudes to the longer base, as shown in the diagram. If the base angles of this trapezoid are congruent, then $\triangle BEA \cong \triangle CFD$ by AAS since ∡A ≅ ∡D, right angles *BEA* and *CFD* are congruent, and $\overline{BE} \cong \overline{CF}$ (parallel lines are everywhere equidistant). Therefore, $\overline{AB} \cong \overline{CD}$ by CPCTC and *ABCD* is an isosceles trapezoid. The congruent-diagonals part of the theorem can be established by first proving $\triangle BED \cong \triangle CFA$ in order to obtain

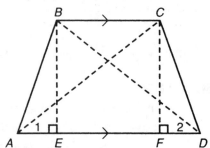

∡1 ≅ ∡2, which can then be used to help prove $\triangle ABD \cong \triangle DCA$. By CPCTC, $\overline{AB} \cong \overline{DC}$. You are asked to supply the details of this proof in Exercise 33 at the end of this chapter.

SUMMARY OF PROPERTIES OF DIAGONALS OF SPECIAL QUADRILATERALS

Special	Diagonals Are Always		Diagonals Always Bisect	
Quadrilateral	Congruent	Perpendicular	Each Other	Vertex Angles
Parallelogram	No	No	Yes	No
Rectangle	Yes	No	Yes	No
Rhombus	No	Yes	Yes	Yes
Square	Yes	Yes	Yes	Yes
Trapezoid	No	No	No	No
Isosceles Trapezoid	Yes	No	No	No

REVIEW EXERCISES FOR CHAPTER 9

1. In rhombus $RSTW$, diagonal $\overline{RT}$ is drawn. If m $\angle RST = 108$, find m $\angle SRT$.

2. In parallelogram $MATH$ the measure of angle T exceeds the measure of angle H by 30. Find the measure of each angle of the parallelogram.

3. In parallelogram $TRIG$, m$\angle R = 2x + 19$ and m$\angle G = 4x - 17$. Find the measure of each angle of the parallelogram.

4. Which statement about a diagonal of a parallelogram is *always* true?
 (1) It bisects the other diagonal of the parallelogram.
 (2) It bisects an angle of the parallelogram.
 (3) It is congruent to the other diagonal of the parallelogram.
 (4) It is perpendicular to the other diagonal of the parallelogram.

5. The length of the median drawn to the hypotenuse of a right triangle is represented by the expression $3x - 7$, while the length of the hypotenuse is represented by $5x - 4$. Find the length of the median.

6. In triangle RST, E is the midpoint of $\overline{RS}$ and F is the midpoint of $\overline{ST}$. If $EF = 5y - 1$ and $RT = 7y + 10$, find the lengths of $\overline{EF}$ and $\overline{RT}$.

7. In trapezoid $BYTE$, $\overline{BE} \parallel \overline{YT}$ and median $\overline{LM}$ is drawn.
 (a) $LM = 35$. If the length of $\overline{BE}$ exceeds the length of $\overline{YT}$ by 13, find the lengths of the bases.
 (b) If $YT = x + 9$, $LM = x + 15$, and $BE = 2x - 5$, find the lengths of $\overline{YT}$, $\overline{LM}$, and $\overline{BE}$.

8. In parallelogram $RSTW$ diagonals $\overline{RT}$ and $\overline{SW}$ intersect at point A. If $SA = x - 13$ and $AW = 2x - 37$, find SW.

9. The lengths of the sides of a triangle are 9, 40, and 41. Find the perimeter of the triangle formed by joining the midpoints of the sides.

10. GIVEN: $\square ABCD$, $\overline{AE} \cong \overline{CF}$.
 PROVE: $\measuredangle ABE \cong \measuredangle CDF$.

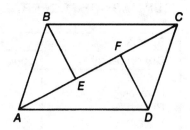

11. GIVEN: $\square ABCD$,
 $\overline{EF} \cong \overline{HG}$.
 PROVE: $\overline{AF} \cong \overline{CG}$.

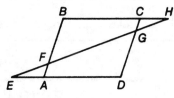

12. GIVEN: $\square ABCD$,
 B is the midpoint of $\overline{AE}$.
 PROVE: $\overline{EF} \cong \overline{FD}$.

13. GIVEN: Rectangle $ABCD$,
 M is the midpoint of $\overline{BC}$.
 PROVE: $\triangle AMD$ is isosceles.

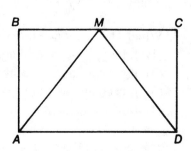

14. GIVEN: Rhombus $ABCD$.
 PROVE: $\triangle ASC$ is isosceles.

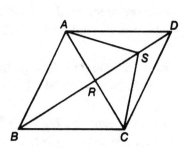

15. GIVEN: Rectangle $ABCD$,
 $\overline{BE} \cong \overline{CE}$.
 PROVE: $\overline{AF} \cong \overline{DG}$.

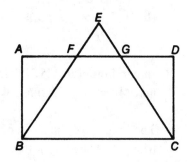

16. GIVEN: *ABCD* is a parallelogram,
 AD > *DC*.
 PROVE: m∡*BAC* > m∡*DAC*.

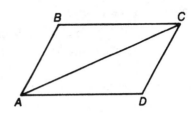

17. GIVEN: *ABCD* is a parallelogram,
 $\overline{BR}$ bisects ∡*ABC*,
 $\overline{DS}$ bisects ∡*CDA*.
 PROVE: *BRDS* is a parallelogram.

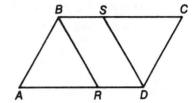

Use the accompanying diagram to solve Exercises 18 and 19.

18. GIVEN: □*BMDL*,
 $\overline{AL} \cong \overline{CM}$.
 PROVE: *ABCD* is a parallelogram.

19. GIVEN: □*ABCD*,
 ∡*ABL* ≅ ∡*CDM*.
 PROVE: *BLDM* is a parallelogram.

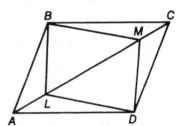

20. GIVEN: □*ABCD* is a rhombus,
 $\overline{BL} \cong \overline{CM}, \overline{AL} \cong \overline{BM}$.
 PROVE: *ABCD* is a square.

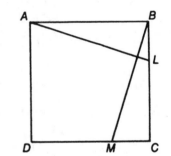

21. GIVEN: □*ABCD*, m∡2 > m∡1.
 PROVE: □*ABCD* is *not* a rectangle.

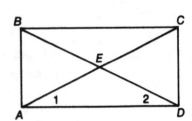

22. GIVEN: $\overline{DE} \cong \overline{DF}$,
 D, E, and *F* are the midpoints
 of $\overline{AC}, \overline{AB}$, and $\overline{BC}$, respectively.
 PROVE: △*ABC* is isosceles.

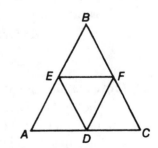

23. GIVEN: $\square RSTW$;
 in $\triangle WST$, B and C are midpoints.
 PROVE: $WACT$ is a parallelogram.

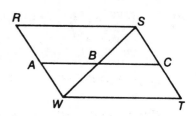

24. GIVEN: Isosceles trapezoid $RSTW$.
 PROVE: $\triangle RPW$ is isosceles.

25. GIVEN: Trapezoid $ABCD$,
 $\overline{EF} \cong \overline{EG}$,
 $\overline{AF} \cong \overline{DG}$,
 $\overline{BG} \cong \overline{CF}$.
 PROVE: Trapezoid $ABCD$ is isosceles.

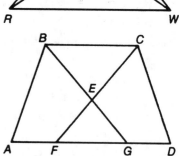

26. GIVEN: Trapezoid $ABCD$ with median $\overline{LM}$,
 P is the midpoint of $\overline{AD}$,
 $\overline{LP} \cong \overline{MP}$.
 PROVE: Trapezoid $ABCD$ is isosceles.

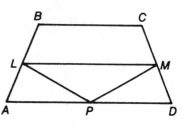

27. GIVEN: Isosceles trapezoid $ABCD$,
 $\angle BAK \cong \angle BKA$.
 PROVE: $BKDC$ is a parallelogram.

28. GIVEN: D is the midpoint of $\overline{AB}$,
 F is the midpoint of $\overline{AC}$,
 $\overline{BE} \cong \overline{CG}$, $\angle B \cong \angle C$,
 $\angle 1$ is supplementary to $\angle 2$,
 $\overline{FD} \cong \overline{FG}$.
 PROVE: (a) $\triangle DEB \cong \triangle FGC$.
 (b) Quadrilateral $DEGF$ is a square.

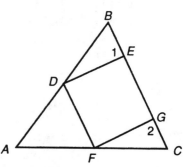

29. Prove that in a rhombus the longer diagonal lies opposite the larger angle of the rhombus. (*Hint:* Given rhombus $ABCD$ with diagonals $\overline{AC}$ and $\overline{DB}$ intersecting at point E, assume $m\angle CDA > m\angle BAD$. Prove $AC > BD$. Work with $\triangle AED$ and first establish that $AE > DE$.)

30. Prove parallel lines are everywhere equidistant. (*Hint:* Given any pair of parallel lines, select two distinct points on one of the lines and from each point draw a segment that is perpendicular to the other parallel line. Show that the lengths of these perpendicular segments are equal.)

31. Prove that the quadrilateral formed by joining consecutively the midpoints of the sides of a parallelogram is a parallelogram.

32. Prove that, if the midpoints of the sides of a rectangle are joined consecutively, the resulting quadrilateral is a rhombus.

33. Prove that, if the diagonals of a trapezoid are congruent, the trapezoid is isosceles.

RATIO, PROPORTION, AND SIMILARITY

WHAT YOU WILL LEARN

Congruent figures have the same shape and exactly the same dimensions. We now consider *similar* figures, which are figures with the same shape but not necessarily the same dimensions. In this chapter you will learn:

- the ways that proportions arise in geometric situations;
- facts about figures that have the same shape but whose corresponding dimensions are in proportion rather than equal;
- the way to prove two triangles are similar;
- the way to prove proportions and products of line segments equal by using similar triangles;
- the special properties of perimeters, altitudes, and medians of similar triangles.

SECTIONS IN THIS CHAPTER

Ratio and Proportion

Proportions in a Triangle

Defining Similar Polygons

Proving Triangles Similar

Proving Lengths of Sides of Similar Triangles in Proportion

Proving Products of Segment Lengths Equal

Ratio and Proportion

RATIOS

Allan is 30 years old and Bob is 10 years old. How do their ages compare? Obviously, Allan is 20 years older than Bob. It is sometimes desirable, however, to compare two numbers by determining how many times larger (or smaller) one number is compared to a second number. This can be accomplished by dividing the first number by the second number:

$$\frac{\text{Allan's age}}{\text{Bob's age}} = \frac{30}{10} = \frac{3}{1} \quad \text{or} \quad 3$$

Allan is 3 times as old as Bob. The result of dividing two numbers is called a *ratio*.

> ### DEFINITION OF RATIO
> The *ratio* of two numbers a and b ($b \neq 0$) is the quotient of the numbers. The numbers a and b are referred to as the *terms* of the ratio.

The ratio of two numbers, a and b, may be written in a variety of ways. For example,

$$\frac{a}{b} \quad a \div b \quad a \text{ to } b \quad a:b$$

We will normally express a ratio by using either the first or the last of these forms.

In writing the ratio of two numbers, it is usually helpful to express the ratio (fraction) in *simplest* form. For example, the ratio of 50 to 100 is expressed as follows:

$$\frac{50}{100} = \frac{1}{2} \quad \text{or} \quad 1:2 \text{ (read as "1 is to 2")}$$

In forming the ratio of two numbers, each number may be expressed in different units of measurement. For example, if a person travels 120 miles in a car in 3 hours, then the ratio of the distance traveled to the time traveled is

$$\frac{120 \text{ miles}}{3 \text{ hours}} = 40 \frac{\text{miles}}{\text{hour}}$$

The value 40 miles/hour is the average *rate* of speed during the trip.

If, however, a ratio is formed in order to determine how many times larger or smaller one value is than another, both quantities must be expressed in the same unit of measurement. For example, if the length of $\overline{AB}$ is 2 feet and the length of $\overline{XY}$ is 16 inches, then, to determine how many times larger $\overline{AB}$ is compared to $\overline{XY}$, we must convert one of the units of measurement into the other. In this example it is convenient to express feet in terms of inches. Since 2 feet is equivalent to 24 inches, we may write

$$\frac{AB}{XY} = \frac{24}{16} = \frac{3}{2}$$

The ratio of the length of $\overline{AB}$ to the length of $\overline{XY}$ is $3:2$. Since the decimal representation of $\frac{3}{2}$ is 1.5, we may say that the length of $\overline{AB}$ is 1.5 times the length of $\overline{XY}$.

EXAMPLE 10.1

If the measure of angle A is 60 and angle B is a right angle, find the ratio of the measure of angle A to the measure of angle B.

SOLUTION

$$\frac{m\measuredangle A}{m\measuredangle B} = \frac{60}{90} = \frac{2}{3} \quad \text{or} \quad 2:3$$

EXAMPLE 10.2

Find each of the following ratios using the figures provided:

a $AB:XY$ b $BC:YZ$ c $XZ:AC$

 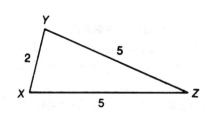

SOLUTION

a $\frac{8}{2} = 4:1$ b $\frac{7}{5} = 7:5$ c $\frac{5}{10} = 1:2$

EXAMPLE 10.3

The measures of a pair of consecutive angles of a parallelogram are in the ratio of $1:8$. Find the measure of the smaller of these angles.

SOLUTION

Let x = measure of smaller angle.
Then $8x$ = measure of larger angle.

Since consecutive angles of a parallelogram are supplementary,

$$x + 8x = 180$$
$$9x = 180$$
$$\frac{9x}{9} = \frac{180}{9}$$
$$x = 20$$

The measure of the smaller angle is 20.

EXAMPLE 10.4

In an isosceles triangle, the ratio of the measure of the vertex angle to the measure of a base angle is $2:5$. Find the measure of each angle of the triangle.

SOLUTION

Let $2x$ = measure of vertex angle.
Then $5x$ = measure of one base angle,
 and $5x$ = measure of the other base angle.

Since the sum of the measures of the angles of a triangle is 180,

$$2x + 5x + 5x = 180$$
$$12x = 180$$
$$x = \frac{180}{12} = 15$$
$$2x = 2(15) = 30$$
$$5x = 5(15) = 75$$

The measures of the three angles of the triangle are 30, 75, and 75.

PROPORTIONS

The ratio $\frac{24}{16}$ may be simplified and written as $\frac{3}{2}$:

$$\frac{24}{16} = \frac{3}{2}$$

An equation that states that two ratios are equal is called a *proportion*. The preceding proportion may also be written in the form $24:16 = 3:2$.

> ### DEFINITION OF PROPORTION
> A *proportion* is an equation that states that two ratios are equal:
> $$\frac{a}{b} = \frac{c}{d} \quad \text{or} \quad a:b = c:d \quad \text{(provided } b \neq 0 \text{ and } d \neq 0\text{)}$$

Each term of a proportion is given a special name according to its position in the proportion.

First proportional ⟶ ⎤ ⎡⟵ Third proportional
$$\frac{a}{b} = \frac{c}{d}$$
Second proportional ⟶ ⎦ ⎣⟵ Fourth proportional

The pair of terms that form the first and fourth proportionals are referred to as the *extremes* of a proportion; the second and third proportionals of a proportion are called the *means* of a proportion.

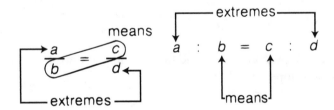

> ### DEFINITION OF THE TERMS OF A PROPORTION
> In the proportion $\frac{a}{b} = \frac{c}{d}$, a is called the *first* proportional, b is called the *second* proportional, c is called the *third* proportional, and d is called the *fourth* proportional. The pair of terms a and d are referred to as the *extremes* of the proportion; the pair of terms b and c are referred to as the *means* of the proportion.

Returning to the proportion $\frac{24}{16} = \frac{3}{2}$, we see that 24 and 2 (the outermost terms) are the extremes. The two innermost terms, 16 and 3, are the means. Notice that the *cross-products* of a proportion are equal:

$$\frac{24}{16} \rightleftharpoons = \rightleftharpoons \frac{3}{2}$$
$$16 \times 3 = 24 \times 2$$
$$48 = 48$$

THEOREM 10.1 EQUAL CROSS-PRODUCTS THEOREM

In a proportion the product of the means is equal to the product of the extremes. (Forming this product is sometimes referred to as cross-multiplying.)

EXAMPLE 10.5

Find the first proportional if the remaining terms of a proportion are 3, 14, and 21.

SOLUTION

Let x = first proportional. Then

$$\frac{x}{3} = \frac{14}{21}$$

By Theorem 10.1,

$$21x = 42$$
$$\frac{21x}{21} = \frac{42}{21}$$
$$x = 2$$

The first proportional is 2.

Sometimes we may be able to work with smaller numbers by simplifying an arithmetic ratio in the original proportion before cross-multiplying: If in Example 10.5 we write $\frac{14}{21}$ as $\frac{2}{3}$, then:

$$\frac{x}{3} = \frac{14}{21}$$
$$\frac{x}{3} = \frac{2}{3}$$
$$3x = 6$$
$$x = 2$$

EXAMPLE 10.6

The first term of a proportion is 2 and the second and third terms are both 8. Find the fourth proportional.

SOLUTION

Let x = fourth proportional. Then,

$$\frac{2}{8} = \frac{8}{x}$$

Simplify *before* applying Theorem 10.1:

$$\frac{1}{4} = \frac{8}{x}$$

Cross-multiply:

$$x = 32$$

The fourth proportional is 32.

In Example 10.6, the means were equal since the second and third terms of the proportion were both equal to 8. Whenever the means of a proportion are identical, then the value that appears in the means is referred to as the *mean proportional* (or *geometric mean*) between the first and fourth terms of the proportion. In Example 10.6, 8 is said to be the mean proportional between 2 and 32.

DEFINITION OF MEAN PROPORTIONAL

If the second and third terms of a proportion are the same, then either term is referred to as the *mean proportional* or *geometric mean* between the first and fourth terms of the proportion:

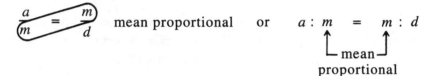

mean proportional or $a : m = m : d$

└── mean ──┘
proportional

EXAMPLE 10.7 Find the mean proportional between each pair of extremes.
a. 3 and 27 **b.** 5 and 7

SOLUTION

a. Let m = mean proportional between 3 and 27. Then

$$\frac{3}{m} = \frac{m}{27}$$
$$m^2 = 3(27)$$
$$= 81$$
$$m = \sqrt{81} = 9$$

b. Let m = mean proportional between 5 and 7.

$$\frac{5}{m} = \frac{m}{7}$$
$$m^2 = 35$$
$$m = \sqrt{35}$$

It is sometimes useful to be able to determine whether a pair of ratios are in proportion. Two ratios are in proportion if *the product of the means of the resulting proportion is equal to the product of the extremes*. Are the ratios $\frac{2}{5}$ and $\frac{12}{30}$ in proportion? We write a tentative proportion and then determine whether the cross-products are equal:

$$\frac{2}{5} \stackrel{?}{=} \frac{12}{30}$$
$$5 \times 12 \stackrel{?}{=} 30 \times 2$$
$$60 = 60$$

Therefore, $\dfrac{2}{5}$ and $\dfrac{12}{30}$ are in proportion.

Now we will repeat this procedure, this time to investigate whether the ratios $\dfrac{8}{12}$ and $\dfrac{6}{8}$ are in proportion:

$$\dfrac{8}{12} \overset{?}{=} \dfrac{6}{8}$$
$$12 \times 6 \overset{?}{=} 8 \times 8$$
$$72 \neq 64$$

This result implies that $\dfrac{8}{12}$ and $\dfrac{6}{8}$ are *not* in proportion.

Some algebraic properties of proportions are worth remembering.

- **PROPERTY 1** If the numerators and denominators of a proportion are switched, then an equivalent proportion results.

 If $\dfrac{a}{b} = \dfrac{c}{d}$, then $\dfrac{b}{a} = \dfrac{d}{c}$

 (provided a, b, c, and d are nonzero numbers).

- **PROPERTY 2** If either pair of opposite terms of a proportion are interchanged, then an equivalent proportion results.

 a If $\dfrac{a}{b} = \dfrac{c}{d}$, then $\dfrac{d}{b} \searrow \dfrac{c}{a}$.

 b If $\dfrac{a}{b} = \dfrac{c}{d}$, then $\dfrac{a}{c} \nearrow \dfrac{b}{d}$.

- **PROPERTY 3** If the denominator is added to or subtracted from the numerator on each side of the proportion, then an equivalent proportion results.

 a If $\dfrac{a}{b} = \dfrac{c}{d}$, then $\dfrac{a+b}{b} = \dfrac{c+d}{d}$.

 b If $\dfrac{a}{b} = \dfrac{c}{d}$, then $\dfrac{a-b}{b} = \dfrac{c-d}{d}$.

- **PROPERTY 4** If the product of two nonzero numbers equals the product of another pair of nonzero numbers, then a proportion may be formed by making the factors of one product the extremes, and making the factors of the other product the means. For example, if $R \times S = T \times W$, then we may

 a make R and S the *extremes:* $\dfrac{R}{T} = \dfrac{W}{S}$

 or

 b make R and S the *means:* $\dfrac{T}{R} = \dfrac{S}{W}$.

Proportions in a Triangle

In Chapter 9 we saw that a line passing through the midpoints of two sides of a triangle was parallel to the third side (and one-half of its length). Suppose we draw a line parallel to a side of a triangle so that it intersects the other two sides, but not necessarily at their midpoints. Many such lines can be drawn, as shown in Figure 10.1.

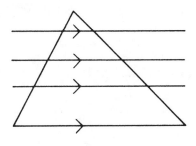

FIGURE 10.1

We will consider one of these lines and the segments that it forms on the sides of the triangle, as shown in Figure 10.2.

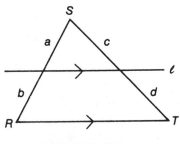

FIGURE 10.2

It will be convenient to postulate that line ℓ divides $\overline{RS}$ and $\overline{ST}$ in such a way that the lengths of corresponding segments on each side have the same ratio:

$$\frac{a}{b} = \frac{c}{d} \quad \text{or} \quad \frac{a}{RS} = \frac{c}{ST} \quad \text{or} \quad \frac{b}{RS} = \frac{d}{ST}$$

If any of the above ratios holds, then the line segments are said to be *divided proportionally*. We notice that each of these ratios has the form

$$\frac{\text{Upper segment of side } \overline{RS}}{\text{Lower segment of side } \overline{RS}} = \frac{\text{upper segment of side } \overline{ST}}{\text{lower segment of side } \overline{ST}}$$

or

$$\frac{\text{Upper segment of side } \overline{RS}}{\text{Whole side } (\overline{RS})} = \frac{\text{upper segment of side } \overline{ST}}{\text{whole side } (\overline{ST})}$$

or

$$\frac{\text{Lower segment of side } \overline{RS}}{\text{Whole side } (\overline{RS})} = \frac{\text{lower segment of side } \overline{ST}}{\text{whole side } (\overline{ST})}$$

Keep in mind that the algebraic properties of proportions allow these three proportions to be expressed in equivalent forms. For example, the numerator and denominator of each fraction may be interchanged (that is, each ratio may be *inverted*).

POSTULATE 10.1

A line parallel to one side of a triangle and intersecting the other two sides *divides these sides proportionally.*

In triangle RST, line segment $\overline{EF}$ is parallel to side $\overline{RT}$, intersecting side $\overline{RS}$ at point E and side $\overline{TS}$ at point F.

a. If $SE = 8$, $ER = 6$, $FT = 15$, find SF.
b. If $SF = 4$, $ST = 12$, $SR = 27$, find SE.
c. If $SE = 6$, $ER = 4$, $ST = 20$, find FT.

SOLUTION

a.
$$\frac{SE}{ER} = \frac{SF}{FT}$$
$$\frac{8}{6} = \frac{SF}{15}$$
$$\frac{4}{3} = \frac{SF}{15}$$
$$3(SF) = 60$$
$$SF = \frac{60}{3} = 20$$

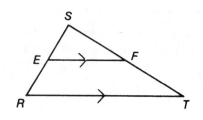

b.
$$\frac{SE}{SR} = \frac{SF}{ST}$$
$$\frac{SE}{27} = \frac{4}{12}$$
$$\frac{SE}{27} = \frac{1}{3}$$
$$3(SE) = 27$$
$$SE = \frac{27}{3} = 9$$

c.
$$\frac{ER}{RS} = \frac{FT}{ST}$$
$$\frac{4}{4+6} = \frac{FT}{20}$$
$$\frac{4}{10} = \frac{FT}{20}$$
$$\frac{2}{5} = \frac{FT}{20}$$
$$5(FT) = 40$$
$$FT = \frac{40}{5} = 8$$

The converse of Postulate 10.1 is also true. If a line is drawn so that the ratio of the segment lengths it cuts off on one side of a triangle is equal to the ratio of the segment lengths it cuts off on a second side of a triangle, then the line must be parallel to the third side of the triangle.

POSTULATE 10.2

A line that divides two sides of a triangle *proportionally* is parallel to the third side of the triangle.

Determine whether $\overline{AB} \parallel \overline{KJ}$ if:
a. $KA = 2$, $AL = 5$, $JB = 6$, and $BL = 15$.
b. $AL = 3$, $KL = 8$, $JB = 10$, and $JL = 16$.
c. $AL = 5$, $KA = 9$, $LB = 10$, and $JB = 15$.

SOLUTION

In each instance, write a tentative proportion and determine whether the proportion is true. If it is true, then $\overline{AB}$ is parallel to $\overline{KJ}$. The proportion written must conform to the meaning of divided proportionally.

a. On the basis of the information provided, use the proportion $\dfrac{KA}{AL} = \dfrac{JB}{BL}$ (equivalent proportions may also be formed). Determine whether this proportion is true using the numbers provided:

$$\frac{2}{5} \stackrel{?}{=} \frac{6}{5}$$
$$2 \times 15 = 5 \times 6$$
$$30 = 30$$

Therefore, $\overline{AB} \parallel \overline{KJ}$.

b. Use the proportion $\dfrac{AL}{KL} = \dfrac{BL}{JL}$ (other proportions can also be used). Since $BL = 16 - 10 = 6$;

$$\frac{3}{8} \stackrel{?}{=} \frac{6}{16}$$

$\overline{AB}$ is parallel to $\overline{KJ}$ since $3 \times 16 = 8 \times 6$.

c. Use the proportion $\dfrac{AL}{KA} = \dfrac{LB}{JB}$:

$$\frac{5}{9} \stackrel{?}{=} \frac{10}{15}$$

$\overline{AB}$ is *not* parallel to $\overline{KJ}$ since $5 \times 15 \neq 9 \times 10$.

GIVEN: Quadrilateral $RSTW$ with $\overline{KJ}$ and $\overline{LM}$ drawn,

$$\frac{RK}{KS} = \frac{RJ}{JW},$$
$$\frac{TL}{LS} = \frac{TM}{MW}.$$

PROVE: $\overline{KJ} \parallel \overline{LM}$.

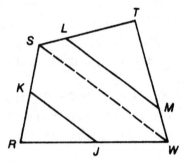

SOLUTION

PLAN: Draw $\overline{SW}$. Applying Postulate 10.2 gives $\overline{KJ} \parallel \overline{SW}$ and $\overline{LM} \parallel \overline{SW}$. Hence, $\overline{KJ} \parallel \overline{LM}$.

PROOF:

Statements	Reasons
1. Quadrilateral $RSTW$ with $\overline{KJ}$ and $\overline{LM}$ drawn.	1. Given.
2. Draw $\overline{SW}$. In $\triangle SRW$:	2. Two points determine a line.
3. $\dfrac{RK}{KS} = \dfrac{RJ}{JW}$.	3. Given.
4. $\overline{KJ} \parallel \overline{SW}$.	4. A line that divides two sides of a triangle proportionally is parallel to the third side of the triangle.
In $\triangle WTS$:	
5. $\dfrac{TL}{LS} = \dfrac{TM}{MW}$.	5. Given.
6. $\overline{LM} \parallel \overline{SW}$.	6. Same as reason 4.
7. $\overline{KJ} \parallel \overline{LM}$.	7. Two lines parallel to the same line are parallel to each other.

Defining Similar Polygons

Compare the three triangles in Figure 10.3. Triangles I and III have exactly the same size and shape since they agree in three pairs of angles *and* in three pairs of sides. Triangle I is *congruent* to triangle III.

Triangles II and III have three pairs of congruent corresponding angles, but *each* side of triangle II is twice the length of the corresponding side of triangle III. As a result, triangles II and III have the same shape, although not the same size. Polygons that have the same shape are said to be *similar*.

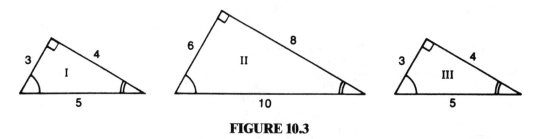

FIGURE 10.3

The concept of similarity is frequently encountered in everyday life. When a photograph is enlarged, the original and enlarged objects are *similar* since they have exactly the same shape. In designing a blueprint, everything must be drawn to scale so that the figures in the blueprint are *similar* to the actual figures.

DEFINITION OF SIMILAR POLYGONS
Two polygons are *similar* if their vertices can be paired so that corresponding angles are congruent *and* the ratios of the lengths of all corresponding sides are equal.

The definition of *similar* polygons assumes that there exists a one-to-one correspondence between the vertices of the polygons. This means that each vertex of the first polygon is matched with exactly one vertex of the second polygon and vice versa.

Congruent polygons have the same size *and* shape, while *similar* polygons have only the same shape. The symbol for similarity is ~. The expression △*ABC* ~ △*RST* is read as "Triangle *ABC* is similar to triangle *RST*." If it is known that two polygons are similar, it may be concluded that each pair of corresponding angles are congruent *and* the ratios of the lengths of all pairs of corresponding sides are equal.

FIGURE 10.4

In Figure 10.4, since quadrilateral *ABCD* is similar to quadrilateral *JKLM*, then the following relationships must hold:

Corresponding Angles are Congruent	Lengths of Corresponding Sides are in Proportion
$\angle A \cong \angle J$ $\angle B \cong \angle K$ $\angle C \cong \angle L$ $\angle D \cong \angle M$	$\dfrac{AB}{JK} = \dfrac{BC}{KL} = \dfrac{CD}{LM} = \dfrac{AD}{JM}$

FIGURE 10.5

For example, let's find the measures of the parts of quadrilateral *JKLM* in Figure 10.5. In Figure 10.4 quadrilaterals *ABCD* and *JKLM* were given to be similar. Therefore:

$$m\angle J = m\angle A = 65$$
$$m\angle K = m\angle B = 85$$
$$m\angle L = m\angle C = 120$$
$$m\angle M = m\angle D = 90$$

The ratio of the lengths of each pair of corresponding sides must be equal to

$$\frac{JM}{AD} = \frac{5}{10} = \frac{1}{2}$$

The ratio of the length of any side of quadrilateral *JKLM* to the length of the corresponding side of quadrilateral *ABCD* must also be 1:2. Hence, *JK* = 4, *KL* = 2, and *LM* = 1.

EXAMPLE 10.11

GIVEN: $\triangle HLX \sim \triangle WKN$.

 a. List the pairs of corresponding congruent angles.

 b. Write an extended proportion that forms the ratios of the lengths of corresponding sides.

 c. If *HL* = 8, *LX* = 14, *HX* = 18, and *WN* = 27, find the lengths of the remaining sides of $\triangle WKN$.

SOLUTION

a. The order in which the vertices of the triangles are written is significant—it defines the pairs of corresponding vertices:

$$H \leftrightarrow W \qquad L \leftrightarrow K \qquad X \leftrightarrow N$$

Hence, $\angle H \cong \angle W$, $\angle L \cong \angle K$, and $\angle X \cong \angle N$.

b. Corresponding sides connect corresponding vertices:

Correspondence	Corresponding Sides
$\triangle HLX \sim \triangle WKN$	$\overline{HL} \leftrightarrow \overline{WK}$
$\triangle HLX \sim \triangle WKN$	$\overline{LX} \leftrightarrow \overline{KN}$
$\triangle HLX \sim \triangle WKN$	$\overline{HX} \leftrightarrow \overline{WN}$

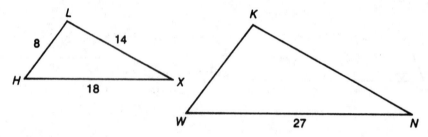

The resulting proportion may be written as follows:

$$\frac{\text{Side of } \triangle HLX}{\text{Corresponding side of } \triangle WKN} = \frac{HL}{WK} = \frac{LX}{KN} = \frac{HX}{WN}$$

After determining the pairs of corresponding vertices, it is probably easier to determine the corresponding sides of two similar figures by drawing a diagram (see the accompanying figure) and then applying the principle that *corresponding sides lie opposite corresponding angles* (vertices). $\overline{HX}$ and $\overline{WN}$ lie opposite corresponding angles L and K; $\overline{HX}$ and $\overline{WN}$ therefore represent a pair of corresponding sides.

c. Since $\overline{HX}$ and $\overline{WN}$ are corresponding sides, the lengths of each pair of corresponding sides must be the same as the ratio of HX to WN:

$$\frac{HX}{WN} = \frac{18}{27} = \frac{2}{3}$$

The length of any side of $\triangle HLX$ to the length of the corresponding side of $\triangle WKN$ must be in the ratio of $2:3$. Using this fact, we can find the lengths of the remaining sides of $\triangle WKN$.

To find WK:

$$\frac{2}{3} = \frac{HL}{WK}$$

$$\frac{2}{3} = \frac{8}{WK}$$

$$2(WK) = 24$$

$$WK = \frac{24}{2} = 12$$

To find KN:

$$\frac{2}{3} = \frac{LX}{KN}$$

$$\frac{2}{3} = \frac{14}{KN}$$

$$2(KN) = 42$$

$$KN = \frac{42}{2} = 21$$

EXAMPLE
10.12

In each part, determine whether the pair of polygons are similar.

a. b. c.

SOLUTION

a. No. The ratios of the lengths of corresponding sides are equal (3 : 1), but corresponding angles are *not* congruent.

b. No. Corresponding angles are congruent (each figure contains four right angles), but the ratios of the lengths of corresponding sides are *not* all equal.

c. Yes. Corresponding angles are congruent and the ratios of the lengths of all pairs of corresponding sides are the same (1 : 2).

If two polygons are similar then the ratio of the lengths of any pair of corresponding sides is called the *ratio of similitude*. If the lengths of a pair of corresponding sides of two similar polygons are 3 and 12, then the ratio of similitude is $\frac{3}{12}$ or 1 : 4.

Are congruent polygons similar? Yes, since they satisfy the two conditions of similarity; all corresponding pairs of angles are congruent and the ratios of the lengths of all pairs of corresponding sides are the same (1 : 1). Are similar polygons also congruent? Generally speaking, no! Similar polygons are congruent only if their ratio of similitude is 1 : 1.

EXAMPLE
10.13

Two quadrilaterals are similar and have a ratio of similitude of 1 : 3. If the lengths of the sides of the smaller quadrilateral are 2, 5, 8, and 12, find the lengths of the sides of the larger quadrilateral.

SOLUTION

Since the ratio of similitude is 1 : 3, the length of each side of the larger quadrilateral is 3 times the length of the corresponding side in the smaller quadrilateral. The lengths of the sides of the larger quadrilateral are 6, 15, 24, and 36.

EXAMPLE
10.14

Two quadrilaterals are similar. The length of the sides of the smaller quadrilateral are 4, 6, 12, and 18. The length of the longest side of the larger quadrilateral is 27. Determine each of the following:

a. The ratio of similitude.

b. The lengths of the remaining sides of the larger quadrilateral.

c. The ratio of the perimeters of the two quadrilaterals.

SOLUTION

a. The longest sides of the two quadrilaterals are 27 and 18. The ratio of similitude (larger to smaller quadrilateral) is $\dfrac{27}{18}$ or $\dfrac{3}{2}$.

b. Let x = length of a side of the larger quadrilateral. Then:

$\dfrac{3}{2} = \dfrac{x}{4}$ implies $2x = 12$ and $x = 6$

$\dfrac{3}{2} = \dfrac{x}{6}$ implies $2x = 18$ and $x = 9$

$\dfrac{3}{2} = \dfrac{x}{12}$ implies $2x = 36$ and $x = 18$

The sides of the larger quadrilateral are 6, 9, 18, and 27. Alternatively, we could have simply multiplied the length of each side of the smaller quadrilateral by the ratio of similitude $\left(\dfrac{3}{2}\right)$ in order to obtain the length of the corresponding side in the larger quadrilateral.

c. $\dfrac{\text{Perimeter of larger quadrilateral}}{\text{Perimeter of smaller quadrilateral}} = \dfrac{6 + 9 + 18 + 27}{4 + 6 + 12 + 18}$

$= \dfrac{60}{40} = \dfrac{3}{2}$

Notice that in part **c** of Example 10.14 the ratio of the perimeters of the quadrilaterals is the same as the ratio of the lengths of a pair of corresponding sides.

THEOREM 10.2 RATIO OF PERIMETERS THEOREM

The perimeters of a pair of similar polygons have the same ratio as the lengths of any pair of corresponding sides.

Proving Triangles Similar

To prove triangles *congruent*, we did not have to show that three pairs of angles were congruent *and* three pairs of sides were congruent. Instead, we used shortcut methods that depended on showing that a particular set of *three* parts of one triangle were congruent to the corresponding parts of the second triangle.

The definition of *similarity* requires that, to show two triangles are similar, we need to demonstrate that three pairs of corresponding angles are congruent *and* that the ratios of the lengths of three pairs of corresponding sides are the same. A formidable task! Fortunately, there is shortcut method that can be used to prove a pair of *triangles* similar.

Let's try the following experiment. Suppose we wish to draw a triangle that has at least two angles with the same measures as two angles of triangle *ABC* (Figure 10.6). We may begin by drawing a line ℓ and choosing *any* two distinct points of line ℓ, say *D* and *E* (Figure 10.7). At points *D* and *E* we will use a protractor to draw angles having the same measures as angles *A* and *B*. The point at which the two rays having *D* and *E* as end points meet will be called *F*.

FIGURE 10.6

FIGURE 10.7

How do the measures of angles *C* and *F* compare? These angles must be congruent since, if two angles of a triangle are congruent to two angles of another triangle, then the third pair of angles must also be congruent (the sum of their measures must equal 180 in each case). Another interesting thing happens. If we actually measured the sides of each triangle and then compared the ratios $\frac{AB}{DE}, \frac{BC}{EF}$, and $\frac{AC}{DF}$, we would find that they are equal! Hence, $\triangle ABC$ must be similar to $\triangle DEF$ since the requirements of the definition of similarity are satisfied. We began by making two angles of $\triangle DEF$ congruent to two angles of $\triangle ABC$. This forced the third pair of angles to be congruent and the lengths of the corresponding sides to be in proportion.

THEOREM 10.3 THE ANGLE-ANGLE (AA) THEOREM OF SIMILARITY

If two angles of one triangle are congruent to two angles of another triangle, then the triangles are similar.

Theorem 10.3 provides a simple method of proving triangles similar. All we need do is show that *two* angles of one triangle are congruent to *two* angles of another triangle. The proof of this theorem is the topic of Exercise 16 at the end of the chapter.

GIVEN: $\overline{CB} \perp \overline{BA}$,
 $\overline{CD} \perp \overline{DE}$.

PROVE: $\triangle ABC \sim \triangle EDC$.

SOLUTION

PLAN: Use the AA Theorem. The two triangles include right and vertical angles that yield two pairs of congruent angles.

PROOF:

Statements	Reasons
1. $\overline{CB} \perp \overline{BA}$ and $\overline{CD} \perp \overline{DE}$.	1. Given.
2. Angles *ABC* and *EDC* are right angles.	2. Perpendicular lines intersect to form right angles.
3. $\angle ABC \cong \angle EDC$. (Angle)	3. All right angles are congruent.
4. $\angle ACB \cong \angle ECD$. (Angle)	4. Vertical angles are congruent.
5. $\triangle ABC \sim \triangle EDC$.	5. AA Theorem.

Proving Lengths of Sides of Similar Triangles in Proportion

SIDES OF SIMILAR TRIANGLES

We have learned that congruent triangles may be used to establish that a pair of angles or segments are congruent. In like fashion, we can establish that the lengths of four segments are in proportion by first showing that a pair of triangles that contain these segments as corresponding sides are similar. We may then apply the reverse of the definition of similarity and conclude that the lengths of the segments are in proportion, using as a reason: *The lengths of corresponding sides of similar triangles are in proportion.*

GIVEN: $\overline{AB} \parallel \overline{DE}$.

PROVE: $\dfrac{EC}{BC} = \dfrac{ED}{AB}$.

SOLUTION

PLAN: **1** Select the triangles that contain these segments as sides. Read across the proportion:

$$\triangle ECD$$
$$\downarrow \qquad \downarrow$$
$$\frac{EC}{BC} = \frac{ED}{AB}$$
$$\uparrow \qquad \uparrow$$
$$\triangle BCA$$

2 Mark the diagram with the Given and all pairs of corresponding congruent angle pairs.

3 Write the proof.

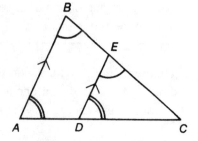

PROOF:

Statements	Reasons
1. $\overline{AB} \parallel \overline{DE}$.	1. Given.
2. $\angle CED \cong \angle CBA$. (Angle) $\angle CDE \cong \angle CAB$. (Angle)	2. If two lines are parallel, then their corresponding angles are congruent.
3. $\triangle ECD \sim \triangle BCA$.	3. AA Theorem.
4. $\dfrac{EC}{BC} = \dfrac{ED}{AB}$.	4. The lengths of corresponding sides of similar triangles are in proportion.

- In this proof notice that the statement "The lengths of corresponding sides of similar triangles are in proportion" is analogous to the statement "Corresponding sides of *congruent* triangles are congruent," which was abbreviated as CPCTC. We will *not* introduce an abbreviation for the expression that appears as reason 4 in the preceding proof.
- Statement 3 of the proof establishes that a *line intersecting two sides of a triangle and parallel to the third side forms two similar triangles.*

GIVEN: $\overline{AC} \perp \overline{CB}, \overline{ED} \perp \overline{AB}.$
PROVE: $EB{:}AB = ED{:}AC.$

SOLUTION

PLAN: **1** Rewrite the Prove in fractional form: $\dfrac{EB}{AB} = \dfrac{ED}{AC}$.

2 Determine the pair of triangles that must be proved similar.

$$\triangle EBD$$

$$\frac{EB}{AB} = \frac{ED}{AC}$$

$$\triangle ABC$$

3 Mark the diagram with the Given and all pairs of corresponding congruent angle pairs.

4 Write the proof.

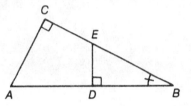

PROOF:

Statements	Reasons
1. $\overline{AC} \perp \overline{CB}, \overline{ED} \perp \overline{AB}.$	1. Given.
2. Angles C and EDB are right angles.	2. Perpendicular lines intersect to form right angles.
3. $\angle C \cong \angle EDB.$ (Angle)	3. All right angles are congruent.
4. $\angle B \cong \angle B.$ (Angle)	4. Reflexive property of congruence.
5. $\triangle EBD \sim \triangle ABC.$	5. AA Theorem.
6. $EB{:}AB = ED{:}AC$	6. The lengths of corresponding sides of similar triangles are in proportion.

SUMMARY

To prove that the lengths of segments are in proportion:
1. Use the proportion provided in the Prove to help identify the triangles that contain the desired segments as sides.
2. Mark the diagram with the Given as well as with any additional information that may be deduced (for example, vertical angles, right angles, congruent alternate interior angles).
3. Prove the triangles similar.
4. Write the desired proportion using, as a reason: "The lengths of corresponding sides of similar triangles are in proportion."

ALTITUDES AND MEDIANS OF SIMILAR TRIANGLES

In similar triangles the lengths of altitudes or medians drawn to corresponding sides have the same ratio as the lengths of any pair of corresponding sides.

THEOREM 10.4 RATIO OF ALTITUDES AND MEDIANS IN SIMILAR TRIANGLES

If a pair of triangles are similar, then the lengths of a pair of corresponding altitudes or medians have the same ratio as the lengths of any pair of corresponding sides.

EXAMPLE 10.18

$\triangle RST \sim \triangle KLM$. The length of altitude $\overline{SA}$ exceeds the length of altitude $\overline{LB}$ by 5. If $RT = 9$ and $KM = 6$, find the length of each altitude.

SOLUTION

Let x = length of altitude $\overline{LB}$.
Then $x + 5$ = length of altitude $\overline{SA}$.

Applying Theorem 10.4 gives

$$\frac{x+5}{x} = \frac{9}{6}$$
$$9x = 6(x + 5)$$
$$= 6x + 30$$
$$3x = 30$$
$$x = 10$$

Altitude $LB = x = 10$
Altitude $SA = x + 5 = 15$

SUMMARY

The lengths of corresponding *altitudes*, the lengths of corresponding *medians*, and the *perimeters* of similar triangles have the same ratio as the lengths of any pair of corresponding sides:

$$\frac{\text{Perimeter of } \triangle ABC}{\text{Perimeter of } \triangle RST} = \frac{BX}{SY} = \frac{BM}{SL} = \frac{AB}{RS} = \frac{BC}{ST} = \frac{AC}{RT}$$

Proving Products of Segment Lengths Equal

We know that, if $\dfrac{A}{B} = \dfrac{C}{D}$, then $A \times D = B \times C$. The reason is that in a proportion the product of the means equals the product of the extremes. Instead of generating a product from a proportion, we sometimes need to be able to take a product and determine the related proportion that would yield that product.

Suppose the lengths of four segments are related in such a way that

$$KM \times LB = LM \times KD$$

What proportion gives this result when the products of its means and extremes are set equal to each other? A true proportion may be derived from the product by designating a pair of terms appearing on the same side of the equal sign as the extremes (say, *KM* and *LB*). The pair of terms on the opposite side of the equal sign then becomes the means (*LM* and *KD*):

$$\dfrac{KM}{LM} = \dfrac{KD}{LB}$$

An equivalent proportion results if *KM* and *LB* are made the means rather than the extremes.

These ideas are needed in problems in which similar triangles are used to prove products of segment lengths equal. As an illustration, let's look at the following problem.

GIVEN: $\square ABCD$.
PROVE: $KM \times LB = LM \times KD$.

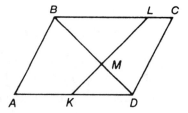

We can identify the needed proportion and the pair of triangles that we will have to prove similar by *working backwards*, from the products in the Prove:

1. Express the product as an equivalent proportion:

$$\dfrac{KM}{LM} = \dfrac{KD}{LB}$$

2. From the proportion (and, in some problems, in conjunction with the Given), determine the pair of triangles to be proved similar:

$$\triangle KMD$$
$$\dfrac{KM}{LM} = \dfrac{KD}{LB}$$
$$\triangle LMB$$

3. Mark the diagram with the Given and decide how to show that the triangles are similar.
 STRATEGY: Use the AA Theorem.

4. Write the formal two-column proof. The steps in the proof should reflect the logic of this analysis, proceeding from step 4 back to step 1 (proving the triangles similar, forming the appropriate proportion, and, lastly, writing the product):

PROOF: Statements	Reasons
1. $\square ABCD$.	1. Given.
2. $\overline{AD} \parallel \overline{BC}$.	2. Opposite sides of a parallelogram are parallel.
3. $\angle 1 \cong \angle 2$, $\angle 3 \cong \angle 4$.	3. If two lines are parallel, then their alternate interior angles are congruent.
4. $\triangle KMD \sim \triangle LMB$.	4. AA Theorem.
5. $\dfrac{KM}{LM} = \dfrac{KD}{LB}$.	5. The lengths of corresponding sides of similar triangles are in proportion.
6. $KM \times LB = LM \times KD$.	6. In a proportion, the product of the means equals the product of the extremes.

In this proof, notice that:

• Compared with our previous work with similar triangles, the only new step is the last statement/reason of the proof. To prove products of segment lengths equal, we must first prove that a related proportion is true; to prove the proportion is true, we must first establish that the triangles that contain these segments as sides are similar.

• In our analysis of the equality $KM \times LB = LM \times KD$, suppose we formed the proportion

$$\frac{KM}{KD} = \frac{LM}{LB}$$

Reading across the top (K-M-L), we do not find a set of letters that correspond to the vertices of a triangle. When this happens, we switch the terms in either the means or the extremes of the proportion:

$$\frac{KM}{LM} \rightleftharpoons \frac{KD}{LB}$$

Reading across the top (K-M-D) now gives us the vertices of one of the desired triangles.

REVIEW EXERCISES FOR CHAPTER 10

1. Find the measure of the largest angle of a triangle if the measures of its interior angles are in the ratio $3:5:7$.

2. Find the measure of the vertex angle of an isosceles triangle if the measures of the vertex angle and a base angle have the ratio $4:3$.

3. The measures of a pair of consecutive angles of a parallelogram have the ratio $5:7$. Find the measure of each angle of the parallelogram.

4. Solve for x.

 (a) $\dfrac{2}{6} = \dfrac{8}{x}$ (b) $\dfrac{2}{x} = \dfrac{x}{50}$ (c) $\dfrac{2x-5}{3} = \dfrac{9}{4}$

 (d) $\dfrac{4}{x+3} = \dfrac{1}{x-3}$ (e) $\dfrac{3}{x} = \dfrac{x-4}{7}$

5. Find the mean proportional between each pair of extremes.

 (a) 4 and 16 (b) $3e$ and $12e^3$

 (c) $\dfrac{1}{2}$ and $\dfrac{1}{8}$ (d) 6 and 9

6. In each part, determine whether the pairs of ratios are in proportion:

 (a) $\dfrac{1}{2}$ and $\dfrac{9}{18}$ (b) $\dfrac{12}{20}$ and $\dfrac{3}{5}$

 (c) $\dfrac{4}{9}$ and $\dfrac{12}{36}$ (d) $\dfrac{15}{25}$ and $\dfrac{20}{12}$

7. In $\triangle BAG$, $\overline{LM}$ intersects side $\overline{AB}$ at L and side $\overline{AG}$ at M so that $\overline{LM}$ is parallel to $\overline{BG}$. Write at least three different true proportions. (Do *not* generate equivalent proportions by inverting the numerator and denominator of each ratio.)

8. For each of the following segment lengths, determine whether $\overline{TP} \parallel \overline{BC}$.

 (a) $AT = 5, TB = 15, AP = 8, PC = 24$.
 (b) $TB = 9, AB = 18, AP = 6, PC = 6$.
 (c) $AT = 4, AB = 12, AP = 6, AC = 15$.
 (d) $AT = 3, TB = 9, PC = 4, AC = 12$.
 (e) $AT = 1/3 \cdot AB$ and $PC = 2 \cdot AP$.

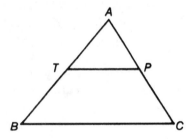

9. If $\overline{KW} \parallel \overline{EG}$, find the length of each indicated segment.

 (a) $HE = 20, KE = 12, WG = 9, HG = ?$
 (b) $KH = 7, KE = 14, HG = 12, HW = ?$
 (c) $HW = 4, WG = 12, HE = 28, KH = ?$
 (d) $KH = 9, KE = 12, HG = 42, WG = ?$
 (e) $KH = 2x - 15, KE = x, HW = 1, HG = 4$.
 Find KH and KE.

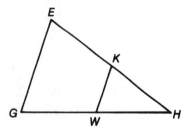

10. $\triangle GAL \sim \triangle SHE$. Name three pairs of congruent angles and three equal ratios.

11. The ratio of similitude of two similar polygons is $3:5$. If the length of the shortest side of the smaller polygon is 24, find the length of the shortest side of the larger polygon.

12. $\triangle ZAP \sim \triangle MYX$. If $ZA = 3$, $AP = 12$, $ZP = 21$, and $YX = 20$, find the lengths of the remaining sides of $\triangle MYX$.

13. Quadrilateral $ABCD \sim$ quadrilateral $RSTW$. The lengths of the sides of quadrilateral $ABCD$ are 3, 6, 9, and 15. If the length of the longest side of quadrilateral $RSTW$ is 20, find the perimeter of $RSTW$.

14. The longest side of a polygon exceeds twice the length of the longest side of a similar polygon by 3. If the ratio of similitude of the polygons is $4:9$, find the length of the longest side of each polygon.

15. $\triangle RST \sim \triangle JKL$.

 (a) $\overline{RA}$ and $\overline{JB}$ are medians to sides $\overline{ST}$ and $\overline{KL}$, respectively. $RS = 10$ and $JK = 15$. If the length of JB exceeds the length of RA by 4, find the lengths of medians $\overline{JB}$ and $\overline{RA}$.
 (b) $\overline{SH}$ and $\overline{KO}$ are altitudes to sides $\overline{RT}$ and $\overline{JL}$, respectively. If $SH = 12$, $KO = 15$, $LK = 3x - 2$, and $TS = 2x + 1$, find the lengths of $\overline{LK}$ and $\overline{TS}$.
 (c) The perimeter of $\triangle RST$ is 25 and the perimeter of $\triangle JKL$ is 40. If $\overline{ST} = 3x + 1$ and $\overline{KL} = 4x + 4$, find the lengths of ST and KL.
 (d) The ratio of similitude of $\triangle RST$ to $\triangle JKL$ is $3:x$. The length of altitude $\overline{SU}$ is $x - 4$ and the length of altitude $\overline{KV}$ is 15. Find the length of altitude $\overline{SU}$.

16. Supply the missing reasons in Part I of the following proof of the AA Theorem of Similarity (Theorem 10.3). Supply statements and reasons for Part II of the proof.

 GIVEN: $\angle R \cong \angle L$, $\angle T \cong \angle P$.
 PROVE: $\triangle RST \sim \triangle LMP$.
 PLAN: $\angle S \cong \angle M$. Show sides are in proportion.

 PROOF:

Statements	Reasons
Part I. To show that $$\frac{SR}{ML} = \frac{ST}{MP}$$	
1. Assume $SR > ML$. On $\overline{SR}$ choose point A such that $SA = ML$.	1. At a given point on a line, a segment may be drawn equal in length to a given segment.
2. Through point A draw a line parallel to $\overline{RT}$, intersecting side $\overline{ST}$ at point B.	2. ?
3. $\dfrac{SR}{SA} = \dfrac{ST}{SB}$.	3. ?

4. $\dfrac{SR}{ML} = \dfrac{ST}{SB}.$ | 4. ?

Show $\triangle ASB \cong \triangle LMP$ in order to obtain $SB = MP$.

5. $\angle SAB \cong \angle R.$ | 5. ?
6. $\angle R \cong \angle L$ and $\angle T \cong \angle P.$ | 6. ?
7. $\angle SAB \cong \angle L.$ | 7. ?
8. $\angle S \cong \angle M.$ | 8. ?
9. $\triangle ASB \cong \triangle LMP.$ | 9. ?
10. $SB = MP.$ | 10. ?
11. $\dfrac{SR}{ML} = \dfrac{ST}{MP}.$ | 11. ?

Part II. Show that

$$\frac{ST}{MP} = \frac{RT}{LP}.$$

HINT: Locate a point C on side $\overline{RT}$ such that $TC = PL$, and use the procedure followed in statements 2 through 11. Since all pairs of corresponding angles are congruent and $\dfrac{SR}{ML} = \dfrac{ST}{MP} = \dfrac{RT}{LP}$, the reverse of the definition of similar polygons is satisfied. Thus, $\triangle RST \sim \triangle LMP$.

17. GIVEN: $\overline{XW} \cong \overline{XY}$,
 $\overline{HA} \perp \overline{WY}, \overline{KB} \perp \overline{WY}.$
 PROVE: $\triangle HWA \sim \triangle KYB.$

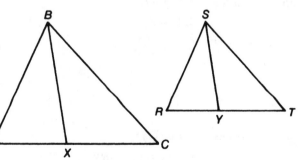

18. GIVEN: $\triangle ABC \sim \triangle RST$,
 $\overline{BX}$ bisects $\angle ABC$,
 $\overline{SY}$ bisects $\angle RST$.
 PROVE: $\triangle BXC \sim \triangle SYT.$

Use the accompanying diagram for Exercises 19 and 20.

19. GIVEN: $\overline{AC} \perp \overline{BD}$ and $\overline{DE} \perp \overline{AB}.$
 PROVE: $\triangle EFA \sim \triangle CFD.$

20. GIVEN: $\overline{AE} \cong \overline{AF}$ and $\overline{DF} \cong \overline{DC}$.
 PROVE: $\dfrac{AF}{EF} = \dfrac{DF}{FC}$.

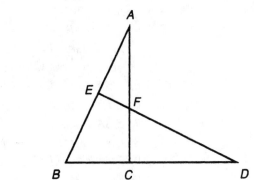

Use the accompanying diagram for Exercises 21 to 23.

21. GIVEN: $\triangle MCT \sim \triangle BAW$,
 $\overline{SW}$ bisects $\measuredangle AWM$.
 PROVE: $\triangle MCT \sim \triangle BCW$.

22. GIVEN: $\overline{AW} \parallel \overline{ST}$,
 $\overline{MS} \cong \overline{MW}$,
 $\overline{WC} \cong \overline{WA}$.
 PROVE: $\triangle BCW \sim \triangle BTS$.

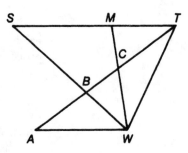

23. GIVEN: $\overline{WB} \cong \overline{WC}$, $\overline{ST} \parallel \overline{AW}$,
 $\overline{AT}$ bisects $\measuredangle STW$.
 PROVE: $\triangle ABW \sim \triangle TCW$.

24. GIVEN: $\overline{HW} \parallel \overline{TA}$,
 $\overline{HY} \parallel \overline{AX}$.
 PROVE: $\dfrac{AX}{HY} = \dfrac{AT}{HW}$.

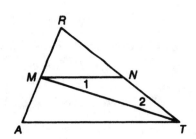

25. GIVEN: $\overline{MN} \parallel \overline{AT}$,
 $\measuredangle 1 \cong \measuredangle 2$.
 PROVE: $\dfrac{NT}{AT} = \dfrac{RN}{RT}$.

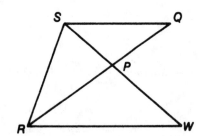

26. GIVEN: $\overline{SR} \cong \overline{SQ}$,
 $\overline{RQ}$ bisects $\measuredangle SRW$.
 PROVE: $\dfrac{SQ}{RW} = \dfrac{SP}{PW}$.

Use the accompanying diagram for Exercises 27 and 28.

27. GIVEN: $\overline{MC} \perp \overline{JK}$, $\overline{PM} \perp \overline{MQ}$,
$\overline{TP} \cong \overline{TM}$.

PROVE: $\dfrac{PM}{MC} = \dfrac{PQ}{MK}$.

28. GIVEN: T is the midpoint of $\overline{PQ}$,
$\overline{MP} \cong \overline{MQ}$, $\overline{JK} \parallel \overline{MQ}$.

PROVE: $\dfrac{PM}{JK} = \dfrac{TQ}{JT}$.

29. GIVEN: $\overline{EF}$ is the median of trapezoid $ABCD$.
PROVE: $EI \times GH = IH \times EF$.

30. GIVEN: $\overline{RS} \perp \overline{ST}$,
$\overline{SW} \perp \overline{RT}$.
PROVE: $(ST)^2 = TW \times RT$.

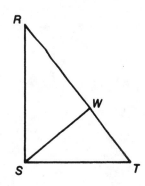

31. GIVEN: $\overline{AF}$ bisects $\angle BAC$,
$\overline{BH}$ bisects $\angle ABC$,
$\overline{BC} \cong \overline{AC}$.
PROVE: $AH \times EF = BF \times EH$.

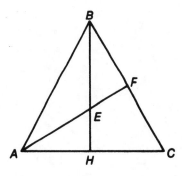

32. GIVEN: $\overline{XY} \parallel \overline{LK}$,
$\overline{XZ} \parallel \overline{JK}$.
PROVE: $JY \times ZL = XZ \times KZ$.

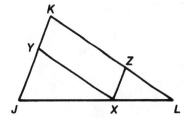

33. GIVEN: Quadrilateral *ABCD*,
 $\overline{AB} \perp \overline{BC}, \overline{AB} \perp \overline{AD}$,
 $\overline{AC} \perp \overline{CD}$.
 PROVE: $BC \times AD = (AC)^2$.

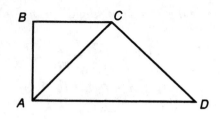

34. GIVEN: $\triangle ABC$ with $\overline{CDA}, \overline{CEB}, \overline{AFB}$,
 $\overline{DE} \parallel \overline{AB}, \overline{EF} \parallel \overline{AC}$,
 $\overline{CF}$ intersects $\overline{DE}$ at *G*.
 PROVE: **(a)** $\triangle CAF \sim \triangle FEG$.
 (b) $DG \times GF = EG \times GC$.

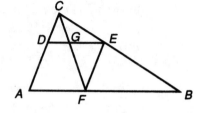

THE RIGHT TRIANGLE

WHAT YOU WILL LEARN

Right triangles have special properties that are developed by applying the properties of similar triangles. In this chapter you will learn:

- the parts of a right triangle that are in proportion when an altitude is drawn to the hypotenuse of a right triangle;
- the use of the Pythagorean Theorem to find the length of any side of a right triangle when the lengths of the other two sides are known;
- the way to prove the Pythagorean Theorem;
- the special relationships between the sides in a right triangle whose acute angles measure 30 and 60;
- the special relationships between the sides in a right triangle whose acute angles measure 45 and 45;
- the use of trigonometry to find the measures of the angles or sides of a triangle.

SECTIONS IN THIS CHAPTER

Proportions in a Right Triangle

The Pythagorean Theorem

Special Right-Triangle Relationships

Trigonometric Ratios

Indirect Measurement in a Right Triangle

Proportions in a Right Triangle

Let's begin by reviewing some important terminology and introducing some new notation. In a right triangle, the side opposite the right angle is called the *hypotenuse*; each of the remaining sides is called a *leg*. It is sometimes convenient to refer to the length of a side of a triangle by using the lowercase letter of the vertex that lies opposite the side. See Figure 11.1.

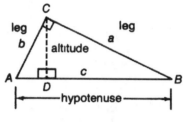

FIGURE 11.1

If we draw an altitude to the hypotenuse of the right triangle shown in Figure 11.1, we notice that altitude $\overline{CD}$ divides hypotenuse $\overline{AB}$ into two segments, $\overline{AD}$ and $\overline{BD}$. Hypotenuse segment $\overline{AD}$ is adjacent to leg $\overline{AC}$, while hypotenuse segment $\overline{BD}$ is adjacent to leg $\overline{BC}$. How many triangles do we see in Figure 11.1? There are *three* different triangles: the original right triangle ($\triangle ACB$) and the two smaller right triangles formed by the altitude on the hypotenuse ($\triangle ADC$ and $\triangle CDB$). As a result of the similarity relationships that exist between these pairs of triangles, a set of proportions can be written. Theorem 11.1 summarizes these relationships.

THEOREM 11.1 PROPORTIONS IN A RIGHT TRIANGLE THEOREM

If in a right triangle the altitude to the hypotenuse is drawn, then:

- The altitude separates the original triangle into two new triangles that are similar to the original triangle and to each other (see Figure 11.1):

$$\triangle ADC \sim \triangle ACB$$
$$\triangle CDB \sim \triangle ACB$$
$$\triangle ADC \sim \triangle CDB$$

- The length of each leg is the mean proportional between the length of the hypotenuse segment adjacent to the leg and the length of the entire hypotenuse.

 Since $\triangle ADC \sim \triangle ACB$,

$$\frac{AD}{AC} = \frac{AC}{AB}$$

Since $\triangle CDB \sim \triangle ACB$,

$$\frac{BD}{BC} = \frac{BC}{AB}$$

- The length of the altitude is the mean proportional between the lengths of the segments it forms on the hypotenuse.

 Since $\triangle ADC \sim \triangle CBD$,

$$\frac{AD}{CD} = \frac{CD}{DB}$$

OUTLINE OF PROOF: $\triangle ADC \sim \triangle ACB$ since $\angle A \cong \angle A$ and $\angle ADC \cong \angle ACB$. $\triangle CDB \sim \triangle ACB$ since $\angle B \cong \angle B$ and $\angle CDB \cong \angle ACB$. $\triangle ADC \sim \triangle CDB$ since, if each triangle is similar to the same triangle, the triangles must be similar to each other. The indicated proportions follow from the principle that the lengths of corresponding sides of similar triangles are in proportion.

EXAMPLE 11.1

Find the value of x.

a.

b.

c.
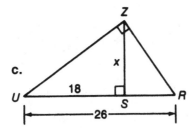

SOLUTION

a. $\dfrac{AD}{AC} = \dfrac{AC}{AB} \quad \left(\dfrac{\text{hyp segment}}{\text{leg}} = \dfrac{\text{leg}}{\text{hyp}}\right)$

$\dfrac{x}{10} = \dfrac{10}{25}$

$25x = 100$

$x = \dfrac{100}{25} = 4$

b. $\dfrac{RW}{RS} = \dfrac{RS}{RT} \quad \left(\dfrac{\text{hyp segment}}{\text{leg}} = \dfrac{\text{leg}}{\text{hyp}}\right)$

$\dfrac{4}{x} = \dfrac{x}{16}$

$x^2 = 64$

$x = \sqrt{64} = 8$

c. $\dfrac{US}{ZS} = \dfrac{ZS}{RS}$ $\left(\dfrac{\text{hyp segment 1}}{\text{altitude}} = \dfrac{\text{altitude}}{\text{hyp segment 2}} \right)$

$\dfrac{18}{x} = \dfrac{x}{8}$ NOTE : $RS = 26 - 18 = 8$

$x^2 = 144$

$x = \sqrt{144} = 12$

EXAMPLE 11.2

In right triangle JKL, $\angle K$ is a right angle. Altitude $\overline{KH}$ is drawn so that the length of $\overline{JH}$ exceeds the length of $\overline{HL}$ by 5. If $KH = 6$, find the length of the hypotenuse.

SOLUTION

Let x = length of $\overline{LH}$.
Then $x + 5$ = length of $\overline{JH}$.

$\dfrac{x}{6} = \dfrac{6}{x + 5}$ $\left(\dfrac{\text{hyp segment 1}}{\text{altitude}} = \dfrac{\text{altitude}}{\text{hyp segment 2}} \right)$

$x(x + 5) = 36$

$x^2 + 5x = 36$

$x^2 + 5x - 36 = 36 - 36$

$x^2 + 5x - 36 = 0$

$(x + 9)(x - 4) = 0$

$x + 9 = 0$ or $x - 4 = 0$

$x = -9$ $x = 4$
(Reject since a $LH = x = 4$
length cannot be a $JH = x + 5 = 9$
negative number.) $JL = LH + JH = 4 + 9 = 13$

SUMMARY

Proportions in a right triangle

- $\dfrac{x}{b} = \dfrac{b}{c}$ and $\dfrac{y}{a} = \dfrac{a}{c}$

- $\dfrac{x}{h} = \dfrac{h}{y}$

The Pythagorean Theorem

One of the most famous and useful theorems in mathematics provides a means for finding the length of *any* side of a *right* triangle, given the lengths of the other two sides. The sides are related by the equation

$$(\text{Hypotenuse})^2 = (\text{leg 1})^2 + (\text{leg 2})^2$$

This relationship, known as the *Pythagorean Theorem*, is named in honor of the Greek mathematician Pythagoras, who is believed to have presented the first proof of this theorem in about 500 B.C. Since that time, many different proofs have been offered. Here are the formal statement of the theorem and one such proof.

THEOREM 11.2 THE PYTHAGOREAN THEOREM

The square of the length of the hypotenuse of a right triangle is equal to the sum of the squares of the lengths of the legs.

GIVEN: Right triangle *ACB* with right angle *C*.
PROVE: $c^2 = a^2 + b^2$.

PLAN: Draw the altitude from *C* to hypotenuse $\overline{AB}$, intersecting $\overline{AB}$ at point *D*. For convenience, refer to $\overline{AD}$ as *x* and $\overline{DB}$ as $c - x$. Apply Theorem 11.1.

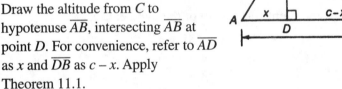

PROOF:

Statements	Reasons
1. Right $\triangle ACB$ with right angle *C*.	1. Given.
2. Draw the altitude from vertex *C* to hypotenuse $\overline{AB}$, intersecting $\overline{AB}$ at point *D*.	2. From a point not on a line, exactly one perpendicular may be drawn to the line.
Apply Theorem 11.1: 3. $\dfrac{x}{b} = \dfrac{b}{c}$ and $\dfrac{c-x}{a} = \dfrac{a}{c}$.	3. If in a right triangle the altitude to the hypotenuse is drawn, then the length of each leg is the mean proportional between the length of the hypotenuse segment adjacent to the leg and the length of the entire hypotenuse.
4. $cx = b^2$ and $c(c - x) = a^2$, which may be written as $c^2 - cx = a^2$.	4. In a proportion, the product of the means equals the product of the extremes.
5. $\begin{aligned} c^2 - cx &= a^2 \\ + \quad cx &= b^2 \\ \hline c^2 \qquad &= a^2 + b^2 \end{aligned}$	5. Addition property.

**EXAMPLE
11.3**

Find the value of x.

a.

b.

c.

d.

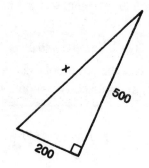

SOLUTION

a. Since x represents the length of the hypotenuse.

$$x^2 = 3^2 + 4^2$$
$$= 9 + 16$$
$$= 25$$
$$x = \sqrt{25} = 5$$

b. Since x represents the length of a leg and the hypotenuse is 13,

$$13^2 = x^2 + 5^2$$
$$169 = x^2 + 25$$
$$169 - 25 = x^2 + 25 - 25$$
$$144 = x^2$$

or

$$x^2 = 144$$
$$x = \sqrt{144} = 12$$

c. Since x represents the length of a leg and the hypotenuse is 7,

$$7^2 = x^2 + 3^2$$
$$49 = x^2 + 9$$
$$49 - 9 = x^2 + 9 - 9$$
$$40 = x^2$$

or

$$x^2 = 40$$
$$x = \sqrt{40}$$

To express a radical in simplest form, write the number underneath the radical sign as the product of two numbers, one of which is the *highest* perfect square factor of the number. Next, distribute the radical sign and simplify:

$$x = \sqrt{40} = \sqrt{4} \times \sqrt{10} = 2\sqrt{10}$$

d. Squaring 200 and 500 would be quite cumbersome. Since each is divisible by 100, work with the numbers 2 and 5. Using these numbers, find the length of the hypotenuse. *Multiply* the answer obtained by 100 to compensate for dividing the original lengths by 100.

$$c^2 = 2^2 + 5^2$$
$$= 4 + 25$$
$$= 29$$
$$c = \sqrt{29}$$

Find the actual value, x, by multiplying the value obtained by 100:

$$x = 100c = 100\sqrt{29}$$

PYTHAGOREAN TRIPLES

In the solutions for Example 11.3, parts **a** and **b**, whole-number values were obtained for the missing sides. Any set of three whole numbers x, y, and z is called a *Pythagorean triple* if the numbers satisfy the equation

$$z^2 = x^2 + y^2$$

The set of numbers $\{3, 4, 5\}$ is an example of a Pythagorean triple. The sets $\{5, 12, 13\}$ and $\{8, 15, 17\}$ are also Pythagorean triples. There are many others.

If $\{x, y, z\}$ is a Pythagorean triple, then so is the set that includes any whole number multiple of *each* member of this set. Thus, the set $\{6, 8, 10\}$ is a Pythagorean triple since each member was obtained by multiplying the corresponding member of the Pythagorean triple $\{3, 4, 5\}$ by 2:

$$\{6, 8, 10\} = \{2 \cdot 3, 2 \cdot 4, 2 \cdot 5\}$$

The following table shows additional examples.

Pythagorean Triple	Multiple of a Pythagorean Triple	Multiplying Factor
$\{3, 4, 5\}$	$\{15, 20, 25\}$	5
$\{5, 12, 13\}$	$\{10, 24, 26\}$	2
$\{8, 15, 17\}$	$\{80, 150, 170\}$	10

You should *memorize* the sets $\{3, 4, 5\}$, $\{5, 12, 13\}$, and $\{8, 15, 17\}$ as Pythagorean triples and be able to recognize their multiples. The need for this will become apparent as you work on the next set of examples.

 EXAMPLE 11.4 The base (length) of a rectangle is 12 and its altitude (height) is 5. Find the length of a diagonal.

SOLUTION

Triangle ABC is a 5-12-13 right triangle. Diagonal $AC = 13$. If you didn't see this pattern, you could apply the Pythagorean Theorem:

$$x^2 = 5^2 + 12^2$$
$$= 25 + 144 = 169$$
$$x = \sqrt{169} = 13$$

EXAMPLE 11.5 The diagonals of a rhombus are 18 and 24. Find the length of a side of the rhombus.

SOLUTION

Recall that the diagonals of a rhombus bisect each other and intersect at right angles.

$\triangle AED$ is a multiple of a 3-4-5 right triangle. Each member of the triple is multiplied by 3. Hence, the length of side $\overline{AD}$ is $3 \cdot 5$ or 15.

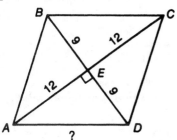

EXAMPLE 11.6 The length of each leg of an isosceles trapezoid is 17. The lengths of its bases are 9 and 39. Find the length of an altitude.

SOLUTION

Drop two altitudes, one from each of the upper vertices.

Quadrilateral *BEFC* is a rectangle (since $\overline{BE}$ and $\overline{CF}$ are congruent, parallel, and intersect $\overline{AD}$ at right angles). Hence $BC = EF = 9$. Since right $\triangle AEB \cong$ right $\triangle DFC$, $AE = DF = 15$. $\triangle AEB$ is an 8-15-17 right triangle. The length of an altitude is 8.

EXAMPLE 11.7 Find the length of a side of a square if a diagonal has a length of 8.

SOLUTION

There are no Pythagorean triples here. Since a square is equilateral, we may represent the lengths of sides $\overline{AB}$ and $\overline{AD}$ as x and apply the Pythagorean Theorem in right $\triangle BAD$:

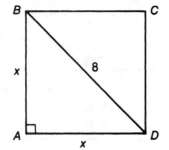

$$x^2 + x^2 = 8^2$$
$$2x^2 = 64$$
$$x^2 = 32$$
$$x = \sqrt{32} = \sqrt{16} \cdot \sqrt{2} = 4\sqrt{2}$$

The length of a side is $4\sqrt{2}$.

Pythagorean Triples

Some commonly encountered Pythagorean triples are:

$$\{3n, 4n, 5n\}$$
$$\{5n, 12n, 13n\}$$
$$\{8n, 15n, 17n\}$$

where n is any positive integer ($n = 1, 2, 3, \ldots$). There are many other families of Pythagorean triples, including $\{7, 24, 25\}$ and $\{9, 40, 41\}$.

CONVERSE OF THE PYTHAGOREAN THEOREM

If the lengths of the sides of a triangle are known, then we can determine whether the triangle is a right triangle by applying the *converse* of the Pythagorean Theorem.

THEOREM 11.3 CONVERSE OF THE PYTHAGOREAN THEOREM

If in a triangle the square of the length of a side is equal to the sum of the squares of the lengths of the other two sides, then the triangle is a right triangle.

Suppose a, b, and c represent the lengths of the sides of a triangle and c is the largest of the three numbers. If $c^2 = a^2 + b^2$, then the triangle is a right triangle.

For example, to determine whether the triangle whose sides have lengths of 11, 60, and 61 is a right triangle, we test whether the square of the largest of the three numbers is equal to the sum of the squares of the other two numbers:

$$61^2 \overset{?}{=} 11^2 + 60^2$$
$$3721 \overset{?}{=} 121 + 3600$$
$$3721 \overset{\checkmark}{=} 3721 \qquad \text{Hence, the triangle is a right triangle.}$$

Special Right-Triangle Relationships

THE 30-60 RIGHT TRIANGLE

Suppose the length of each side of the equilateral triangle in Figure 11.2 is represented by $2s$. If you drop an altitude from the vertex of this triangle, several interesting relationships materialize:

FIGURE 11.2

- Since $\triangle BDA \cong \triangle BDC$, $\overline{BD}$ bisects $\angle ABC$ and also bisects side $\overline{AC}$. Hence, $m\angle ABD = 30$. The acute angles of right triangle ADB measure 60 and 30. Also, $AD = \frac{1}{2}(AC) = \frac{1}{2}(2s) = s$.

- The length of leg $\overline{BD}$ may be found using the Pythagorean Theorem:

$$(BD)^2 + (AD)^2 = (AB)^2$$
$$(BD)^2 + (s)^2 = (2s)^2$$
$$(BD)^2 + s^2 = 4s^2$$
$$(BD)^2 = 3s^2$$
$$BD = \sqrt{3s^2} = \sqrt{3} \cdot \sqrt{s^2} = \sqrt{3}\, s$$

Let's summarize the information we have gathered about $\triangle ADB$. First, $\triangle ADB$ is referred to as a *30-60 right triangle* since these numbers correspond to the measures of its acute angles. In a 30-60 right triangle the following relationships hold:

30-60 RIGHT-TRIANGLE SIDE RELATIONSHIPS

- The length of the *shorter* leg (the side opposite the 30 degree angle) is one-half the length of the hypotenuse:

$$AD = \frac{1}{2} AB$$

- The length of the *longer* leg (the side opposite the 60 degree angle) is one-half the length of the hypotenuse multiplied by $\sqrt{3}$:

$$BD = \frac{1}{2} \times AB \times \sqrt{3}$$

- The length of the longer leg is equal to the length of the shorter leg multiplied by $\sqrt{3}$:

$$BD = AD \cdot \sqrt{3}$$

Fill in the following table:

EXAMPLE **11.8**

	RS	ST	RT
a.	?	?	12
b.	4	?	?
c.	?	$7\sqrt{3}$	?

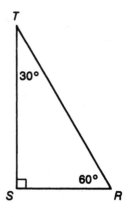

SOLUTION

a. $RS = \dfrac{1}{2}(RT) = \dfrac{1}{2}(12) = 6$

$ST = RS(\sqrt{3}) = 6\sqrt{3}$

b. $ST = RS(\sqrt{3}) = 4\sqrt{3}$

$RT = 2(RS) = 2(4) = 8$

c. $RS = 7$

$RT = 2(RS) = 2(7) = 14$

In □$RSTW$, m∡R = 30 and RS = 12. Find the length of an altitude.

SOLUTION

$$SH = \frac{1}{2}(12) = 6$$

In △JKL, m∡K = 120 and JK = 10. Find the length of the altitude drawn from vertex J to side $\overline{LK}$ (extended if necessary).

SOLUTION

Since ∡K is obtuse, the altitude falls in the exterior of the triangle, intersecting the extension of $\overline{KL}$, say at point H. △JHK is a 30-60 right triangle. Hence,

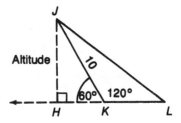

$$JH = \frac{1}{2} \cdot 10 \cdot \sqrt{3} = 5\sqrt{3}$$

THE 45-45 RIGHT TRIANGLE

Another special right triangle is the isosceles right triangle. Since the legs of an isosceles right triangle are congruent, the angles opposite must also be congruent. This implies that the measure of each acute angle of an isosceles right triangle is 45. A *45-45 right triangle* is another name for an isosceles right triangle.

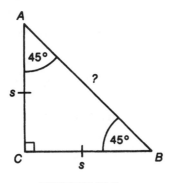

FIGURE 11.3

To determine the relationships between the lengths of the sides of a 45-45 right triangle, we will represent the length of each leg by the letter s (Figure 11.3).

Then the hypotenuse $\overline{AB}$ may be expressed in terms of s by applying the Pythagorean Theorem:

$$(AB)^2 = s^2 + s^2$$
$$= 2s^2$$
$$AB = \sqrt{2s^2} = \sqrt{2} \times \sqrt{s^2}$$
$$= \sqrt{2} \times s$$

Thus, the length of the hypotenuse is $\sqrt{2}$ times the length of a leg:

$$\text{Hypotenuse} = \sqrt{2} \times \text{leg}$$

By dividing each side of this equation by $\sqrt{2}$ we may find an expression for the length of a leg in terms of the length of the hypotenuse:

$$\text{Leg} = \frac{\text{Hypotenuse}}{\sqrt{2}}$$

Rationalizing the denominator, we obtain:

$$\text{Leg} = \frac{\text{Hypotenuse}}{\sqrt{2}} \times \frac{\sqrt{2}}{\sqrt{2}} = \frac{\text{Hypotenuse}}{2} \times \sqrt{2}$$

45-45 RIGHT TRIANGLE SIDE RELATIONSHIPS

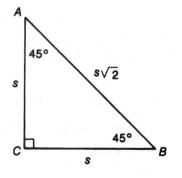

- The *lengths of the legs* are equal:

 $$AC = BC$$

- The *length of the hypotenuse* is equal to the length of either leg multiplied by $\sqrt{2}$:

 $$AB = AC \text{ (or } BC) \cdot \sqrt{2}$$

- The *length of either leg* is equal to one-half the length of the hypotenuse multiplied by $\sqrt{2}$:

 $$AC \text{ (or } BC) = \frac{1}{2} AB \cdot \sqrt{2}$$

EXAMPLE 11.11

In isosceles trapezoid *ABCD*, the measure of a lower base angle is 45 and the length of upper base $\overline{BC}$ is 5. If the length of an altitude is 7, find the lengths of the legs, $\overline{AB}$ and $\overline{DC}$.

SOLUTION

Drop altitudes from *B* and *C*, forming two congruent 45-45 right triangles. $AE = BE = 7$. Also, $FD = 7$. $AB = AE \times \sqrt{2} = 7\sqrt{2}$. The length of each leg is $7\sqrt{2}$.

Trigonometric Ratios

DEFINING TRIGONOMETRIC RATIOS

Each of the triangles pictured in Figure 11.4 is a 30-60 right triangle. The lengths of the sides in each of these right triangles must obey the relationships developed in the section on special right-triangle relationships.

In particular, we may write:

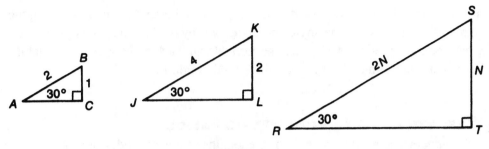

FIGURE 11.4

In △*ABC*	In △*JKL*	In △*RST*
$\dfrac{BC}{AB} = \dfrac{1}{2}$	$\dfrac{KL}{JK} = \dfrac{2}{4} = \dfrac{1}{2}$	$\dfrac{ST}{RS} = \dfrac{N}{2N} = \dfrac{1}{2}$

In each of these triangles the ratio of the length of the leg opposite the 30° angle to the length of the hypotenuse is 1 : 2 or 0.5. This ratio will hold in *every* 30-60 right triangle no matter how large or small the triangle is. For other choices of congruent acute angles in right triangles, will this type of ratio also be constant? As the set of triangles in Figure 11.5 illustrates, the answer is yes! In each right triangle, the ratio of the length of the leg opposite the angle whose measure is *x* degrees to the length of the hypotenuse must be the same.

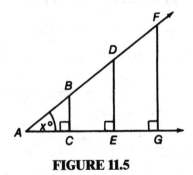

FIGURE 11.5

The ratios $\dfrac{BC}{AB}, \dfrac{DE}{AD}, \dfrac{FG}{AF}$ are equal since they represent the ratios of the lengths of corresponding sides of similar triangles. You should observe that Figure 11.5 includes three right triangles each of which is similar to the other two; that is, △*ABC* ~ △*ADE* ~ △*AFG* by the AA Theorem of similarity since:

- $\angle A \cong \angle A \cong \angle A$.
- Each triangle includes a right angle, and all right angles are congruent.

We are then entitled to write the following extended proportion:

$$\frac{BC}{AB} = \frac{DE}{AD} = \frac{FG}{AF} = \frac{\text{length of leg opposite angle } A}{\text{length of hypotenuse}}$$

For any given acute angle, this ratio is the same regardless of the size of the right triangle. It will be convenient to refer to this type of ratio by a special name—the *sine ratio*. The sine of an acute angle of a right triangle is the ratio formed by taking the length of the leg opposite the angle and dividing it by the length of the hypotenuse. In a right triangle, other ratios may also be formed with each being given a special name. These names are summarized in the following definitions.

DEFINITIONS OF TRIGONOMETRIC RATIOS

A *trigonometric ratio* is the ratio of the lengths of *any* two sides of a *right* triangle with respect to a given angle. Three commonly formed trigonometric ratios are called the *sine, cosine,* and *tangent* ratios:

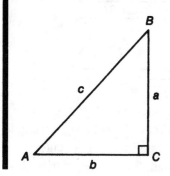

$$\text{Sine of } \angle A = \frac{\text{length of leg opposite } \angle A}{\text{length of hypotenuse}} = \frac{BC}{AB}$$

$$\text{Cosine of } \angle A = \frac{\text{length of leg adjacent } \angle A}{\text{length of hypotenuse}} = \frac{AC}{AB}$$

$$\text{Tangent of } \angle A = \frac{\text{length of leg opposite } \angle A}{\text{length of leg adjacent } \angle A} = \frac{BC}{AC}$$

WORKING WITH TRIGONOMETRIC RATIOS

When working with trigonometric ratios, keep in mind the following:

- Sine, cosine, and tangent may be abbreviated as *sin, cos,* and *tan*, respectively.
- The trigonometric ratios may be taken with respect to either of the acute angles of the triangle:

$$\sin A = \frac{a}{c} \qquad \sin B = \frac{b}{c}$$
$$\cos A = \frac{b}{c} \qquad \cos B = \frac{a}{c}$$
$$\tan A = \frac{a}{b} \qquad \tan B = \frac{b}{a}$$

- Since $\sin A$ and $\cos B$ are both equal to $\dfrac{a}{c}$, they are equal to each other. Similarly, $\cos A$ and $\sin B$ are equal since both are equal to $\dfrac{b}{c}$. Since angles A and B are complementary, the sine of an angle is equal to the cosine of the angle's complement. For example, $\sin 60° = \cos 30°$.
- The definitions of the three trigonometric ratios should be memorized.

EXAMPLE 11.12

In right triangle *ABC*, $\angle C$ is a right angle. If $AC = 4$ and $BC = 3$, find the value of $\tan A$, $\sin A$, and $\cos A$.

SOLUTION

Since $\triangle ABC$ is a 3-4-5 right triangle, $AB = 5$.

$$\tan A = \frac{\text{leg opposite } \angle A}{\text{leg adjacent } \angle A} = \frac{3}{4} = 0.75$$

$$\sin A = \frac{\text{leg opposite } \angle A}{\text{hypotenuse}} = \frac{3}{5} = 0.6$$

$$\cos A = \frac{\text{leg adjacent } \angle A}{\text{hypotenuse}} = \frac{4}{5} = 0.8$$

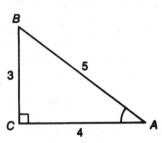

EXAMPLE 11.13 Express the values of sin 60°, cos 60°, and tan 60° in either radical or decimal form.

SOLUTION

$$\sin 60° = \frac{\sqrt{3} \times n}{2n} = \frac{\sqrt{3}}{2}$$

$$\cos 60° = \frac{n}{2n} = \frac{1}{2} = 0.5$$

$$\tan 60° = \frac{\sqrt{3} \times n}{n} = \sqrt{3}$$

USING A SCIENTIFIC CALCULATOR TO FIND TRIGONOMETRIC VALUES

In Example 11.13 we were able to use special right-triangle relationships to find the exact values of the sine, cosine, and tangent of 60°. Suppose that we need to find the value of the sine, cosine, or tangent of 38°. One approach would be to use a protractor to construct a right triangle with an acute angle of 38° and then, with a ruler, measure the lengths of the three sides of the right triangle. Using these lengths, we could then calculate ratios that would give us an approximate value for the sine, cosine, or tangent of 38°. Unfortunately, this procedure is time consuming, prone to error, and usually does not give results that are sufficiently close to the actual values.

Scientific calculators provide a fast and easy way to obtain values for the sine, cosine, and tangent of a given angle. When you first turn on a scientific calculator, the unit of angle measurement is typically set to DEGrees. If you do not see DEG in small type in the display window, press the MODE key or DGR key until DEG is displayed. To evaluate the sine, cosine, or tangent of $n°$, enter n and then press the key labeled with the name of the appropriate trigonometric function. Since most scientific calculators will display a decimal value with at least eight-place decimal accuracy, you will usually need to round off the answer that appears in the display window. As a general practice, you should round off answers correct to four decimal places.

For example, to evaluate sin 38° correct to *four decimal places*, follow this procedure:

STEP 1: Enter 38.

STEP 2: Press the SIN key.

STEP 3: Round off the value that appears in the display window. Thus, correct *to four decimal places*, sin 38° = 0.6157.

Not all calculators work in the same way. For example, some scientific calculators require that you first press the �it SIN ██ key, enter 38, and then press the ██ = ██ key.

Now consider the opposite situation. Suppose you know that cos x = 0.8387. How can you find the measure of angle x correct to the nearest degree? A scientific calculator has either an INVerse function key, a 2nd function key, or a SHIFT key that allows you to find the degree measure of an angle when the value of a trigonometric function of that angle is known. For example, to find x when *cos x* = 0.8387, follow these steps:

STEP 1: Enter .8387.
STEP 2: Press the ██ INV ██ or ██ 2nd ██ or ██ SHIFT ██ key, depending on the calculator you are using.
STEP 3: Press the ██ COS ██ key.
STEP 4: Round off the value that appears in the display window to the nearest degree. Thus, correct to the *nearest degree, x* = 42.

If this procedure does not work with the calculator you are using, try Steps 2 and 3 first, then enter the angle and press the ██ = ██ key. For example, if your calculator has a SHIFT key, use this key sequence:

$$\boxed{\text{SHIFT}} \rightarrow \boxed{\text{COS}} \rightarrow .8387 \rightarrow \boxed{=}$$

 Use your calculator to find the value of each of the following:
EXAMPLE 11.14 **a.** tan 27° **b.** sin 44° **c.** cos 35°

SOLUTION
 a. tan 27° = 0.5095 **b.** sin 44° = 0.6947 **c.** cos 35° = 0.8192

 Use your calculator to find the measure of angle x if:
EXAMPLE 11.15 **a.** sin x = 0.7071 **b.** tan x = 0.6009 **c.** cos x = 0.7986

SOLUTION
 a. x = 45° **b.** x = 31° **c.** x = 37°

 Use your calculator to find the measure of angle A, correct to the *nearest degree*, if:
EXAMPLE 11.16 **a.** tan A = 0.7413 **b.** cos A = 0.8854 **c.** sin A = 0.6599

SOLUTION
 a. A = 37° **b.** A = 28° **c.** A = 41°

You should use your calculator to verify that for angles from 0° to 90°:

- As an angle increases in value, its sine and tangent ratios increase, while its cosine ratio decreases.
- The minimum value of sine, cosine, and tangent is 0. (*Note:* sin 0° = 0, cos 90° = 0, tan 0° = 0)
- The maximum value of cosine and sine is 1. (*Note:* cos 0° = 1 and sin 90° = 1.) As the measure of an acute angle of a right triangle approaches 90°, the value of its tangent ratio gets larger and larger, with *no* upper limit. We say that the tangent ratio is *unbounded* or undefined at 90°.

Indirect Measurement in a Right Triangle

Trigonometric ratios may be used to arrive at the measure of a side or angle of a right triangle that may be difficult, if not impossible, to calculate by direct measurement. For example, consider a plane that takes off from a runway, and climbs while maintaining a constant angle with the horizontal ground. Suppose that, at the instant of time when the plane has traveled 1000 meters, its altitude is 290 meters. Using our knowledge of trigonometry, we can approximate the measure of the angle at which the plane has risen with respect to the horizontal ground.

A right triangle may be used to represent the situation we have just described. The hypotenuse corresponds to the path of the rising plane, the vertical leg of the triangle represents the plane's altitude, and the acute angle formed by the hypotenuse and the horizontal leg (the ground) is the desired angle whose measure we must determine.

FIGURE 11.6

To find the value of x in Figure 11.6, we must first determine the appropriate trigonometric ratio. The sine ratio relates the three quantities under consideration:

$$\sin x = \frac{\text{leg opposite } \sphericalangle}{\text{hypotenuse}}$$
$$= \frac{290}{1000} = 0.2900$$

To find the value of x to the nearest degree, we use a scientific calculator, which gives us $x = 17°$, correct to the *nearest degree*.

Find m$\sphericalangle R$ to the *nearest degree*.

EXAMPLE 11.17

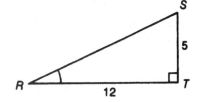

SOLUTION
- Decide which trigonometric ratio to use and write the corresponding equation:

$$\tan R = \frac{\text{leg opposite } \sphericalangle R}{\text{leg adjacent } \sphericalangle R}$$
$$\tan R = \frac{5}{12}$$

- Express the ratio in decimal form. Using a scientific calculator, you obtain the answer, rounded off to four decimal places: 0.4167. Hence,

$$\tan R = 0.4167$$

- Again use a scientific calculator to find that R measures 23°, correct to the *nearest degree*.

The next two examples illustrate that, given the length of a side of a right triangle and the measure of an acute angle, the length of either of the two remaining sides of the triangle can be found.

EXAMPLE 11.18

Find the value of x to the *nearest tenth*.

SOLUTION

- Decide which trigonometric ratio to use, and then write the corresponding equation:

$$\cos J = \frac{\text{leg adjacent} \measuredangle j}{\text{hypotenuse}}$$

$$\cos 40° = \frac{x}{20}$$

- Use a scientific calculator to obtain $40° = 0.7660$. Hence,

$$0.7660 = \frac{x}{20}$$

- Solve the equation. It may help to think of the equation as a proportion:

$$\frac{0.7660}{1} = \frac{x}{20}$$

Solve by cross-multiplying:

$$x = 20(0.7660) = 15.32$$

- Round off the answer to the desired accuracy. This example asks for the answer correct to the *nearest tenth*. Hence:

$$x = 15.3$$

EXAMPLE 11.19

Find the value of x to the *nearest tenth*.

SOLUTION

$$\tan \measuredangle R = \frac{\text{leg opposite} \measuredangle R}{\text{leg adjacent} \measuredangle R}$$

$$\tan 75° = \frac{28}{x}$$

$$3.7321 = \frac{28}{x}$$

$$3.7321x = 28$$

$$\frac{3.7321x}{3.7321} = \frac{28}{3.7321}$$

$$x = 7.502 = 7.5 \text{ to the } \textit{nearest tenth}$$

EXAMPLE 11.20

The lengths of the diagonals of a rhombus are 12 and 16. Find to the *nearest degree* the measures of the angles of the rhombus.

SOLUTION

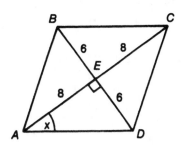

- In $\triangle AED$:

$$\tan x = \frac{\text{side opposite}}{\text{side adjacent}} = \frac{ED}{AE}$$

$$= \frac{6}{8} = 0.7500$$

$x = 37°$ to the *nearest degree*

- Since the diagonals of a rhombus bisect its angles,

$$m\angle BAD = 2x = 2(37) = 74$$

- Since consecutive angles of a rhombus are supplementary,

$$
\begin{aligned}
m\angle ABC &= 180 - m\angle BAD \\
&= 180 - 74 \\
&= 106
\end{aligned}
$$

- Since opposite angles of a rhombus are equal in measure:

$$m\angle BCD = m\angle BAD = 74$$
$$m\angle ADC = m\angle ABC = 106$$

REVIEW EXERCISES FOR CHAPTER 11

In Exercises 1 to 3, find the values of r, s, *and* t.

1.

2.

3.

In Exercises 4 to 9, in right triangle JKL, angle JKL is the right angle and $\overline{KH} \perp \overline{JL}$.

4. If $JH = 4$ and $HL = 16$, find KH.

5. If $JH = 5$ and $HL = 4$, find KL.

6. If $JH = 8$, $JL = 20$, find KH.

7. If $KL = 18$, $JL = 27$, find JK.

8. If $JK = 14$, $HL = 21$, find JH.

9. If $KH = 12$, $JL = 40$, find JK (assume $\overline{JK}$ is the shorter leg of right $\triangle JKL$).

10. The altitude drawn to the hypotenuse of a right triangle divides the hypotenuse into segments such that their lengths are in the ratio of $1:4$. If the length of the altitude is 8, find the length of:
 (a) Each segment of the hypotenuse. (b) The longer leg of the triangle.

In Exercises 11 to 15, find the value of x.

11.

12.

13.

14.

15.

16. If the lengths of the diagonals of a rhombus are 32 and 24, find the perimeter of the rhombus.

17. If the perimeter of a rhombus is 164 and the length of the longer diagonal is 80, find the length of the shorter diagonal.

18. Find the length of the altitude drawn to a side of an equilateral triangle whose perimeter is 30.

19. The length of the base of an isosceles triangle is 14. If the length of the altitude drawn to the base is 5, find the length of each of the legs of the triangle.

20. The measure of the vertex angle of an isosceles triangle is 120 and the length of each leg is 8. Find the length of:
 (a) The altitude drawn to the base. **(b)** The base.

21. If the perimeter of a square is 24, find the length of a diagonal.

22. If the length of a diagonal of a square is 18, find the perimeter of the square.

23. Find the length of the altitude drawn to side $\overline{AC}$ of $\triangle ABC$ if $AB = 8$, $AC = 14$, and $m\angle A$ equals:
 (a) 30 **(b)** 120 **(c)** 135

24. The lengths of the bases of an isosceles trapezoid are 9 and 25. Find the lengths of the altitude and each of the legs if the measure of each lower base angle is:
 (a) 30 **(b)** 45 **(c)** 60

25. Find the value of x, y, and z.

26. The lengths of two adjacent sides of a parallelogram are 6 and 14. If the measure of an included angle is 60, find the length of the shorter diagonal of the parallelogram.

27. The length of each side of a rhombus is 10 and the measure of an angle of the rhombus is 60. Find the length of the longer diagonal of the rhombus.

28. In right $\triangle ABC$, angle C is the right angle, $AC = 7$, and $BC = 24$. Find the value of the sine of the largest acute angle of the triangle.

29. In right $\triangle RST$, angle T is the right angle. If $\sin R = \dfrac{9}{41}$, find the values of $\cos R$ and $\tan R$.

30. The lengths of a pair of adjacent sides of a rectangle are 10 and 16. Find, correct to the *nearest degree*, the angle a diagonal makes with the longer side.

31. The measure of a vertex angle of an isosceles triangle is 72. If the length of the altitude drawn to the base is 10, find to the *nearest whole number* the length of the base and the length of each leg of the triangle.

32. The shorter diagonal of a rhombus makes an angle of 78° with a side of the rhombus. If the length of the shorter diagonal is 24, find to the *nearest tenth* the lengths of:
 (a) A side of the rhombus. **(b)** The longer diagonal.

33. Find the values of x and y, correct to the *nearest tenth*.

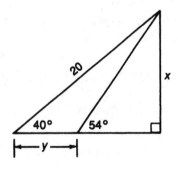

34. *ABCD* is a trapezoid. If $AB = 14$, $BC = 10$, and $m\angle BCD = 38$, find the lengths of $\overline{AD}$ and $\overline{CD}$, correct to the *nearest tenth*.

CUMULATIVE REVIEW EXERCISES: CHAPTERS 8–11

1. Which set of numbers may be the lengths of the sides of a triangle?
 (1) {5,4,1} (3) {5,4,9}
 (2) {5,4,6} (4) {5,4,10}

2. In quadrilateral $ABCD$, $\overline{AB} \cong \overline{DC}$ and $\overline{AB} \parallel \overline{DC}$. Which statement *must* be true?
 (1) $\overline{BD} \cong \overline{AC}$ (3) $\overline{AC} \cong \overline{AD}$
 (2) $\overline{AB} \cong \overline{BC}$ (4) $\overline{AD} \cong \overline{BC}$

3. The lengths of the sides of a triangle are 5, 7, and 8. If the length of the longest side of a similar triangle is 24, find the perimeter of the larger triangle.

4. In $\triangle ABC$, $AB = 8$ and $BC = 6$. Which statement is *always* true?
 (1) $AC > 14$ (3) $AC < 2$ and $AC > 14$
 (2) $AC < 2$ (4) $AC > 2$ and $AC < 14$

5. In equilateral triangle ABC, $AB = 16$. Find the perimeter of the triangle formed by connecting the midpoints of the sides of triangle ABC.

6. The altitude to the hypotenuse of right triangle ABC divides the hypotenuse into segments whose lengths are represented by $4x$ and x. If the length of the altitude is 2, find x.

7. What is the perimeter of a rhombus whose diagonals are 12 and 16?

8. In the accompanying diagram of $\triangle ABC$, $BC > AB$, and $\overline{AB} \cong \overline{BD}$. Which statement is true?
 (1) $BC > BD$
 (2) $AB > BC$
 (3) $m\angle BDA > m\angle BAD$
 (4) $m\angle BDA < m\angle BCA$

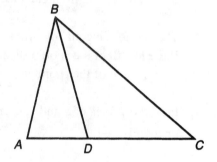

9. Which statement *must* be true?
 (1) If a parallelogram is not a rectangle, then it is not a square.
 (2) If a parallelogram is not a square, then it is not a rectangle.
 (3) All rectangles are squares.
 (4) If a parallelogram is a rectangle, then it is a square.

10. In the accompanying diagram, parallel lines $\overleftrightarrow{AB}$ and $\overleftrightarrow{CD}$ are cut by transveral $\overleftrightarrow{AC}$. Segments $\overline{BC}$ and $\overline{AD}$ intersect at E, and $m\angle BAC < m\angle ACD$. If $\overline{CB}$ bisects $\angle ACD$ and AD bisects $\angle BAC$, then which statement is true?
 (1) $AC = AB$
 (2) $AD < CD$
 (3) $AE < CE$
 (4) $\angle DAC > \angle BCD$

11. In $\triangle ABC$, $\overline{AB} \perp \overline{BC}$ and m$\angle CAB = 45$. If $AC = 12$, find AB in radical form.

12. In triangle ABC, $\overline{DE}$ is drawn parallel to $\overline{AB}$ and intersects $\overline{AC}$ and $\overline{BC}$ at D and E, respectively. If $CD = 4$, $DA = 2$, and $BE = 3$, find CE.

13. In rhombus $ABCD$, the bisectors of $\angle B$ and $\angle C$ *must* be
 (1) parallel
 (2) oblique
 (3) perpendicular
 (4) congruent

14. If the midpoints of the sides of a quadrilateral are joined consecutively, the figure formed *must* be
 (1) equiangular
 (2) equilateral
 (3) a trapezoid
 (4) a parallelogram

15. What is the length of an altitude of an equilateral triangle whose perimeter is 12?

16. The legs of a right triangle have lengths of 5 and 12. Expressed as a fraction, what is the cosine of the larger acute angle?

17. GIVEN: $\overline{CD}$ intersects $\overline{EMB}$ at A, $\overline{CB}$ is drawn, $\overline{ED} \perp \overline{EB}$, $\overline{CD} \perp \overline{CB}$, and $\overline{CM}$ is the perpendicular bisector of $\overline{AB}$.
 PROVE: $\triangle CMB \sim \triangle DEA$

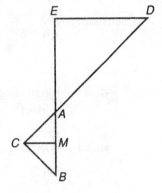

18. GIVEN: $\triangle ABC$, $\overline{CD}$ intersects $\overline{AB}$ at D, and $\overline{CD}$ bisects $\angle BCA$.
 PROVE: $CA > DA$

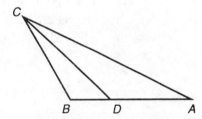

19. GIVEN: $\overline{BC} \parallel \overline{AD}$. $\triangle ABC$ is *not* isosceles.
 PROVE: $\overline{AC}$ does *not* bisect $\angle BAD$.

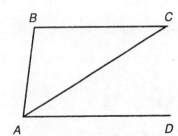

20. GIVEN: $\triangle ABC$ with $\overline{AB} \cong \overline{BC}$. Points D, E, and F
 are on $\overline{AB}$, $\overline{BC}$, and $\overline{AC}$, respectively, so that
 $\overline{ED} \perp \overline{DF}$, $\overline{DF} \perp \overline{AC}$, and $\overline{FE} \perp \overline{BC}$.

 PROVE: **(a)** $\triangle FEC \sim \triangle EDF$
 (b) $\triangle EDF \sim \triangle DFA$

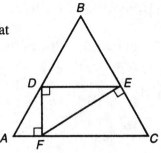

21. GIVEN: Trapezoid $ABCD$ with diagonal $\overline{AFC} \cong$
 diagonal $\overline{BFD}$; through C a line is drawn
 parallel to $\overline{BD}$ and intersecting $\overline{AB}$ at E.

 PROVE: **(a)** $BECD$ is a parallelogram.
 (b) $\overline{AC} \cong \overline{CE}$.
 (c) $\triangle AFB$ is isosceles.

22. GIVEN: Rectangle $ABCD$, $\overline{DFEC}$, $\overline{AGE}$, $\overline{BGF}$,
 $\overline{DF} \cong \overline{EC}$.

 PROVE: **(a)** $\angle 1 \cong \angle 2$.
 (b) $\angle 3 \cong \angle 4$.
 (c) $\overline{AG} \cong \overline{GB}$.

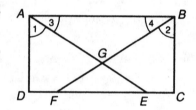

23. GIVEN: Quadrilateral $ABCD$, diagonals $\overline{AC}$
 and $\overline{BD}$, $\angle 1 \cong \angle 2$, $\overline{BD}$ bisects $\overline{AC}$
 at E.

 PROVE: $ABCD$ is a parallelogram.

24. GIVEN: V is a point on $\overline{ST}$ such that $\overline{RVW}$
 bisects $\angle SRT$, $\overline{TW} \cong \overline{TV}$.

 PROVE: $RW \times SV = RV \times TW$.

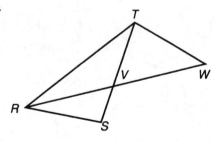

CIRCLES AND ANGLE MEASUREMENT

WHAT YOU WILL LEARN

Curved figures are encountered in geometry as well as in everyday life. Imagine riding on a bicycle or in an automobile that didn't have circular wheels. It's difficult to think of a machine that doesn't include some circular parts. In this chapter you will learn:

- the names of the various parts of a circle;
- the way in which the measure of an angle whose sides intersect a circle depends on the location of the vertex of the angle;
- the way to find the measure of an angle by using the measure(s) of its intercepted arc(s).

The Parts of a Circle

If we take a compass and using a fixed setting, draw a closed figure (see Figure 12.1), the resulting figure is a *circle*. The small puncture hole or impression that the metal compass point makes on the paper is called the *center* of the circle. The compass-setting distance is called the *radius* of the circle. Since the compass setting (that is, the radius) remained the same while we drew the circle with the compass, we may define a circle as follows:

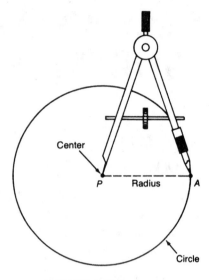

FIGURE 12.1

DEFINITION OF A CIRCLE

A *circle* is the set of all points in a plane having the same distance from a fixed point. The fixed point is called the *center* of the circle. The distance between the center of the circle and any point of the circle is called the *radius* (plural: radii) of the circle.

A capital letter is used to designate the center of a circle. If the center of a circle is designated by the letter P, the circle is referred to as *circle P*. The shorthand notation $\odot P$, where the symbol that precedes the letter P is a miniature circle, is commonly used.

From the definition of a circle, it follows that *all radii of the same circle are congruent*. Two different circles are congruent if their radii are congruent. In circle O, shown in Figure 12.2a, several radii are drawn. Each of these radii must have the same length so that $\overline{OA} \cong \overline{OB} \cong \overline{OC} \cong \overline{OD} \cong \ldots$ *Circle O* and *circle P* (Figure 12.2b) are congruent (written as $\odot O \cong \odot P$) if $\overline{OA} \cong \overline{PX}$.

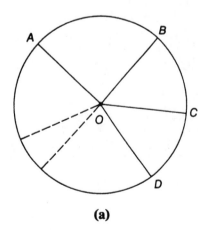

(a) (b)

FIGURE 12.2

Several basic terms with which you should be familiar are associated with circles. These terms are defined below and illustrated in Figure 12.3(*a*) and Figure 12.3(*b*).

- A *chord* is a line segment whose end points lie on the circle.
- A *diameter* is a chord that passes through the center of the circle.

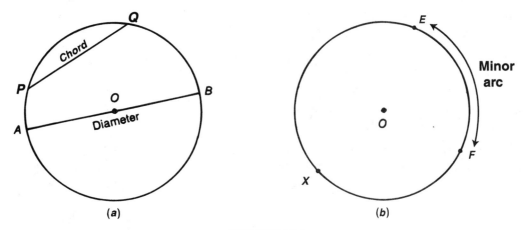

FIGURE 12.3

- An *arc* is a curved $\overline{PQ}$ portion of a circle. In Figure 12.3*a*, diameter $\overline{AB}$ divides the circle into two arcs, each of which is called a *semicircle*. In Figure 12.3*b*, E and F are the end points of two arcs. The shorter path from E to F represents a *minor arc*. A minor arc is less than a semicircle. The notation $\overparen{EF}$ is read "arc *EF*." The long way around from E to F is a *major arc*. A major arc is greater than a semicircle. Does $\overparen{EF}$ name a major arc or a minor arc whose end points are E and F? By using three letters to name a major arc, as in $\overparen{EXF}$, we can avoid any possible confusion. Thus, $\overparen{EF}$ names the *minor* arc whose end points are E and F, while $\overparen{EXF}$ refers to the *major* arc whose end points are E and F.

In Figure 12.3*a*, chord $\overline{PQ}$ cuts off or *determines* an arc of the circle. Referring to Figure 12.4, we can say that chord $\overline{PQ}$ *intercepts* $\overparen{PQ}$ or *has* $\overparen{PQ}$ as its arc.

In Figure 12.3*b* one and only one chord can be drawn that joins the end points of $\overparen{EF}$. Referring to Figure 12.5, we can say that arc *EF* has $\overline{EF}$ as its chord or that arc *EF* *determines* chord $\overline{EF}$.

A diameter of a circle consists of two radii that lie on the same straight line. In Figure 12.3*a* diameter AB = radius OA + radius OB. Hence, the length of a diameter of a circle is twice the length of a radius of the circle. If, for example, the length of a diameter of a circle is 12, then the length of a radius is 6.

FIGURE 12.4

FIGURE 12.5

In circle O, radius $OA = 3n - 10$ and radius $OB = n + 2$. Find the length of a diameter of circle O.

SOLUTION

Since all radii of a circle are congruent,

$$OA = OB$$
$$3n - 10 = n + 2$$
$$3n = n + 12$$
$$2n = 12$$
$$n = 6$$
$$OB = OA = n + 2 = 6 + 2 = 8$$

The length of a diameter of $\odot O = 2 \times 8 = 16$

GIVEN: $\overline{AB}$ and $\overline{CD}$ are diameters of circle O.
PROVE: $\overline{AD} \cong \overline{BC}$.

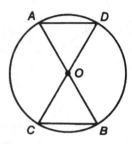

SOLUTION

PLAN: Show $\triangle AOD \cong \triangle BOC$.

PROOF:

Statements		Reasons
1. $\overline{AB}$ and $\overline{CD}$ are diameters of $\odot O$.		1. Given.
2. $\overline{OA} \cong \overline{OB}$.	(S)	2. All radii of a circle are congruent.
3. $\angle AOD \cong \angle BOC$.	(A)	3. Vertical angles are congruent.
4. $\overline{OD} \cong \overline{OC}$.	(S)	4. Same as reason 2.
5. $\triangle AOD \cong \triangle BOC$.		5. SAS Postulate.
6. $\overline{AD} \cong \overline{BC}$.		6. CPCTC.

Consider circle O and a point P. If the length of $\overline{OP}$ is less than the length of the radius of circle O, point P must be located within the boundaries of the circle. If the length of $\overline{OP}$ is greater than the length of the radius of the circle, then point P must fall outside the circle. Suppose the radius of circle O is 5 inches. If the length of $\overline{OP}$ is 6 inches, then point P must lie in the exterior (outside) of the circle. If the length of $\overline{OP}$ is 2 inches, then point P must lie in the interior (inside) of the circle. If the length of $\overline{OP}$ is 5 inches, then point P is a point on the circle.

DEFINITIONS OF INTERIOR AND EXTERIOR OF A CIRCLE
- The *interior of a circle* is the set of all points whose distance from the center of the circle is less than the length of the radius of the circle.
- The *exterior of a circle* is the set of all points whose distance from the center of the circle is greater than the length of the radius of the circle.

See Figure 12.6.

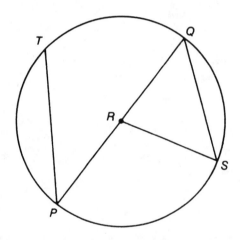

Point	Location
P	interior of ⊙ O
Q	on ⊙ O
R	exterior of ⊙ O

FIGURE 12.6

$\overline{PQ}$ is a diameter of the accompanying circle.

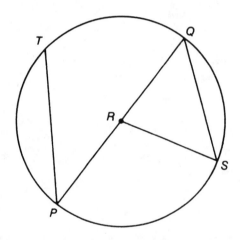

a. What point is the center of the circle?

b. Which segments are chords?

c. Which segments are radii?

d. Name all congruent segments.

e. If $RS = 6$, find PQ.

f. If $PQ = 15$, find RQ.

g. For each of the following conditions, determine whether point X lies in the interior of the circle, the exterior of the circle, or on the circle:

 (i) $RQ = 5$, and $RX = 4$.

 (ii) $RP = 6$, and $RX = 8$.

 (iii) $PQ = 14$, and $RX = 7$.

h. If $m\angle QRS = 60$ and $RQ = 9$, find QS.

i. If $m\angle QRS = 90$ and $QS = 15$, find RQ and RS.

j. If $m\angle QRS = 90$ and $RS = 6$, find RQ and QS.

SOLUTION

a. Point R.

b. $\overline{PT}$, $\overline{PQ}$, and $\overline{QS}$.

c. $\overline{RP}$, $\overline{RQ}$, and $\overline{RS}$.

d. $\overline{RP} \cong \overline{RQ} \cong \overline{RS}$ since all radii of a circle are congruent.

e. Diameter $PQ = 2(6) = 12$.

f. Radius $RQ = \dfrac{1}{2}(15) = 7.5$.

g. (i) X is located in the interior of the circle.

 (ii) X is located in the exterior of the circle.

 (iii) X is located on the circle.

h. $RQ = RS$, which implies $m\angle RSQ = m\angle RQS$ (base angles of an isosceles triangle are equal in measure). Since the vertex angle has measure 60, each base angle has a measure of one-half of 120 or 60. Triangle RQS is equiangular. An equiangular triangle is also equilateral. Hence, $QS = 9$.

i. Triangle QRS is a 45-45 right triangle.

$$RS = RQ = \frac{\sqrt{2}}{2} \times 15 = 7.5\sqrt{2}$$

j. Triangle QRS is a 45-45 right triangle.

$$RQ = RS = 6$$

$$QS = RS \times \sqrt{2} = 6\sqrt{2}$$

Arcs and Central Angles

The *length* of an arc of a circle is expressed in linear units of measurement such as inches, centimeters, and feet. Degrees are used to represent the *measure* of an arc. A degree is a unit of measurement obtained by dividing one complete revolution (that is, a circle) into 360 equal parts, and then referring to each part as a *degree*. A circle contains 360 degrees, so the measure of an arc of a circle must be some fractional part of 360 degrees.

Since the measure of a circle is 360 degrees, the measure of a semicircle (Figure 12.7*a*) is 180 degrees. An arc whose degree measure is less than 180 is called a *minor arc*. A *major arc* has a degree measure greater than 180 but less than 360. See Figure 12.7*b*.

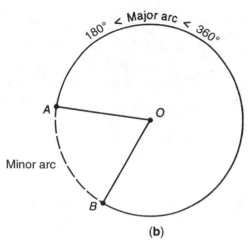

FIGURE 12.7

The measure of an arc of a circle may also be related to an angle whose vertex is the center of the circle and whose sides intercept arcs of the circle. In Figure 12.8 angle *AOB* is called a *central angle*. The minor arc that lies in the interior of the angle is *defined* to have the same measure as its central angle, $\angle AOB$. For example, if $m\angle AOB = 83$, then $m\overarc{AB} = 83$.

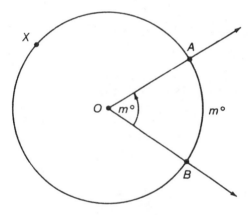

FIGURE 12.8

At times we will need to refer to a central angle and *its intercepted arc*. Although the sides of a central angle determine a minor *and* a major arc, we will refer to the minor arc as *the* intercepted arc.

SUMMARY

- A *semicircle* is an arc of a circle whose end points are the end points of a diameter of the circle; its measure is 180 degrees.
- A *central angle* is an angle whose vertex is at the center of the circle.
- A *minor arc* of a circle is an arc that lies in the interior of the central angle that intercepts the arc. The measure of a minor arc is the same as the measure of its central angle.

- A *major arc* of a circle is an arc that lies in the exterior of the central angle that intercepts the arc. The degree measure of a major arc is equal to 360 minus the measure of the central angle that determines the end points of the arc.
- The *measure* of an arc, as with angle measurement, is denoted by preceding the name of the arc by the letter "m": m$\widehat{AB}$ = 67 is read as "the measure of arc *AB* is 67." When this notation is used, the degree symbol is omitted.

a. Name four minor arcs of circle *O*.

b. Name major $\widehat{TEJ}$ in three different ways. (Use three letters.)

c. Name four different major arcs of circle *O*.

d. If m∡*JOT* = 73 and $\overline{NT}$ is a diameter of circle *O*, find:

 (i) m$\widehat{JT}$ **(ii)** m$\widehat{JN}$

 (iii) m$\widehat{JET}$ **(iv)** m$\widehat{TEN}$

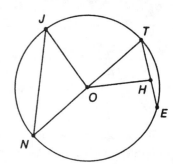

SOLUTION

a. $\widehat{JT}$, $\widehat{JN}$, $\widehat{ET}$, and $\widehat{EN}$.

b. $\widehat{JET}$, $\widehat{TNJ}$, and $\widehat{JNT}$.

c. $\widehat{NTE}$, $\widehat{TNE}$, $\widehat{NTJ}$, and $\widehat{TNJ}$.

d. **(i)** m$\widehat{JT}$ = m∡*JOT* = 73

 (ii) m$\widehat{JN}$ = m∡*JON* = 180 — 73 = 107

 (iii) m$\widehat{JET}$ = 360 — m∡*JOT* = 360 — 73 = 287

 (iv) m$\widehat{TEN}$ = 180

In circle *O* of Figure 12.9, the measures of arcs *AC*, *CB*, and *AB* may be determined as follows:

$$m\widehat{AC} = m\angle AOC = 30$$

$$m\widehat{CB} = m\angle BOC = 40$$

$$m\widehat{AB} = m\angle AOB = 70$$

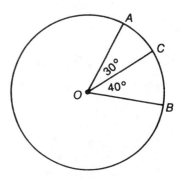

FIGURE 12.9

Notice that m$\widehat{AB}$ = m$\widehat{AC}$ + m$\widehat{CB}$. This result holds provided that arcs *AC* and *CB* are consecutive, having exactly one point (*C*) in common. This concept of arc addition is analogous to the concept of betweenness of points on line segments.

The measures of arcs can be added or subtracted in the same way that measures of angles and line segments are combined.

Are arcs that have the same degree referring to measure congruent (that is, equal in length)? If we are referring to the same circle, the answer is yes. But what if we are comparing two arcs having the same measure in two different circles? We will consider two circles that have the same center (called *concentric* circles).

In Figure 12.10, although m $\widehat{AB}$ = 40 and m $\widehat{XY}$ = 40, arcs *AB* and *XY* are *not* congruent. This observation will help us to frame the definition of congruent arcs.

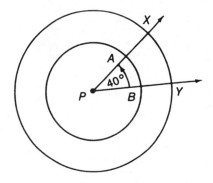

FIGURE 12.10

DEFINITION OF CONGRUENT ARCS

Congruent arcs are arcs in the *same circle* or in *congruent circles* that have the same degree measure.

It follows from the definition of congruent arcs that, in the same circle or in congruent circles, *congruent central angles intercept congruent arcs* and *congruent arcs have congruent central angles*. See Figure 12.11.

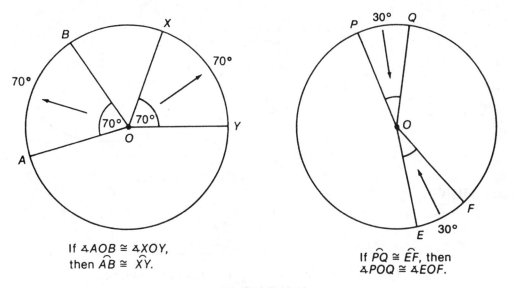

If ∡*AOB* ≅ ∡*XOY*, then $\widehat{AB}$ ≅ $\widehat{XY}$.

If $\widehat{PQ}$ ≅ $\widehat{EF}$, then ∡*POQ* ≅ ∡*EOF*.

FIGURE 12.11

THEOREM 12.1 ≅ CENTRAL ANGLES IMPLIES ≅ ARCS AND VICE VERSA

In the same circle or congruent circles:
- Congruent central angles intercept congruent arcs.
- Congruent arcs have congruent central angles.

If point M is located on $\overset{\frown}{AB}$ so that $\overset{\frown}{AM} \cong \overset{\frown}{MB}$, then M is the midpoint of $\overset{\frown}{AB}$, as shown in Figure 12.12.

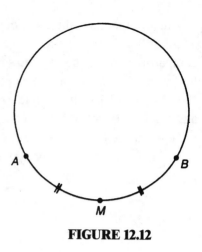

FIGURE 12.12

DEFINITION OF MIDPOINT OF AN ARC

The *midpoint of an arc* is the point of the arc that divides the arc into two congruent arcs.

EXAMPLE 12.5

GIVEN: B is the midpoint of $\overset{\frown}{AC}$.
PROVE: $\triangle AOB \cong \triangle COB$.

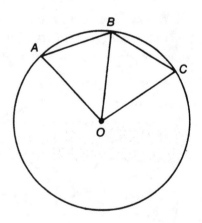

SOLUTION

PLAN: Prove $\triangle AOB \cong \triangle COB$ by SAS.

PROOF:

Statements	Reasons
1. $\overline{OA} \cong \overline{OC}$. (Side)	1. Radii of the same circle are congruent.
2. B is the midpoint of $\overset{\frown}{AC}$.	2. Given.
3. $\overset{\frown}{AB} \cong \overset{\frown}{BC}$.	3. A midpoint of an arc divides the arc into two congruent arcs.
4. $\angle AOB \cong \angle COB$. (Angle)	4. In the same circle, congruent arcs have congruent central angles.
5. $\overline{OB} \cong \overline{OB}$. (Side)	5. Reflexive property of congruence.
6. $\triangle AOB \cong \triangle COB$.	6. SAS Postulate.

**EXAMPLE
12.6**

GIVEN: $\overset{\frown}{AB} \cong \overset{\frown}{CD}$.
PROVE: $\overline{AB} \cong \overline{CD}$.

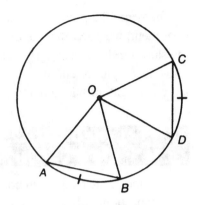

SOLUTION

PLAN: Show $\triangle AOB \cong \triangle DOC$.

PROOF:

Statements	Reasons
1. $\overline{AO} \cong \overline{DO}$. (Side)	1. Radii of the same circle are congruent.
2. $\overset{\frown}{AB} \cong \overset{\frown}{CD}$.	2. Given.
3. $\angle AOB \cong \angle DOC$. (Angle)	3. In the same circle, congruent arcs have congruent central angles.
4. $\overline{OB} \cong \overline{OC}$. (Side)	4. Same as reason 1.
5. $\triangle AOB \cong \triangle DOC$.	5. SAS Postulate.
6. $\overline{AB} \cong \overline{CD}$.	6. CPCTC.

Example 12.6 establishes the following theorem:

THEOREM 12.2 CONGRUENT ARCS IMPLY CONGRUENT CHORDS

In the same circle or in congruent circles, congruent arcs have congruent chords.

In Figure 12.13, if circle O is congruent to circle P and $\overset{\frown}{AB} \cong \overset{\frown}{CD}$, then Theorem 12.2 entitles us to conclude that $\overline{AB} \cong \overline{CD}$.

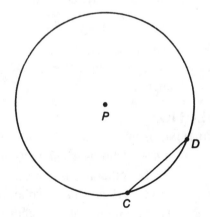

FIGURE 12.13

Theorem 12.2 provides a convenient method for proving a pair of chords are congruent; we show that the chords to be proved congruent cut off congruent arcs in the same circle or in congruent circles. The converse of Theorem 12.2 states that, if we know that a pair of chords are congruent, we may conclude that their intercepted arcs are congruent, provided that we are working in the same circle or in congruent circles.

THEOREM 12.3 CONGRUENT CHORDS IMPLY CONGRUENT ARCS

In the same circle or in congruent circles, congruent chords intercept congruent arcs.

EXAMPLE 12.7

GIVEN: $\overline{CA} \cong \overline{DB}$.
PROVE: $\angle A \cong \angle B$.

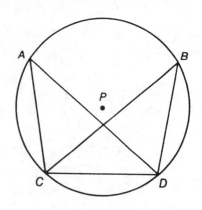

SOLUTION

PLAN: Show $\triangle ACD \cong \triangle BDC$ by SSS Postulate.

PROOF:

Statements	Reasons
1. $\overline{CD} \cong \overline{CD}$. (Side)	1. Reflexive property of congruence.
2. $\overline{CA} \cong \overline{DB}$. (Side)	2. Given.
3. $m\overset{\frown}{CA} = m\overset{\frown}{DB}$.	3. In the same circle, congruent chords intercept equal arcs.
4. $m\overset{\frown}{AB} = m\overset{\frown}{AB}$.	4. Reflexive property of equality.
5. $m\overset{\frown}{ACD} = m\overset{\frown}{BDC}$	5. Arc addition.
6. $\overline{AD} \cong \overline{BC}$. (Side)	6. In the same circle, equal arcs have congruent chords.
7. $\triangle ACD \cong \triangle BDC$.	7. SSS Postulate.
8. $\angle A \cong \angle B$.	8. CPCTC.

EXAMPLE 12.8

Prove that, if two chords are the perpendicular bisectors of each other, then each chord is a diameter.

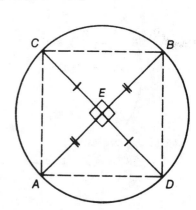

SOLUTION

OUTLINE OF PROOF

GIVEN: Chords $\overline{AB}$ and $\overline{CD}$ are $\perp$ to each other and bisect each other.
PROVE: $\overline{AB}$ and $\overline{CD}$ are diameters.
PLAN: Show that each chord divides the circle into two congruent arcs.

PLAN: To prove that $\overline{AB}$ is a diameter, show that $m\overset{\frown}{ACB} = m\overset{\frown}{ADB}$.

STEPS: • Draw $\overline{AC}$, $\overline{AD}$, $\overline{BC}$, and $\overline{BD}$.
 • Prove $\triangle AEC \cong \triangle AED$ by *SAS*. $\overline{AC} \cong \overline{AD}$, which implies that $m\overset{\frown}{AC} = m\overset{\frown}{AD}$.
 • Prove $\triangle BEC \cong \triangle BED$ by *SAS*. $\overline{BC} \cong \overline{BD}$, which implies that $m\overset{\frown}{BC} = m\overset{\frown}{BD}$.
 •
$$\begin{aligned} m\overset{\frown}{AC} &= m\overset{\frown}{AD} \\ + \quad m\overset{\frown}{BC} &= m\overset{\frown}{BD} \\ \hline m\overset{\frown}{ACB} &= m\overset{\frown}{ADB} \end{aligned}$$
, which implies that $\overline{AB}$ is a diameter since it divides the circle into two equal arcs.
 • Use a similar approach to show that $m\overset{\frown}{CAD} = m\overset{\frown}{CBD}$, which establishes that $\overline{CD}$ is a diameter.

SUMMARY

 • To prove *arcs* are congruent in the same circle or congruent circles, show that one of the following statements is true:
 1. The central angles that intercept the arcs are congruent.
 2. The chords that cut off the arcs are congruent.
 • To prove *central angles* are congruent in the same circle or in congruent circles, show that their intercepted arcs are congruent.
 • To prove *chords* are congruent in the same circle or in congruent circles, show that the chords have congruent arcs.
 • To prove a *chord is a diameter* show that the chord divides the circle into two congruent arcs.

Diameters and Chords

If we draw a chord and then draw a diameter perpendicular to the chord, what appears to be true? The resulting chord segments look as though they are congruent. Also, corresponding pairs of arcs seem to be congruent. These observations are stated in Theorem 12.4.

THEOREM 12.4 DIAMETER ⊥ CHORD THEOREM

In a circle, a diameter drawn perpendicular to a chord bisects the chord and its arcs.

OUTLINE OF PROOF

GIVEN: $\odot O$ with diameter $\overline{AB} \perp \overline{CD}$.

PROVE: **a.** $\overline{CX} \cong \overline{DX}$.

b. $\widehat{CB} \cong \widehat{DB}$.
$\widehat{CA} \cong \widehat{DA}$.

STEPS: • Draw radii $\overline{OC}$ and $\overline{OD}$.

• Prove $\triangle OXC \cong \triangle OXD$ by Hy-Leg.

• $\overline{CX} \cong \overline{DX}$ and $\angle 1 \cong \angle 2$ by *CPCTC*.

• $\widehat{CB} \cong \widehat{DB}$ since $\cong$ central angles intercept $\cong$ arcs.

• $\angle 3 \cong \angle 4$ since supplements of congruent angles are congruent.

• $\widehat{CA} \cong \widehat{DA}$.

According to Theorem 12.4, any segment that passes through the center of a circle and is perpendicular to a chord bisects the chord.

 The length of a diameter of circle O is 20 and the length of chord $\overline{AB}$ is 16. What is the distance between the chord and the center of the circle?

SOLUTION

We recall that the distance from a point (that is, the center of the circle) to a line segment (that is, the chord) is the length of the perpendicular segment from the point to the line segment. We therefore choose to draw a diameter that is perpendicular to chord $\overline{AB}$.

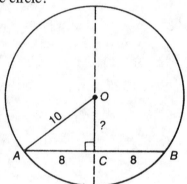

The distance between the chord and the center of the circle is represented by the length of $\overline{OC}$. By drawing radius $\overline{OA}$, we form a right triangle. Since the diameter is 20, the radius is 10. Applying Theorem 12.4, we deduce that the length of $\overline{AC}$ is 8. Hence, $\triangle OCA$ is a right triangle, the lengths of whose sides form a Pythagorean triple (3-4-5). The length of $\overline{OC}$ is 6.

After drawing several pairs of parallel chords in a circle, a pattern emerges, as shown in Figure 12.14.

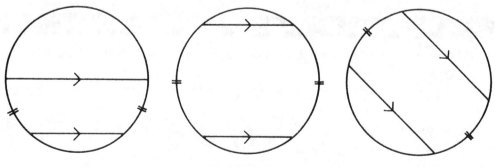

FIGURE 12.14

Notice that the arcs that the parallel chords cut off (and which lie between the chords) appear to be congruent. Here are the formal statement of the corresponding theorem and its proof.

THEOREM 12.5 PARALLEL CHORDS AND CONGRUENT ARCS THEOREM

In a circle, parallel chords cut off equal arcs.

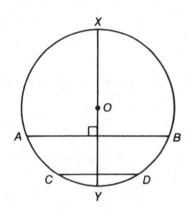

GIVEN: $\odot O$ with $AB \parallel CD$.

PROVE: $m\widehat{AC} = m\widehat{BD}$.

PLAN: Draw diameter $\overline{XY} \perp \overline{AB}$. $\overline{XY}$ will also be $\perp \overline{CD}$. Apply Theorem 12.4.

PROOF:

Statements	Reasons
1. $\odot O$ with $\overline{AB} \parallel \overline{CD}$.	1. Given.
2. Draw $\overline{XY}$ through point O and perpendicular to $\overline{AB}$.	2. Through a point not on a line, exactly one perpendicular may be drawn from the point to the line.
3. $\overline{XY} \perp \overline{CD}$.	3. A line perpendicular to one of two parallel lines is perpendicular to the other line.
4. $\begin{aligned} m\widehat{AY} &= m\widehat{BY} \\ -m\widehat{CY} &= m\widehat{DY}. \end{aligned}$	4. A diameter perpendicular to a chord bisects its arcs.
5. $m\widehat{AC} = m\widehat{BD}$.	5. Arc subtraction.

EXAMPLE 12.10

If chord $\overline{CD}$ is parallel to diameter $\overline{AB}$, and $m\widehat{CD} = 40$, find the measures of arcs AC and BD.

SOLUTION

Since $\overline{CD} \parallel \overline{AB}$, $m\widehat{AC} = m\widehat{BD} = x$

$$
\begin{aligned}
x + 40 + x &= 180 \\
2x + 40 &= 180 \\
2x &= 140 \\
x &= 70 \\
m\widehat{AC} = m\widehat{BD} &= 70
\end{aligned}
$$

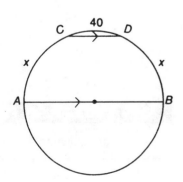

Tangents and Secants

DEFINITIONS AND PROPERTIES OF TANGENTS AND SECANTS

A line may intersect a circle at no points, one point, or two points. See Figure 12.15.

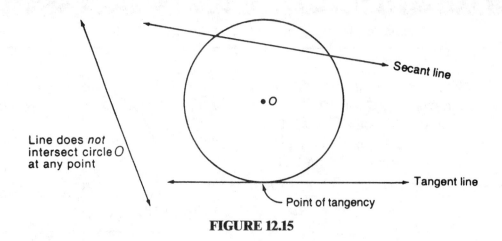

FIGURE 12.15

> ### DEFINITIONS OF TANGENT AND SECANT
> - A *tangent* line is a line that intersects a circle in exactly one point. The point of contact is called the *point of tangency*.
> - A *secant* line is a line that intersects a circle in two different points. (Every secant line includes a chord of the circle.)

In Figure 12.16, lines *j, k*, and ℓ are tangent to circle *O*. Radii have been drawn to their points of tangency. In each case the radius appears to be perpendicular to the tangent.

FIGURE 12.16

THEOREM 12.6 RADIUS ⊥ TANGENT THEOREM

A radius drawn to a point of tangency is perpendicular to the tangent.

OUTLINE OF PROOF

GIVEN: $\overleftrightarrow{AB}$ is tangent to $\odot P$ at point A.

PROVE: $PA \perp \overleftrightarrow{AB}$.

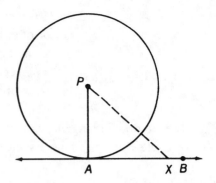

PLAN: Use an indirect proof.

STEPS:
- If we assume $\overline{PA}$ is *not* perpendicular to $\overleftrightarrow{AB}$, then there must be another segment, say PX, which can be drawn from P perpendicular to $\overleftrightarrow{AB}$.
- Since $PX \perp \overleftrightarrow{AB}$, PX represents the *shortest* distance from P to $\overleftrightarrow{AB}$. Thus, PX is *less than PA*.
- But point X lies in the exterior of circle P. Since $\overline{PA}$ is a radius, PX must be *greater than PA*. This contradicts our previous assertion that PX is less than $\overline{PA}$. Our assumption that PA is not perpendicular to $\overleftrightarrow{AB}$ must be false. $\overline{PA}$ is therefore perpendicular to $\overleftrightarrow{AB}$.

The converse of this theorem is given in Theorem 12.7 and may also be proved indirectly.

THEOREM 12.7 RADIUS ⊥ LINE IMPLIES A TANGENT THEOREM

If a radius is perpendicular to a line at the point at which the line intersects a circle, then the line is a tangent.

OUTLINE OF PROOF

GIVEN: $\odot P$ with $\overline{PA} \perp$ line ℓ at point A.

PROVE: ℓ is tangent to $\odot P$.

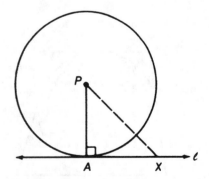

PLAN: Use an indirect proof.

STEPS:
- If ℓ is assumed not to be a tangent line, it must intersect $\odot P$ at another point, say point X.
- Since $\overline{PX}$ is the hypotenuse of right triangle PAX, PX is greater than PA.

- But $\overline{PA}$ is a radius, which implies that point X must lie in the *exterior* of circle P.
- The assumption that there is a second point *on the circle* at which line ℓ intersects is false. Since there is exactly one point at which ℓ intersects the circle, ℓ is a tangent line.

The length of a tangent segment drawn from point P to circle O is 12. If the radius of circle O is 5, find the distance from point P to the center of the circle.

SOLUTION

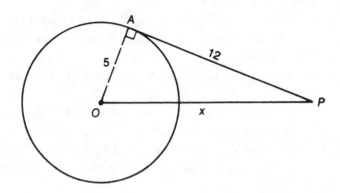

$\overline{PA}$ is a tangent segment since one of its end points, point A, is the point of tangency. Also, PO represents the distance from point P to the center of the circle. Draw a radius to the point of tangency $\overline{OA} \perp \overline{PA}$ and $x = 13$ since a $\triangle OAP$ is a 5-12-*13* right triangle.

INSCRIBED AND CIRCUMSCRIBED POLYGONS

Quadrilateral $ABCD$ in Figure 12.17 is *circumscribed about circle O*. Pentagon $ABCDE$ in Figure 12.18 is *inscribed in* circle O.

FIGURE 12.17

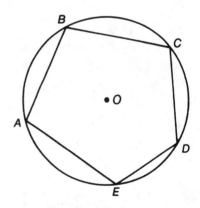

FIGURE 12.18

DEFINITIONS OF CIRCUMSCRIBED AND INSCRIBED POLYGONS
A polygon is *circumscribed about* a circle if each of its sides is tangent to the circle.
A polygon is *inscribed in* a circle if each of its vertices lies on the circle.

EXAMPLE 12.12 A quadrilateral is inscribed in a circle in a way that the sides of the quadrilateral divide the circle into arcs whose measures have the ratio $1:2:3:4$. How many degrees are there in each arc?

SOLUTION

The sum of the measures of the arcs that comprise a circle must equal 360:

$$1x + 2x + 3x + 4x = 360°$$
$$10x = 360°$$
$$x = 36°$$
$$2x = 72°$$
$$3x = 108°$$
$$4x = 144°$$

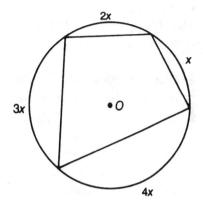

Angle Measurement: Vertex on the Circle

The vertex of an angle whose sides intercept arcs on a circle may be located in any one of the positions shown in Figure 12.19.

At the center of the circle:

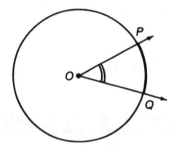

Intercepted arc: $\overset{\frown}{PQ}$

On the circle:

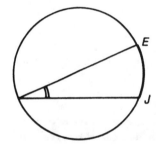

Intercepted arc: $\overset{\frown}{EJ}$

In the interior of the circle but not at the center:

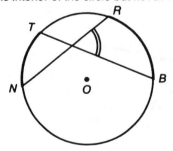

Intercepted arcs: $\overset{\frown}{NT}$ and $\overset{\frown}{RB}$

In the exterior of the circle:

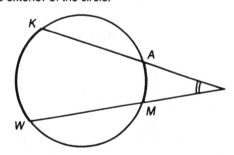

Intercepted arcs: $\overset{\frown}{KW}$ and $\overset{\frown}{AM}$

FIGURE 12.19

In each of these cases a relationship exists between the measures of the intercepted arc or arcs and the measure of the angle. We have already seen, for example, that the measure of any angle whose vertex is at the center of the circle (a central angle) is equal to the measure of its intercepted arc. Let's now consider the situation in which the vertex of the angle is a point on the circle and the sides of the angle are chords (or secants) of the circle. In Figure 12.20, angle *ABC* is called an *inscribed angle* and $\overset{\frown}{AC}$ is its intercepted arc.

DEFINITION OF INSCRIBED ANGLE

An *inscribed angle* is an angle whose vertex lies on a circle and whose sides are chords (or secants) of the circle.

In Figure 12.21, suppose $m\overset{\frown}{AC} = 50$. Then $m\angle AOC = 50$. Since $\angle AOC$ is an exterior angle of $\triangle BOC$, $m\angle AOC = m\angle B + m\angle C$. Since $m\angle B = m\angle C$ (base angles of an isosceles triangle), $m\angle B = 25$. How do the measures of inscribed angle *ABC* and its intercepted arc ($\overset{\frown}{AC}$) compare? Observe that $m\angle ABC = \frac{1}{2}m\overset{\frown}{AC} = \frac{1}{2}(50) = 25$.

FIGURE 12.20

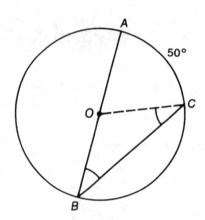

FIGURE 12.21

THEOREM 12.8 INSCRIBED ANGLE THEOREM

The measure of an inscribed angle is equal to one-half the measure of its intercepted arc.

In the accompanying figure,

$$m\angle ABC = \frac{1}{2}x.$$

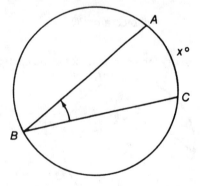

In our example, we assumed that one of the sides of the inscribed angle passed through the center of the circle. In general, this need not be true. The center of the

circle may lie on a side of the inscribed angle, in the interior of the angle, or in the exterior of the angle. The proof of the Inscribed Angle Theorem must treat each of these three cases separately.

The three cases of Theorem 12.8 to be proved are shown in Figure 12.22.

CASE 1: Center lies on a side of the inscribed angle.

CASE 2: Center lies in the interior of the inscribed angle.

CASE 3: Center lies in the exterior of the inscribed angle.

 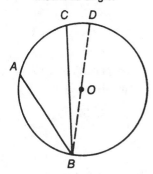

FIGURE 12.22

GIVEN: $\odot O$ with inscribed angle ABC.

PROVE: $m \angle ABC = \dfrac{1}{2} m \widehat{AC}$.

PROOF
OF CASE 1:

PLAN: Use an approach that parallels the solution to the numerical example used to motivate Theorem 12.8.

PROOF: Statements	Reasons
1. Draw radius OC.	1. Two points determine a line.
2. $m \angle 1 = m \angle ABC + m \angle 2$.	2. The measure of an exterior angle of a triangle is equal to the sum of the measures of the two nonadjacent interior angles of the triangle.
3. $\overline{OB} \cong \overline{OC}$.	3. All radii of the same circle are congruent.
4. $m \angle 2 = m \angle ABC$.	4. If two sides of a triangle are congruent, then the angles opposite these sides are equal in measure.
5. $m \angle 1 = m \angle ABC + m \angle ABC$, or $m \angle 1 = 2 \, m \angle ABC$.	5. Substitution.
6. $\dfrac{1}{2} m \angle 1 = m \angle ABC$ or $m \angle ABC = \dfrac{1}{2} m \angle 1$.	6. Division property.
7. $m \angle 1 = m \widehat{AC}$.	7. The measure of a central angle is equal to the measure of its intercepted arc.
8. $m \angle ABC = \dfrac{1}{2} m \widehat{AC}$.	8. Substitution.

OUTLINE OF Draw diameter $\overline{BD}$.
PROOF FOR Apply the result established in Case 1:
CASE 2:

$$\mathrm{m}\angle 1 \ = \frac{1}{2}\mathrm{m}\widehat{AD}$$

$$+ \ \mathrm{m}\angle 2 \ = \frac{1}{2}\mathrm{m}\widehat{DC}$$

$$\overline{\mathrm{m}\angle 1 + \mathrm{m}\angle 2 \ = \frac{1}{2}(\mathrm{m}\widehat{AD} + \mathrm{m}\widehat{DC})}$$

Use substitution:

$$\mathrm{m}\angle ABC = \frac{1}{2}\mathrm{m}\widehat{AC}$$

OUTLINE OF Draw diameter $\overline{BD}$.
PROOF FOR Apply the result established in Case 1:
CASE 3:

$$\mathrm{m}\angle ABD = \frac{1}{2}\mathrm{m}\widehat{ACD}$$

$$-\mathrm{m}\angle CBD = \frac{1}{2}\mathrm{m}\widehat{CD}$$

$$\overline{\mathrm{m}\angle ABD - \mathrm{m}\angle CBD = \frac{1}{2}(\mathrm{m}\widehat{ACD} - \mathrm{m}\widehat{CD})}$$

Use substitution:

$$\mathrm{m}\angle ABC = \frac{1}{2}\mathrm{m}\widehat{AC}$$

EXAMPLE 12.13 In each of the following, find the value of x.

a.

b.

c.

d.

SOLUTION

a. $\widehat{ACB}$ is a semicircle, so $\mathrm{m}\widehat{BC} = 180 - 110 = 70$.
$$x = \frac{1}{2}\mathrm{m}\widehat{BC} = \frac{1}{2}(70)$$
$$= 35$$

b. $\angle AOC$ is a central angle, so $\mathrm{m}\widehat{AC} = 110$.
$$\mathrm{m}\angle ABC = \frac{1}{2}\mathrm{m}\widehat{AC} = \frac{1}{2}(110)$$
$$x = 55$$

c. The measure of the arc intercepted by an inscribed angle must be twice the measure of the inscribed angle. Hence,
$$x = 2(\mathrm{m}\angle ABC) = 2(25)$$
$$= 50$$

d. $\angle AOB$ is a central angle so $\mathrm{m}\widehat{ACB} = 70$.
$$\mathrm{m}\widehat{APB} = 360 - 70 = 290$$
$$x = \frac{1}{2}\mathrm{m}\widehat{APB} = \frac{1}{2}(290)$$
$$= 145$$

A triangle is inscribed in a circle so that its sides divide the circle into arcs whose measures have the ratio of $2:3:7$. Find the measure of the largest angle of the triangle.

SOLUTION

First, determine the degree measures of the arcs of the circle.

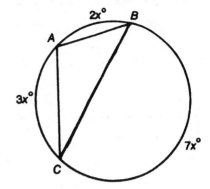

$$m\widehat{AB} + m\widehat{AC} + m\widehat{BC} = 360$$
$$2x + 3x + 7x = 360$$
$$12x = 360$$
$$x = 30$$

The measures of the arcs of the circle are as follows:

$$m\widehat{AB} = 2x = 60$$
$$m\widehat{AC} = 3x = 90$$
$$m\widehat{BC} = 7x = 210$$

Each angle of the triangle is an inscribed angle. The largest angle lies opposite the arc having the greatest measure. Hence,

$$m\angle A = \frac{1}{2}m\widehat{BC} = \frac{1}{2}(210)$$
$$= 105$$

Quadrilateral $ABCD$ is inscribed in circle O. If $m\angle A = x$, and $m\angle B = y$, express the measures of angles of C and D in terms of x and y.

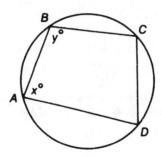

SOLUTION

To find $m\angle C$:

	To find $m\angle D$:
$m\widehat{BCD} = 2(x) = 2x$	$m\widehat{ADC} = 2(y) = 2y$
$m\widehat{BAD} = 360-2(x)$	$m\widehat{ABC} = 360-2y$
$m\angle C = \frac{1}{2}(360-2x)$	$m\angle D = \frac{1}{2}(360-2y)$
$= 180-x$	$= 180-y$

Example 12.15 establishes the following theorem:

THEOREM 12.9 ANGLES OF AN INSCRIBED QUADRILATERAL THEOREM

Opposite angles of an inscribed quadrilateral are supplementary.

$$a + c = 180$$
$$b + d = 180$$

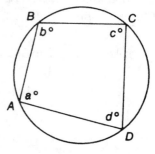

A chord that has as one of its end points the point of tangency of a tangent to a circle forms an angle whose vertex is on the circle. In Figure 12.23, the sides of $\angle ABC$ are chord $\overline{AB}$ and tangent ray $\overrightarrow{BC}$. The end points of the intercepted arc are the vertex of the angle (point B) and the other end point of the chord (point A).

FIGURE 12.23

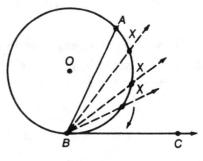

FIGURE 12.24

In Figure 12.24, consider inscribed angle ABX and suppose point X slides down the circle so that $\overrightarrow{BX}$ gets closer and closer to tangent $\overrightarrow{BC}$. At each position, inscribed angle ABX is measured by one-half of its intercepted arc, $\overset{\frown}{AX}$. As point X approaches point B, the measure of $\overset{\frown}{AX}$ gets closer and closer to the measure of $\overset{\frown}{AB}$. Hence, when $\overrightarrow{BX}$ finally coincides with tangent $\overrightarrow{BC}$, the angle formed is measured by one-half the measure of $\overset{\frown}{AB}$.

THEOREM 12.10 CHORD-TANGENT ANGLE THEOREM

The measure of an angle formed by a tangent and a chord drawn to the point of tangency is equal to one-half the measure of the intercepted arc.

In the accompanying figure,

$$m\angle ABC = \frac{1}{2}x$$

OUTLINE OF PROOF

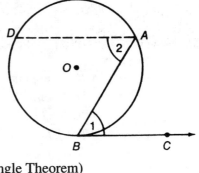

GIVEN: $\overrightarrow{BC}$ is tangent to $\odot O$ at point B.

$\overline{AB}$ is drawn.

PROVE: $m\angle 1 = \frac{1}{2}m\widehat{AB}$.

PLAN: Through point A draw a chord parallel to $\overrightarrow{BC}$, intersecting the circle at point D. Then,

STEPS:
• $m\angle 1 = m\angle 2$
 (Alternate interior angles)

• $m\angle 2 = \frac{1}{2}m\widehat{BD}$ (Inscribed Angle Theorem)

• $m\angle 1 = \frac{1}{2}m\widehat{BD}$ (Substitution)

• $m\widehat{BD} = m\widehat{AB}$ (Since $\overline{AD} \parallel \overrightarrow{BC}$)

• $m\angle 1 = \frac{1}{2}m\widehat{AB}$ (Substitution)

EXAMPLE 12.16

Find the value of x if $m\widehat{ACB} = 250$.

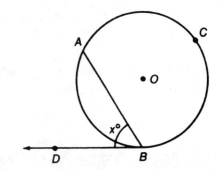

SOLUTION

$m\widehat{AB} = 360 - 250 = 110$

$x = \frac{1}{2}m\widehat{AB} = \frac{1}{2}(110)$

 $= 55$

EXAMPLE 12.17

In circle O, $m\widehat{AC} = 110$. Find the measure of each of the numbered angles.

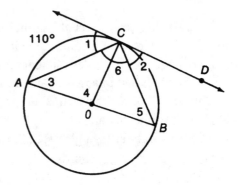

SOLUTION

$m\angle 1 = \frac{1}{2}m\widehat{AC} = \frac{1}{2}(110) = 55$

Since $m\widehat{BC} = 180 - 110 = 70$,

$m\angle 2 = \frac{1}{2}m\widehat{BC} = \frac{1}{2}(70) = 35$

$m\angle 3 = \frac{1}{2}m\widehat{BC} = \frac{1}{2}(70) = 35$

$m\angle 4 = m\widehat{AC} = 110$

$m\angle 5 = \frac{1}{2}m\widehat{AC} = \frac{1}{2}(110) = 55$

$m\angle 6 = \frac{1}{2}m\widehat{AB} = \frac{1}{2}(180) = 90$

NOTE: Angles 1 and 2 are formed by a chord and a tangent; angle 4 is a central angle; and angles 3, 5, and 6 are inscribed angles.

EXAMPLE 12.18 Prove that a tangent drawn parallel to a chord of a circle cuts off congruent arcs between the tangent and the chord.

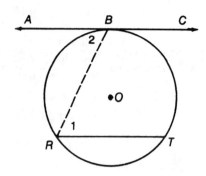

SOLUTION

OUTLINE OF PROOF

GIVEN: $\overleftrightarrow{ABC}$ is tangent to $\odot O$ at point B, $\overleftrightarrow{ABC} \parallel \overline{RT}$.

PROVE: $\overarc{BR} \cong \overarc{BT}$.

STEPS:
- $m\angle 1 = \frac{1}{2} m\overarc{BT}$ and $m\angle 2 = \frac{1}{2} m\overarc{BR}$.

- Since $\overleftrightarrow{ABC} \parallel \overline{RT}$, $m\angle 1 = m\angle 2$ so that $\frac{1}{2} m\overarc{BT} = \frac{1}{2} m\overarc{BR}$.

- Multiplying each side of this equation by 2 yields $m\overarc{BT} = m\overarc{BR}$, or $\overarc{BT} \cong \overarc{BR}$.

Thus, parallel chords, parallel tangents, and a tangent parallel to a chord cut off congruent arcs on a circle.

Angle Measurement: Vertex in the Interior of the Circle

When two chords intersect at a point on the circle, then an inscribed angle is formed. If the chords intersect at a point in the interior of a circle (not at the center), then two pairs of angles whose sides intercept arcs on the circle result, as shown in Figure 12.25. With respect to angle 1 (and its vertical angle 2), the intercepted arcs are the arcs that lie opposite the vertical angle pair: $\overarc{AC}$ and $\overarc{BD}$. With respect to the vertical angle pair formed by angles 3 and 4, the intercepted arcs are $\overarc{AD}$ and $\overarc{BC}$. For convenience we will refer to angles 1, 2, 3, and 4 as *chord-chord* angles since they are formed by intersecting chords.

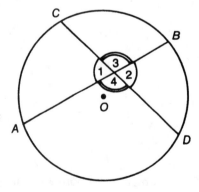

FIGURE 12.25

A chord-chord angle is measured by the *average* of its two intercepted arcs:

$$m\angle 1 = m\angle 2 = \frac{1}{2}(m\widehat{AC} + m\widehat{BD})$$

and

$$m\angle 3 = m\angle 4 = \frac{1}{2}(m\widehat{AD} + m\widehat{BC})$$

THEOREM 12.11 CHORD-CHORD ANGLE THEOREM

The measure of an angle formed by two chords intersecting in the interior of a circle is equal to one-half the sum of the measures of the two intercepted arcs.
 In the accompanying figure,

$$m\angle AEC = \frac{1}{2}(x + y).$$

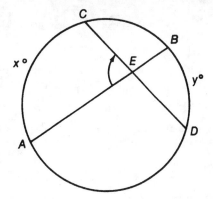

In Theorem 12.11, the sides of the angle may be secants that intersect in the interior of a circle since secants contain chords of the circle.

OUTLINE OF PROOF

GIVEN: In $\odot O$ chords $\overline{AB}$ and $\overline{CD}$ intersect at point E.

PROVE: $m\angle 1 = \dfrac{1}{2}(m\widehat{AC} + m\widehat{BD})$.

PLAN: Draw $\overline{AD}$. Angle 1 is an exterior angle of $\triangle AED$. Hence,

$$m\angle 1 = (m\angle D + m\angle A)$$
$$= \frac{1}{2}m\widehat{AC} + \frac{1}{2}m\widehat{BD}$$
$$= \frac{1}{2}(m\widehat{AC} + m\widehat{BD})$$

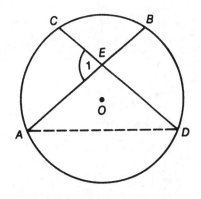

Find the value of x.

(a)

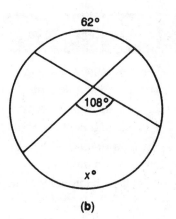

(b)

SOLUTION

a. $x = \frac{1}{2}(65 + 105)$

$= \frac{1}{2}(170) = 85$

b. $108 = \frac{1}{2}(x + 62)$

or

$x + 62 = 2(108) \quad = 216$

$x = 216 - 62 = 154$

Angle Measurement: Vertex in the Exterior of the Circle

If the vertex of an angle whose sides intercept arcs on the circle lies in the exterior of the circle, then the angle may be formed in one of three possible ways, as shown in Figure 12.26.

1. Each side is a secant.

2. One side is a secant and the other side is a tangent.

3. Each side is a tangent.

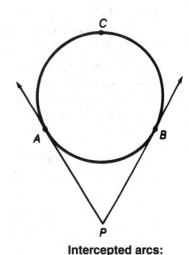

Intercepted arcs:

$\overset{\frown}{BC}$ and $\overset{\frown}{AD}$

Intercepted arcs:

$\overset{\frown}{BC}$ and $\overset{\frown}{AC}$

Intercepted arcs:

$\overset{\frown}{ACB}$ and $\overset{\frown}{AB}$

FIGURE 12.26

In each of these three cases, the angle at vertex P is measured by one-half the *difference* of the intercepted arcs.

THEOREM 12.12 SECANT-SECANT, SECANT-TANGENT, TANGENT-TANGENT THEOREM

The measure of an angle formed by two secants, a secant and a tangent, or two tangents intersecting in the exterior of a circle is equal to one-half the difference of the measures of the intercepted arcs.

In the accompanying figure,

$$m \angle P = \frac{1}{2}(x - y).$$

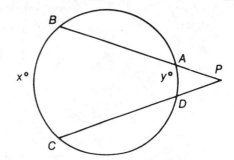

The three cases of Theorem 12.12 to be proved are shown in Figure 12.27.

CASE 1 **Secant-Secant** CASE 2 **Secant-Tangent** CASE 3 **Tangent-Tangent**

 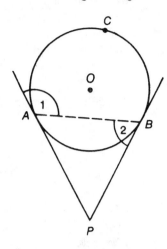

GIVEN: **In ⊙O, secants** GIVEN: **In ⊙O, secant $\overline{PAB}$** GIVEN: **In ⊙O, tangents**
$\overline{PAB}$ and $\overline{PDC}$ **and tangent PC** **$\overline{PA}$ and $\overline{PB}$**
are drawn. **are drawn.** **drawn.**

PROVE: $m \angle P =$ PROVE: $m \angle P =$ PROVE: $m \angle P =$
$\frac{1}{2}(m\widehat{BC} - m\widehat{AD})$ $\frac{1}{2}(m\widehat{BC} - m\widehat{AC})$ $\frac{1}{2}(m\widehat{ACB} - m\widehat{AB})$

FIGURE 12.27

OUTLINE OF PROOF

OF CASE 1: Draw $\overline{BD}$. Angle 1 is an exterior angle of $\triangle DBP$. Hence,

$$m \angle 1 = m \angle 2 + m \angle P$$

or

$$m \angle P = m \angle 1 - m \angle 2$$
$$= \frac{1}{2}m\widehat{BC} - \frac{1}{2}m\widehat{AD}$$
$$= \frac{1}{2}(m\widehat{BC} - m\widehat{AD})$$

The proofs of Cases 2 and 3 follow a similar pattern.

SUMMARY OF ANGLE MEASUREMENT IN CIRCLES

Vertex of Angle	Measure of Angle Equals	Model
1. At center (Central Angle)	The measure of the intercepted arc.	$m\angle AOB = m\widehat{AB}$

(Continued)

SUMMARY OF ANGLE MEASUREMENT IN CIRCLES (*Continued*)

Vertex of Angle	Measure of Angle Equals	Model
2. On circle (Inscribed Angle)	One-half the measure of the intercepted arc.	$m\angle ABC = \frac{1}{2} m\widehat{AC}$
3. On circle (Chord-Tangent Angle)	One-half the measure of the intercepted arc.	$m\angle ABC = \frac{1}{2} m\widehat{AC}$
4. Interior of Circle (Chord-Chord Angle)	One-half the sum of the measures of the intercepted arcs.	$m\angle AED = \frac{1}{2} (m\widehat{AD} + m\widehat{BC})$
5. Exterior of Circle (Secant-Secant, Secant-Tangent, and Tangent-Tangent Angles)	One-half the difference of the measures of the intercepted arcs.	$m\angle APC = \frac{1}{2} (m\widehat{AC} - m\widehat{AB})$

EXAMPLE 12.20

Find the value of x

a.

b.

c.

d.

e.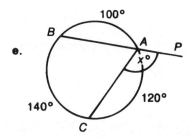

SOLUTION

a. $m\widehat{AC} = 360 - (76 + 84)$

$= 360 - 160$

$= 200$

$x = \frac{1}{2}(m\widehat{AC} - m\widehat{AB})$

$= \frac{1}{2}(200 - 76)$

$= \frac{1}{2}(124) = 62°$

c. $m\angle P = \frac{1}{2}(m\widehat{BD} = m\widehat{AC})$

$36 = \frac{1}{2}(88 - x)$

Multiplying both sides by 2,

$72 = 88 - x$

$x + 72 = 88$

$x = 88 - 72 = 16°$

b. $m\widehat{AB} = 360 - 295 = 65$

$x = \frac{1}{2}(m\widehat{ACB} - m\widehat{AB})$

$= \frac{1}{2}(295 - 65)$

$= \frac{1}{2}(230) = 115°$

d. $m\widehat{AB} = 360 - m\widehat{ACB}$

$m\widehat{AB} = 360 - x$

$m\angle P = \frac{1}{2}(m\widehat{ACB} - m\widehat{AB})$

$54 = \frac{1}{2}[x - (360 - x)]$

Multiplying both sides by 2,

$108 = x - 360 + x$

$108 = 2x - 360$

$468 = 2x$

$x = 234°$

e. A common *error* is to find m∡*CAP* (that is, *x*) by taking $\frac{1}{2}$ m$\widehat{AC}$. This is incorrect since the sides of the angle are a chord and *part of a secant*. Since angles *CAP* and *CAB* are supplementary, we first determine the measure of inscribed angle *CAB*:

$$m\angle CAB = \frac{1}{2}m\widehat{BC} = \frac{1}{2}(140) = 70$$
$$m\angle CAP = 180 - 70$$
$$x = 110°$$

EXAMPLE 12.21 $\overrightarrow{PA}$ is tangent to circle *O* at point *A*. Secant $\overline{PBC}$ is drawn. Chords $\overline{CA}$ and $\overline{BD}$ intersect at point *E*. If m$\widehat{AD}$:m$\widehat{AB}$:m$\widehat{DC}$:m$\widehat{BC}$ = 2:3:4:6, find the measure of each of the numbered angles.

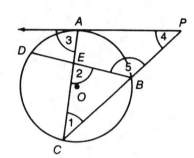

SOLUTION

Let m$\widehat{AD}$ = 2x,
 m$\widehat{AB}$ = 3x,
 m$\widehat{DC}$ = 4x,
 m$\widehat{BC}$ = 6x.

Then $2x + 3x + 4x + 6x = 360$
$$15x = 360$$
$$x = \frac{360}{15} = 24$$

m$\widehat{AD}$ = 2x = 2(24) = 48
m$\widehat{AB}$ = 3x = 3(24) = 72
m$\widehat{DC}$ = 4x = 4(24) = 96
m$\widehat{BC}$ = 6x = 6(24) = 144

- m∡1 = $\frac{1}{2}$m$\widehat{AB}$ = $\frac{1}{2}$(72) = 36

- m∡2 = $\frac{1}{2}$(m$\widehat{AD}$ + m$\widehat{BC}$)
 = $\frac{1}{2}$(48 + 144)
 = $\frac{1}{2}$(192)
 = 96

- m∡3 = $\frac{1}{2}$m$\widehat{AC}$
 = $\frac{1}{2}$(m$\widehat{AD}$ + m$\widehat{DC}$)
 = $\frac{1}{2}$(48 + 96)
 = $\frac{1}{2}$(144)
 = 72

- m∡4 = $\frac{1}{2}$(m$\widehat{AC}$ − m$\widehat{AB}$)
 = $\frac{1}{2}$(144 −72)
 = $\frac{1}{2}$(72)
 = 36

- m∡5 = 180 − m∡*CBD*

Since m∡*CBD* = $\frac{1}{2}$m$\widehat{DC}$
 = $\frac{1}{2}$(96) = 48,
m∡5 = 180 − 48
 = 132

Using Angle-Measurement Theorems

Angle-measurement theorems can be used to help establish that angles, arcs, and chords of a circle are congruent. First, though, we will formally state some observations that you have probably made while working on numerical applications of these theorems.

FIGURE 12.28

In Figure 12.28*a* $\overline{AB}$ is a diameter of circle *O*. Angle *AHB* is said to be *inscribed in* semicircle $\overset{\frown}{AHB}$ since the end points of the chords that form the angle coincide with the end points of a diameter. What is the measure of ∢ *AHB*? Angle $\overset{\frown}{AHB}$ is an inscribed angle having semicircle $\overset{\frown}{AXB}$ as its intercepted arc. Hence, m∢ *AHB* = $\frac{1}{2}$(180) = 90. *An angle inscribed in a semicircle is a right angle.*

In Figure 12.28*b* the measures of inscribed angles 1 and 2 are each equal to one-half of 80° (or 40°). *Inscribed angles that intercept the same arc (or congruent arcs) are congruent.*

The inscribed angles shown in Figure 12.28*c* are given as congruent (each has a measure of 25). It follows that m$\overset{\frown}{AC}$ = 50 and m$\overset{\frown}{DF}$ = 50. *If a pair of inscribed angles are congruent, then their intercepted arcs are congruent.*

Each of the preceding observations was based on a direct application of the Inscribed Angle Theorem (Theorem 12.8) and is therefore referred to as a *corollary*.

COROLLARIES TO THE INSCRIBED ANGLE THEOREM

- An angle inscribed in a semicircle is a right angle.
- If inscribed angles (or angles formed by a tangent and a chord) inter-cept the same or congruent arcs, then they are congruent.
- If inscribed angles are congruent, then their intercepted arcs are congruent.

EXAMPLE 12.22

GIVEN: In $\odot O$, $\overline{CD} \parallel \overline{AB}$. Chords $\overline{AC}$ and $\overline{BD}$ are extended to meet at point P.

PROVE: $\overline{AP} \cong \overline{BP}$.

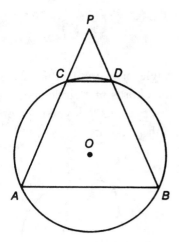

SOLUTION

PLAN: Show that inscribed angles A and B intercept congruent arcs and therefore are congruent. The sides opposite these angles in triangle APB must be congruent by the converse of the Base Angles Theorem.

PROOF:

Statements	Reasons
1. $\overline{CD} \parallel \overline{AB}$.	1. Given.
2. $m\overset{\frown}{AC} = m\overset{\frown}{BD}$.	2. Parallel chords intercept arcs equal in measure.
3. $m\overset{\frown}{ACD} = m\overset{\frown}{BDC}$.	3. Arc Sum Postulate.
4. $m\angle B = \frac{1}{2}m\overset{\frown}{ACD}$, $m\angle A = \frac{1}{2}m\overset{\frown}{BDC}$.	4. The measure of an inscribed angle is one-half the measure of its intercepted arc.
5. $\frac{1}{2}m\overset{\frown}{ACD} = \frac{1}{2}m\overset{\frown}{BDC}$.	5. Halves of equals are equal
6. $m\angle B = m\angle A$.	6. Substitution.
7. $\overline{AP} \cong \overline{BP}$.	7. If two angles of a triangle are equal in measure, then the sides opposite are congruent.

EXAMPLE 12.23

GIVEN: In $\odot O$, $\overline{AB}$ is a diameter. $\overset{\frown}{AC} \cong \overset{\frown}{BD}$.

PROVE: $\triangle ABC \cong \triangle BAD$.

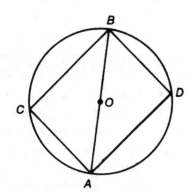

SOLUTION

PLAN: Prove triangles are congruent by Hy-Leg.

PROOF:

Statements	Reasons
1. In $\odot O$, $\overline{AB}$ is a diameter.	1. Given.
2. Angles C and D are right angles.	2. An angle inscribed in a semicircle is a right angle.
3. Triangles ABC and BAD are right triangles.	3. A triangle that contains a right angle is a right triangle.
4. $\overline{AB} \cong \overline{AB}$. (Hy)	4. Reflexive property of congruence.
5. $\overset{\frown}{AC} \cong \overset{\frown}{BD}$.	5. Given.
6. $\overline{AC} \cong \overline{BD}$. (Leg)	6. In a circle, congruent arcs have congruent chords.
7. $\triangle ACB \cong \triangle BDA$.	7. Hy-Leg.

The AAS theorem could have been used in the proof of Example 12.23.

(A) $\angle C \cong \angle D$ (Right angles are congruent.)

(A) $\angle ACB \cong \angle BDA$ (Inscribed angles that intercept congruent arcs are congruent.)

(S) $\overline{AB} \cong \overline{AB}$

REVIEW EXERCISES FOR CHAPTER 12

In Exercises 1 to 10, find the value of x.

1.

$\overline{AB} \parallel \overline{CD}$

2.

3.

4.

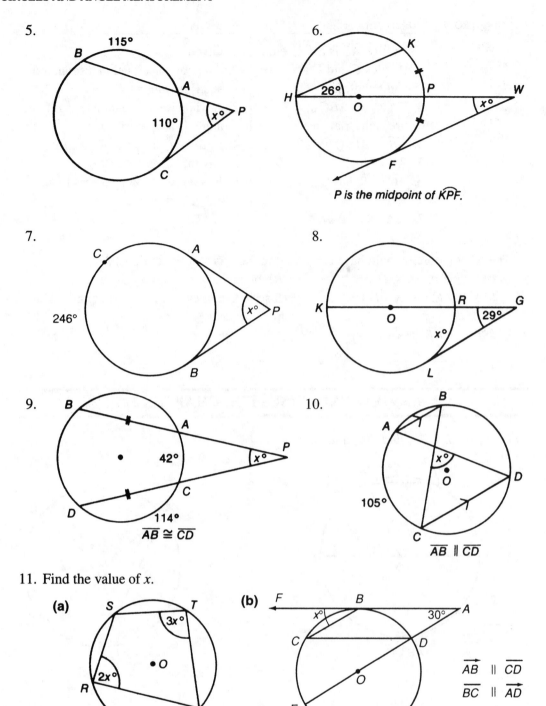

5.

6.

P is the midpoint of $\overarc{KPF}$.

7.

8.

9.

$\overline{AB} \cong \overline{CD}$

10.

$\overline{AB} \parallel \overline{CD}$

11. Find the value of x.

(a)

(b)

$\overrightarrow{AB} \parallel \overline{CD}$
$\overline{BC} \parallel \overrightarrow{AD}$

12. The length of a chord of a circle is 24 and its distance from the center is 5. Find the length of a diameter of the circle.

13. Tangents $\overline{AX}$ and $\overline{AY}$ are drawn to circle P from an exterior point A. Radii $\overline{PX}$ and $\overline{PY}$ are drawn. If $m\angle XPY = 74$, find $m\angle XAY$.

14. The length of tangent segment $\overline{PA}$ drawn from exterior point P to circle O is 24. If the radius of the circle is 7, find the distance from point P to the center of the circle.

15. In circle P, $\overline{ST}$ is a diameter and $\overleftrightarrow{LAB}$ is a
tangent. Point T is the midpoint of $\overset{\frown}{ATK}$. If
$m\angle AKS = 74$, find each of the following:
(a) $m\angle ATS$ (b) $m\angle AST$
(c) $m\angle LAS$ (d) $m\angle TAB$
(e) $m\angle APT$ (f) $m\angle SAT$
(g) $m\angle TSK$ (h) $m\angle KAT$
(i) $m\angle TAP$

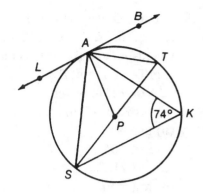

16. Fill in the following table:

	m $\overset{\frown}{BD}$	m $\overset{\frown}{AC}$	m $\angle DEB$	m $\angle ABC$	m $\angle AED$
(a)	118°	62°	?	?	?
(b)	102°	?	?	28°	?
(c)	115°	?	83°	?	?
(d)	?	?	?	41°	64°

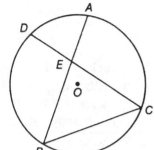

17. Find the values of x, y, and z.

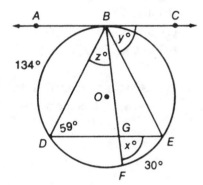

18. A is the midpoint of $\overset{\frown}{DAB}$. Fill in the following table.

	m $\overset{\frown}{ADC}$	m $\overset{\frown}{AB}$	m $\angle CAQ$	m $\angle DBC$	m $\angle AEB$	m $\angle CPQ$
(a)	120	40	?	?	?	?
(b)	132	?	?	?	?	37
(c)	?	64	?	?	82	?
(d)	?	?	71	?	?	49

19. In circle O, tangent $\overrightarrow{PW}$ and secant $\overline{PST}$ are
drawn. Chord $\overline{WA}$ is parallel to chord $\overline{ST}$.
Chords $\overline{AS}$ and $\overline{WT}$ intersect at point B.
If $m\widehat{WA}:m\widehat{AT}:m\widehat{ST} = 1:3:5$, find
each of the following:
 (a) $m\widehat{WA}$, $m\widehat{AT}$, $m\widehat{ST}$, and $m\widehat{SW}$
 (b) $m\angle WTS$
 (c) $m\angle TBS$
 (d) $m\angle TWP$
 (e) $m\angle WPT$

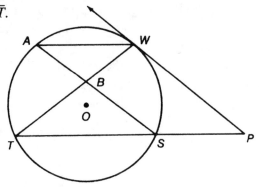

20. In circle P, $\overline{KM}$ is a diameter and $\overline{LFK}$
and $\overline{LHJ}$ are secants. Point F is the
midpoint of $\widehat{KFH}$. If $m\widehat{KJ}:m\widehat{JM} = 5:4$
and $m\angle HEM = 64$, find each of the
following:
 (a) $m\widehat{KF}$, $m\widehat{FM}$, $m\widehat{JK}$, $m\widehat{JM}$, and $m\widehat{HM}$
 (b) $m\angle KPF$
 (c) $m\angle KJH$
 (d) $m\angle KLJ$

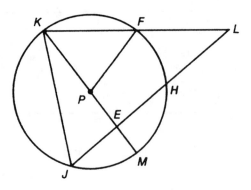

21. GIVEN: In $\odot O$, $\overline{OM} \perp \overline{AB}$.
 PROVE: X is the midpoint of $\widehat{AB}$.

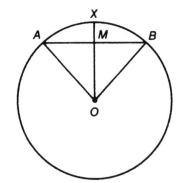

22. GIVEN: In $\odot O$, $\widehat{RST} \cong \widehat{WTS}$.
 PROVE: $\angle RTS \cong \angle WST$.

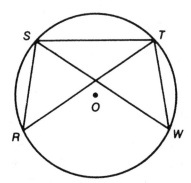

23. GIVEN: In $\odot O$, $OA > AC$.
 PROVE: $\text{m}\widehat{BC} > \text{m}\widehat{AC}$.

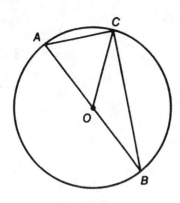

24. GIVEN: In $\odot O$, $\text{m}\widehat{BC} > \text{m}\widehat{AC}$.
 PROVE: $OA > AD$.

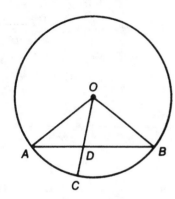

25. GIVEN: $\odot X \cong \odot Y$.
 $\overleftrightarrow{PQ}$ is tangent to $\odot X$ at P
 and tangent to $\odot Y$ at Q.
 PROVE: Point M is the midpoint of $\overline{XY}$.

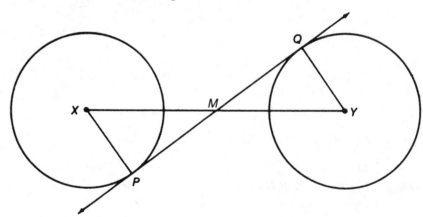

26. GIVEN: $\triangle EFG$ is inscribed in $\odot P$,
 $\overleftrightarrow{AB}$ is tangent to F,
 $\overline{FE} \cong \overline{FG}$.
 PROVE: $\overleftrightarrow{AB} \parallel \overline{EG}$.

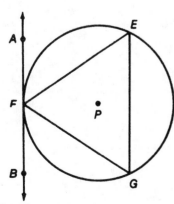

27. GIVEN: In $\odot O$, $\overline{AB}$ is a diameter, point M is the
midpoint of $\overline{BD}$, chords $\overline{BM}$ and $\overline{AC}$
are extended to meet at point D.

PROVE: M is the midpoint of $\overset{\frown}{BMC}$.

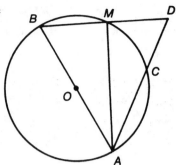

28. GIVEN: $\triangle RST$ is inscribed in $\odot O$ and chords $\overline{SW}$
and $\overline{WK}$ are drawn,
$\overline{SW}$ bisects $\angle RST$, $\overset{\frown}{RW} \cong \overset{\frown}{SK}$.

PROVE: **(a)** $\triangle NWS$ is isosceles.

 (b) $\triangle NTK$ is isosceles.

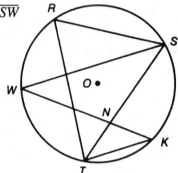

29. GIVEN: Trapezoid $JKLM$ with $\overline{JK} \parallel \overline{LM}$ is inscribed in $\odot O$,
$\mathrm{m}\overset{\frown}{LM} > \mathrm{m}\overset{\frown}{KL}$.

PROVE: $\mathrm{m}\angle MKL > \mathrm{m}\angle JKM$.

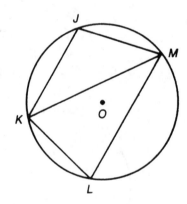

30. GIVEN: $\overline{SR} \cong \overline{TW}$,
$\overline{MA} \cong \overline{MT}$.

PROVE: Quadrilateral $RSTW$ is
a parallelogram.

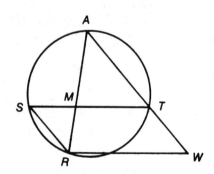

CHORD, TANGENT, AND SECANT SEGMENTS

WHAT YOU WILL LEARN

In Chapter 12 you learned the relationship between the measures of angles and the measures of their intercepted arcs. In this chapter you will learn relationships involving the lengths of chord, tangent, and secant segments. Some of these relationships are not obvious and depend on the properties of similar triangles. In particular, you will learn:

- the conditions under which chords of a circle are congruent and tangent segments drawn to the same circle are congruent;
- the way to draw internally and externally tangent circles;
- the use of angle-measurement theorems in circles to help prove triangles similar;
- the relationship between the four segments formed when two chords intersect inside a circle;
- the relationships between tangent and secant segments drawn to a circle from the same point outside the circle;
- the way to find the circumference of a circle;
- the way to find the length of an arc intercepted by a central angle whose measure is given.

SECTIONS IN THIS CHAPTER
Equidistant Chords
Tangents and Circles
Similar Triangles and Circles
Tangent- and Secant-Segment Relationships
Circumference and Arc Length

Equidistant Chords

In a circle, two chords may have the same length or may have different lengths. In this section we will demonstrate that there is a relationship between how two chords compare in length and the distances of these chords from the center of the circle.

In Figure 13.1, $\overline{OC}$ is drawn perpendicular to chord $\overline{AB}$. The length of $\overline{OC}$ represents the *distance* of chord $\overline{AB}$ from the center (point O) of the circle. If the radius of circle O is 5 and the length of $\overline{AB}$ is 8, what is the length of $\overline{OC}$? If a line passes through the center of a circle and is perpendicular to a chord, then it bisects the chord (see Theorem 12.4). Hence, $AC = 4$. If we now draw radius $\overline{OA}$, a 3-4-5 right triangle is formed, where $OC = 3$. (See Figure 13.2.)

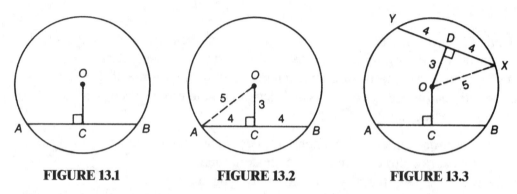

FIGURE 13.1 **FIGURE 13.2** **FIGURE 13.3**

Suppose we now draw another chord in circle O, say $\overline{XY}$, which also has a length of 8. How far is $\overline{XY}$ from the center of the circle? (See Figure 13.3.)

Using a similar analysis, we find that chord $\overline{XY}$ is also 3 units from the center of the circle. What property do chords $\overline{AB}$ and $\overline{XY}$ have in common? What conclusion follows? The answers are stated in Theorem 13.1.

THEOREM 13.1 CONGRUENT CHORDS AND DISTANCE FROM THE CENTER

In the same circle or in congruent circles, congruent chords are equidistant (the same distance) from the center(s) of the circle(s).

OUTLINE OF PROOF

GIVEN: In $\odot O$, $\overline{AB} \cong \overline{XY}$, $\overline{OC} \perp \overline{AB}$, $\overline{OD} \perp \overline{XY}$.

PROVE: $OC = OD$.

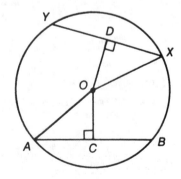

PLAN: Draw $\overline{OA}$ and $\overline{OX}$. Prove $\triangle OCA \cong \triangle ODX$ by Hy-Leg:

STEPS: • $\overline{OA} \cong \overline{OX}$ (Hy)

- $AC = XD$ (leg) since $AC = \frac{1}{2} AB$ and $XD = \frac{1}{2} XY$, and halves of equals (AB and XY) are equal (AC and XD).
- $\overline{OC} \cong \overline{OD}$ by CPCTC from which it follows that $OC = OD$.

EXAMPLE 13.1

GIVEN: In $\odot P$, $\overline{PB} \perp \overline{DE}$, $\overline{PC} \perp \overline{FG}$, $\overline{DE} \cong \overline{FG}$.

PROVE: $\overline{PA}$ bisects $\angle FAD$.

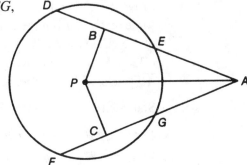

SOLUTION

PLAN: Show right $\triangle PBA \cong$ right $\triangle PCA$.

PROOF:

Statements	Reasons
1. $\overline{PB} \perp \overline{DE}$ and $\overline{PC} \perp \overline{FG}$.	1. Given.
2. $\triangle PBA$ and $\triangle PCA$ are right triangles.	2. Perpendicular lines intersect to form right angles; a triangle that contains a right angle is a right triangle.
3. $\overline{PA} \cong \overline{PA}$. (Hy)	3. Reflexive property of congruence.
4. $\widehat{DE} \cong \widehat{FG}$.	4. Given.
5. $\overline{DE} \cong \overline{FG}$.	5. In a circle, congruent arcs have congruent chords.
6. $PB = PC$.	6. In a circle, congruent chords are equidistant from the center of the circle.
7. $\overline{PB} \cong \overline{PC}$. (Leg)	7. Segments equal in length are congruent.
8. $\triangle PBA \cong \triangle PCA$.	8. Hy-Leg.
9. $\angle PAB \cong \angle PAC$.	9. CPCTC.
10. $\overline{PA}$ bisects $\angle FAD$.	10. If a line divides an angle into two congruent angles, it bisects the angle.

The converse of Theorem 13.1 states that if we know that two chords are the same distance from the center of a circle, then they must be congruent.

THEOREM 13.2 EQUIDISTANT CHORDS

In the same circle or congruent circles, chords equidistant from the center(s) of the circle(s) are congruent.

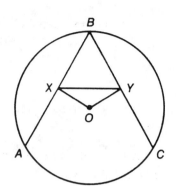

EXAMPLE 13.2

GIVEN: ⊙O, with $\overline{OX} \perp \overline{AB}$, $\overline{OY} \perp \overline{CB}$,
$\angle OXY \cong \angle OYX$.

PROVE: $\widehat{AB} \cong \widehat{CB}$.

SOLUTION

PLAN: Prove that $\overline{AB}$ and $\overline{CB}$ are the same distance from the center of the circle by showing that $OX = OY$. By Theorem 13.2, $\overline{AB} \cong \overline{CB}$ which implies $\widehat{AB} \cong \widehat{CB}$.

PROOF:

Statements	Reasons
1. $\overline{OX} \perp \overline{AB}$ and $\overline{OY} \perp \overline{CB}$.	1. Given.
2. $\angle OXY \cong \angle OYX$.	2. Given.
3. $OX = OY$	3. If two angles of a triangle are congruent, then the sides opposite are equal in length.
4. $\overline{AB} \cong \overline{CB}$.	4. In the same circle, chords equidistant from the center of the circle are congruent.
5. $\widehat{AB} \cong \widehat{CB}$.	5. In the same circle, congruent chords intercept congruent arcs.

Tangents and Circles

TANGENT SEGMENTS

A tangent segment is a segment whose end points are the point of tangency and a fixed point on the tangent. An infinite number of tangent segments can be drawn to a circle. Under what conditions will the lengths of these tangent segments be equal? If two tangent segments are drawn from the *same* exterior point to a circle, they will have the same length.

THEOREM 13.3 CONGRUENT TANGENT SEGMENTS

If two tangent segments are drawn to a circle from the same exterior point, then they are congruent.

OUTLINE OF PROOF

GIVEN: $\overline{PA}$ and $\overline{PB}$ are tangent to ⊙O at points A and B, respectively.

PROVE: $\overline{PA} \cong \overline{PB}$.

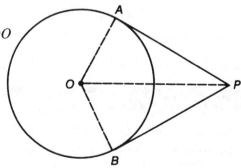

STEPS • Draw $\overline{OP}$ and radii $\overline{OA}$ and $\overline{OB}$.
Angles OAP and OBP are right angles.
• Prove $\triangle OAP \cong \triangle OBP$ by Hy-Leg:
$$\overline{OP} \cong \overline{OP} \text{ (Hy)}$$
$$\overline{OA} \cong \overline{OB} \text{ (Leg)}$$
• By CPCTC, $\overline{PA} \cong \overline{PB}$.

In the proof of Theorem 13.3, notice that since $\triangle OAP \cong \triangle OBP$, $\angle APO \cong \angle BPO$. In other words, $\overline{OP}$ bisects the angle formed by the two tangent segments.

Find the value of x.

a.

b.

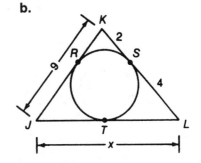

SOLUTION

a. $AC = AB = 3$
$DC = DE = 2$
$x = AC + DC = 5$

b. $KR = KS = 2$
$JR = 9 - KR = 9 - 2 = 7$
$JT = JR = 7$
$LT = LS = 4$
$x = JT + LT = 7 + 4 = 11$

Find the values of x and y.

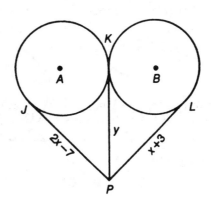

SOLUTION

$PK = PJ$ and $PK = PL$.
Hence, $PJ = PL$.
$$2x - 7 = x + 3$$
$$2x = x + 10$$
$$x = 10$$
$$PL = x + 3 = 13$$
$$y = PK = PL = 13$$

COMMON TANGENTS AND TANGENT CIRCLES

The same line may be tangent to more than one circle (a common tangent), and circles may intersect each other in exactly one point (tangent circles). In order to describe these situations, some terminology must be introduced. In Figure 13.4, line segment $\overline{AB}$ joins the centers of circles A and B and is called the *line of centers* of the two circles. Figure 13.5 illustrates that a line may be tangent to two different circles. A line

which has this property is called a *common tangent*. Lines *j*, *k*, *ℓ*, and *m* are common tangents. In Figure 13.6 two different circles are shown to be tangent to the same line at the same point. These circles are known as *tangent circles*. Circles *A* and *B* are tangent to line *ℓ* at point *P*. Circles *A* and *C* are tangent to line *m* at point *Q*.

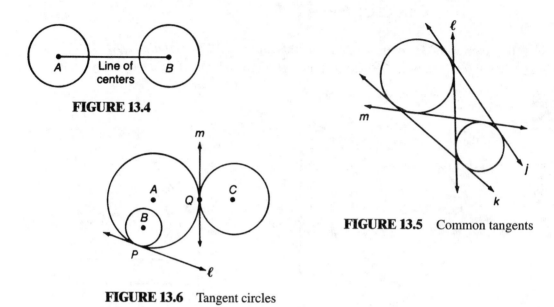

FIGURE 13.4

FIGURE 13.5 Common tangents

FIGURE 13.6 Tangent circles

We may further distinguish between types of common tangents and types of tangent circles. A common tangent may be either a common *internal* or a common *external* tangent, as shown in Figure 13.7.

(a) Common internal tangents

(b) Common external tangents

FIGURE 13.7

In Figure 13.7*a*, lines *ℓ* and *m* are common *internal* tangents since each is tangent to both circles and each intersects the line of centers of the two circles. In Figure 13.7*b*, lines *j* and *k* are common *external* tangents since each is tangent to both circles and each does *not* intersect the line of centers of the two circles.

Tangent circles may be tangent either *internally* or *externally* to each other, as shown in Figure 13.8. In Figure 13.8*a*, circles *A* and *B* are tangent *internally* since they lie on the *same* side of their common tangent. In Figure 13.8*b*, circles *P* and *Q* are tangent *externally* since they lie on *opposite* sides of their common tangent.

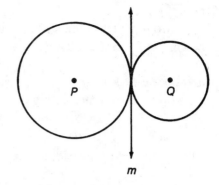

(a) Internally tangent circles **(b)** Externally tangent circles

FIGURE 13.8

Determine the number of common tangents that can be drawn for each of the following situations:

a. Circle A and circle B intersect in two distinct points.

b. Circle A and circle B are externally tangent circles.

SOLUTION

a.

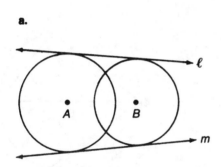

Two common external tangents

b.

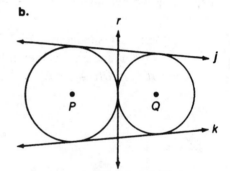

Two common external tangents
and one common internal tangent

Prove that common internal tangent segments drawn to two nonintersecting circles are congruent.

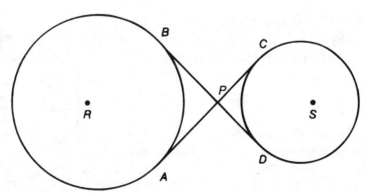

SOLUTION

Draw two nonintersecting circles with their common internal tangent segments shown. Let *P* represent the point at which the tangent segments intersect.

OUTLINE OF PROOF

GIVEN: Nonintersecting circles *R* and *S*, common internal tangent segments $\overline{AC}$ and $\overline{BD}$, intersecting at point *P*.

PROVE: $\overline{AC} \cong \overline{BD}$.

Apply Theorem 13.3:

$$
\begin{aligned}
PA &= PB \\
+ PC &= PD \\
\hline
PA + PC &= PB + PD
\end{aligned}
$$

Since $AC = PA + PC$ and $BD = PB + PD$, it follows that $\overline{AC} \cong \overline{BD}$.

Similar Triangles and Circles

If two triangles overlap a circle, then some of their angles may have the same arc or congruent arcs. This observation may lead to a pair of congruent angles that can be used to help prove that these triangles are similar.

EXAMPLE 13.7

GIVEN: $\overline{EB}$ is tangent to $\odot O$ at *B*, $\overline{AB}$ is a diameter, *B* is the midpoint of $\overset{\frown}{CBD}$.

PROVE: $\triangle ABC \sim \triangle AEB$.

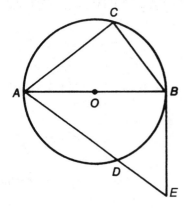

In planning the proof, note in the accompanying figure that:

• Angles *ACB* and *ABE* are congruent since they are both right angles.

• Angles *CAB* and *EAB* are congruent since they are inscribed angles that intercept congruent arcs ($\overset{\frown}{BC} \cong \overset{\frown}{BD}$).

• $\triangle ABC \sim \triangle AEB$ by the AA Theorem of Similarity.

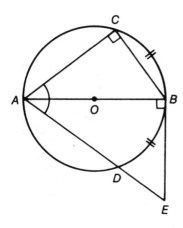

PROOF:

Statements	Reasons
1. B is the midpoint $\overset{\frown}{CBD}$.	1. Given
2. $\overset{\frown}{BC} \cong \overset{\frown}{BD}$.	2. A midpoint of an arc divides the arc into two congruent arcs.
3. $\measuredangle CAB \cong \measuredangle EAB$. (Angle)	3. Inscribed angles of a circle that intercept congruent arcs are congruent.
4. $\measuredangle ACB$ is a right angle.	4. An angle inscribed in a semicircle is a right angle.
5. $\measuredangle ABE$ is a right angle.	5. An angle formed by a radius drawn to the point of tangency is a right angle.
6. $\measuredangle ACB \cong \measuredangle ABE$. (Angle)	6. All right angles are congruent.
7. $\triangle ABC \sim \triangle AEB$.	7. AA Theorem of Similarity.

The properties of similar triangles may be used to establish a special relationship between the segments formed by two chords that intersect in the interior of a circle.

THEOREM 13.4 PRODUCTS OF LENGTHS OF SEGMENTS OF INTERSECTING CHORDS

If two chords intersect in the interior of a circle, then the product of the lengths of the segments of one chord is equal to the product of the lengths of the segments of the other chord. In the accompanying figure,

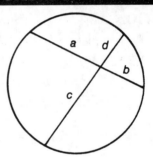

$$a \times b = c \times d.$$

OUTLINE OF PROOF

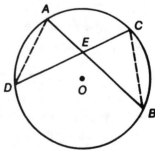

GIVEN: Chords $\overline{AB}$ and $\overline{CD}$ intersect in the interior of $\odot O$ at point E.

PROVE: $AE \cdot EB = CE \cdot ED$.

PLAN: Draw $\overline{AD}$ and $\overline{CB}$. Show $\triangle AED \sim \triangle CEB$.

STEPS: $\triangle AED$ is similar to $\triangle CEB$ since:
- $\measuredangle AED \cong \measuredangle BEC$ (Vertical angles)
- $\measuredangle ADC \cong \measuredangle CBA$ (Inscribed angles that intercept the same arc)
- $\dfrac{AE}{CE} = \dfrac{ED}{EB}$ so $AE \cdot EB = CE \cdot ED$

 **EXAMPLE
13.8** Find the value of x.

a.

b.

c.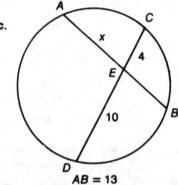

AB = 13

SOLUTION

 a. $AE \cdot EB = CE \cdot ED$ **b.** $(x)(x) = (8)(2)$

 $(x)(12) = (4)(9)$ $x^2 = 6$

 $12x = 36$ $x = \sqrt{16} = 4$

 $x = \dfrac{36}{12} = 3$

 c. If $x = AE$, then $13 - x = EB$.

 $(x)(13 - x) = (10)(4)$

 $13x - x^2 = 40$

Writing the quadratic equation in standard form gives

$$x^2 - 13x + 40 = 0$$
$$(x - 8)(x - 5) = 0$$
$$x - 8 = 0 \quad or \quad x - 5 = 0$$
$$x = 8 \quad or \quad x = 5$$

If $AE = 8$ then $EB = 5$. Alternatively, AE may equal 5, in which case $EB = 8$.

 **EXAMPLE
13.9** A diameter divides a chord of a circle into two segments whose lengths are 7 and 9. If the length of the shorter segment of the diameter is 3, find the length of a radius of the circle.

SOLUTION

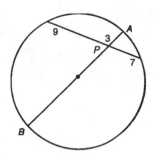

Let $x = PB$.

$(3)(x) = (7)(9)$

$3x = 63$

$x = \dfrac{63}{3} = 21$

Since diameter $AB = 3 + 21 = 24$, the length of a radius of the circle is $\dfrac{1}{2}(24)$ or 12.

Tangent- and Secant-Segment Relationships

In Figure 13.9, $\overline{PAB}$ is called a secant *segment*. Its end points are a point in the exterior of the circle (point P) and the point on the circle furthest from point P at which the secant intersects the circle (point B). The circle divides the secant segment into two segments: an *internal* secant segment (chord $\overline{AB}$) and an *external* secant segment ($\overline{AP}$).

When two secant segments are drawn to a circle from the same exterior point, then a special relationship exists between the lengths of the secant segments and the lengths of their external segments.

FIGURE 13.9

THEOREM 13.5 SECANT-SECANT SEGMENT PRODUCTS

If two secant segments are drawn to a circle from the same exterior point, then the product of the lengths of one secant segment and its *external* segment is equal to the product of the lengths of the other secant segment and its *external* segment.

In the accompanying figure,

$$s_1 \times e_1 = s_2 \times e_2.$$

OUTLINE OF PROOF

GIVEN: $\overline{PB}$ and $\overline{PD}$ are secant segments drawn to $\odot O$.

PROVE: $PB \cdot PA = PD \cdot PC$.

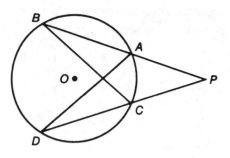

STEPS: • Draw $\overline{AD}$ and $\overline{CB}$.

• $PB \cdot PA = PD \cdot PC$, so $\dfrac{PB}{PD} = \dfrac{PC}{PA}$. Hence, show $\triangle PBC \sim \triangle PDA$.

• $\triangle PBC \sim \triangle PDA$ since $\angle P \cong \angle P$ and $\angle PBC \cong \angle PDA$

Find the value of x.

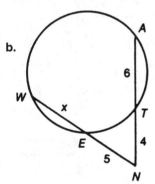

SOLUTION

a. $PA \cdot PB = PC \cdot PD$

$3 \cdot 8 = 2 \cdot x$

$24 = 2x$

$12 = x$

b. $NE \cdot NW = NT \cdot NA$, where

$NE = 5$

$NW = x + 5$

$NT = 4$

$NA - 6 + 4 = 40$

$5(x + 5) = 4 \cdot 10$

$5x + 25 = 40$

$5x = 15$

$x = 3$

When a *tangent* segment and a secant segment are drawn to a circle from the same exterior point, a relationship analogous to the one stated in Theorem 13.5 results. Figure 13.10*a* illustrates that, when secant $\overline{PAB}$ is rotated clockwise, it will eventually be tangent to the circle as in Figure 13.10*b*. This analysis suggests that we replace the

secant-segment length with tangent-segment length $\overline{PA}$ and also replace the external portion of the secant segment with tangent-segment length $\overline{PA}$.

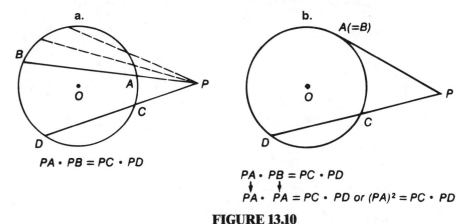

a.

$PA \cdot PB = PC \cdot PD$

b.

$$PA \cdot PB = PC \cdot PD$$
$$PA \cdot PA = PC \cdot PD \text{ or } (PA)^2 = PC \cdot PD$$

FIGURE 13.10

THEOREM 13.6 TANGENT-SECANT SEGMENT PRODUCTS

If a tangent segment and a secant segment are drawn to a circle from the same exterior point, then the square of the length of the tangent segment is equal to the product of the lengths of the secant segment and its external segment.

In the accompanying figure,

$$t^2 = s \times e.$$

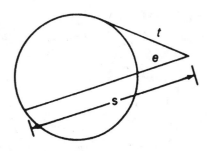

OUTLINE OF PROOF

GIVEN: $\overline{PA}$ is tangent to $\odot O$ at A and $\overline{PC}$ is a secant.

PROVE: $(PA)^2 = PC \cdot PB$.

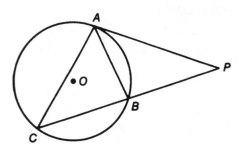

STEPS:
• Draw $\overline{CA}$ and $\overline{BA}$.

• $(PA)^2 = PC \cdot PB$, so $\dfrac{PA}{PC} = \dfrac{PB}{PA}$. Hence, show $\triangle PAB \sim \triangle PCA$.

• $\triangle PAB \sim \triangle PCA$ since:
$$\angle P \cong \angle P$$
$$\angle PCA \cong \angle PAB$$

(NOTE: Both angles are measured by $\dfrac{1}{2} m\overset{\frown}{AB}$.)

EXAMPLE 13.11

Find the value of x.

a.

b.

c.

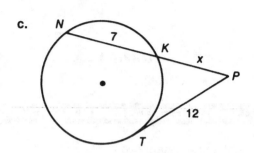

SOLUTION

a.
$$(PA)^2 = PB \cdot PC$$
$$(8)^2 = x \cdot 16$$
$$64 = 16x$$
$$4 = x$$

b.
$$(RG)^2 = RA \cdot RL$$
$$x^2 = 5(15 + 5)$$
$$= 5 \cdot 20$$
$$= 100$$
$$x = \sqrt{100} = 10$$

c.
$$(PT)^2 = PK \cdot PN$$
$$(12)^2 = x(x + 7)$$
$$144 = x^2 + 7x$$

Writing the quadratic equation in standard form gives

$$x^2 + 7x - 144 = 0 \qquad or \qquad x - 9 = 0$$
$$(x + 16)(x - 9) = 0$$
$$x + 16 = 0$$
$$x = -16 \qquad\qquad\qquad x = 9$$

Reject since a length cannot
be a negative number.

Circumference and Arc Length

The distance around a polygon is called the *perimeter* of the polygon. The perimeter of a circle is given a special name, *circumference*.

DEFINITION OF CIRCUMFERENCE

The *circumference* of a circle is the distance around the circle, expressed in linear units of measurement (e.g., inches, centimeters, feet).

If a wheel is rolled along a flat surface, as in Figure 13.11, so that the surface is again tangent to the same point on the circle, then the distance the wheel travels along the surface has the same numerical value as the circumference of the circle.

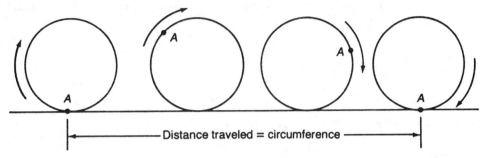

FIGURE 13.11

The longer the diameter of a circle, the larger the circle's circumference. Interestingly, however, if the circumference of a circle is divided by the length of its diameter, then the value obtained will be the same regardless of the size of the circle:

$$\frac{\text{Circumference}}{\text{Diameter}} = 3.1415926\ldots$$

The three dots that follow the decimal number indicate that this value is a nonterminating (never-ending) decimal number. This constant value is referred to as *pi* and is denoted by the Greek letter π. We may therefore write

$$\frac{\text{Circumference}}{\text{Diameter}} = \pi \quad \text{or} \quad \text{Circumference} = \pi \cdot \text{Diameter}$$

where π is *approximately equal to* 3.14 or *approximately equal to* the improper fraction $\frac{22}{7}$.

THEOREM 13.7 CIRCUMFERENCE OF A CIRCLE

The circumference of a circle is equal to the product of π and the length of its diameter: $C = \pi D$.

When working with the constant π, note that:

- $\pi \approx 3.14$ and $\pi \approx \frac{22}{7}$, where the symbol $\approx$ is read as "is approximately equal to."

- Since the length of a diameter is numerically equal to twice the length of the radius, we may write

$$\text{Circumference} = \pi \cdot 2 \cdot \text{radius}$$

In writing a formula that includes a number, it is common practice to place the digit (2 in this case) before any symbols. We therefore write

$$\text{Circumference} = \pi \cdot 2 \cdot \text{radius} \quad \text{or} \quad C = 2\pi r$$

- The approximation to be used for π (3.14 or $\frac{22}{7}$) is usually specified in the statement of the problem. Frequently, you will be asked *not* to substitute an approximation for π, and instead to express the answer in terms of π, as illustrated in Example 13.12, part **c**.

EXAMPLE 13.12 Find the circumference of a circle if:

 a. Diameter = 10 (Use π = 3.14.)

 b. Radius = 14 (Use $\pi = \frac{22}{7}$.)

 c. Diameter = 29 (Express answer in terms of π.)

SOLUTION

 a. $C = \pi D$

 $= 3.14 \times 10 = 31.4$

 b. $C = 2\pi r$

$$= 2 \times \frac{22}{7} \times \frac{\overset{2}{\cancel{14}}}{1} = 88$$

 c. $C = \pi D$

 $= \pi 29$ or 29π

EXAMPLE 13.13 Rectangle *ABCD* has a width of 5 and a length of 12 and is inscribed in circle *O*. Find the circumference of the circle.

SOLUTION

Since $\angle BAD$ is a right angle, diagonal $\overline{BD}$ must coincide with a diameter of the circle. Right $\triangle BAD$ is a 5-12-*13* right triangle where diagonal $BD = 13$. Hence,

$$C = \pi D = 13\pi$$

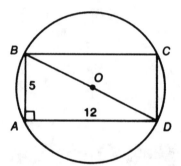

Circumference represents the distance around the *entire* circle. Our next concern is how to determine the length (in linear units) of an *arc* of the circle. Since a circle contains 360°, the circumference of a circle represents the length of a 360° arc of the circle (Figure 13.12). The ratio of the length of an arc to the circumference must be equal to the ratio of the degree measure of the arc to 360°:

$$\frac{\text{Arc measurement}}{\text{Circle measurement}} = \frac{\text{arc length}}{2\pi r} = \frac{n^\circ}{360^0}$$

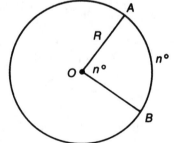

FIGURE 13.12

If you consider the circle to be made of string, then the length of arc *AB* corresponds to the number arrived at by taking the section of the string from *A* to *B*, stretching it out, and then using a ruler to measure its length.

THEOREM 13.8 ARC LENGTH PROPORTION

$$\frac{\text{Length of arc}}{\text{Circumference}} = \frac{\text{degree measure of arc}}{360°}$$

In a circle having a radius of 10, find the length of an arc whose degree measure is 72°. (Leave answer in terms of π.)

SOLUTION

$$\frac{\text{Length of arc}}{\text{Circumference}} = \frac{\text{degree measure of arc}}{360°}$$

$$\frac{L}{2\pi \cdot 10} = \frac{72°}{360°}$$

Simplify each ratio *before* cross-multiplying:

$$\frac{L}{20\pi} = \frac{1}{5}$$

Cross-multiply:

$$5L = 20\pi$$

$$L = 4\pi$$

In a circle a 40° arc has a length of 8π. Determine the radius of the circle.

SOLUTION

$$\frac{\text{Length of arc}}{\text{Circumference}} = \frac{\text{degree measure of arc}}{360°}$$

$$\frac{8\pi}{2\pi r} = \frac{40°}{360°}$$

Simplify each ratio *before* cross-multiplying:

$$\frac{4}{r} = \frac{1}{9}$$

Cross-multiply:

$$r = 36$$

Right triangle *ABC* is inscribed in circle *O* so that $\overline{AB}$ is a diameter and has a length of 27. If $m\angle CAB = 50$, find the length of $\overset{\frown}{AC}$, expressed in terms of π.

SOLUTION

In order to find the degree measure of $\overset{\frown}{AC}$, draw $\overline{OC}$ and determine the measure of central angle AOC. Since $\overline{OA} \cong \overline{OC}$, m$\angle OCA$ = m$\angle CAB$ = 50. The fact that the sum of the angles of a triangle is 180 means that m$\angle AOC$ = 80. Since a central angle and its intercepted arc have the same measure, m$\overset{\frown}{AC}$ = 80.

$$\frac{\text{Length of arc}}{\text{Circumference}} = \frac{\text{degree measure of arc}}{360°}$$

$$\frac{L}{27\pi} = \frac{80°}{360°}$$

$$= \frac{2}{9}$$

$$9L = 54\pi$$

$$L = 6\pi$$

REVIEW EXERCISES FOR CHAPTER 13

1. For each of the given situations, determine the number of common internal tangents and the number of common external tangents that can be drawn.
 (a) Circle P lies in the exterior of circle Q and has no points in common with circle Q.
 (b) Circle P lies in the interior of circle Q and has no points in common with circle Q.
 (c) Circle P and circle Q are tangent internally.

2. Find the value of x.

(a)

(b)

Use the accompanying figure for Exercises 3 to 6.

3. If $RK = 5$, $KH = 9$, $PK = 15$, find KN.

4. If $PK = 27$, $KN = 3$, and K is the midpoint of RH, find RK.

5. If $RH = 16$, $RK = 4$, $PK = 8$, find PN.

6. If $RH = 22$, $PK = 7$, and $KN = 3$, find RK.

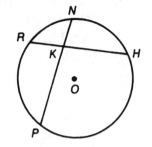

Use the accompanying figure for Exercises 7 to 11.

7. If $PW = 5$, $PG = 8$, $PH = 2$, find KH.

8. If $GW = 7$, $PW = 3$, $PK = 15$, find PH.

9. If W is the midpoint of $\overline{GP}$ and $PH = 5$, $KH = 35$, find PG.

10. If $PW = 6$, $WG = 9$, $PH = 9$, find KH.

11. If $GW = 11$, $PH = 8$, $KH = 2$, find PW.

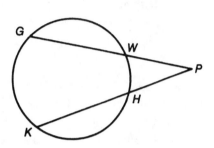

Use the accompanying figure for Exercises 12 to 14.

12. If $RV = 9$, $RM = 3$, find RT.

13. If $MT = 24$, $RM = 1$, find RV.

14. If $RV = 8$, $RM = 4$, find MT.

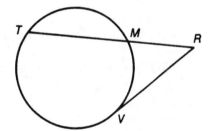

15. $RE = 2$, $RA = 14.5$, $ZG = 6$, $ZF = 8$, $SF = x$, E is the midpoint of $\overline{BG}$. Find SZ.

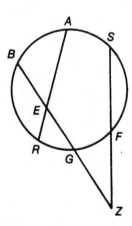

16. (a) If $NA = 4$, $JA = 8$, $WA = 11$, find OK.
 (b) If A is the midpoint of $\overline{JW}$, $AK = 32$,
 $NA = 18$, find JW and OA.
 (c) $JA = 12$, $AW = 9$, and AK is 3
 times the length of $\overline{NA}$. Find OK.

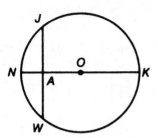

17. A circle divides a secant segment into an internal segment having a length of 8 and an external segment having a length of 2. Find the length of a tangent segment drawn to the circle from the same exterior point.

18. From a point 2 units from a circle, a tangent segment is drawn. If the radius of the circle is 8, find the length of the tangent segment.

Unless otherwise specified, answers to Exercises 19 to 25 may be left in terms of π.

19. Find the circumference of a circle if:
 (a) Diameter = 21 (Use $\pi = 22/7$.)
 (b) Radius = 50 (Use $\pi = 3.14$.)
 (c) Radius = 17 (Express answer in terms of π.)

20. Find the radius of a circle if its circumference is:
 (a) 32π (b) 15π (c) 23

21. Find the circumference of a circle that is inscribed in a square whose side is 9.

22. Find the circumference of a circle that is circumscribed about a square whose perimeter is 36.

23. Fill in the following table:

	Radius of Circle	Degree Measure of Arc	Length of Arc
(a)	9	120°	?
(b)	12	?	3π
(c)	?	72°	4π

24. A square having a side of 12 inches is inscribed in a circle. Find the length of an arc of the circle intercepted by one of the square's sides.

25. In the pulley system of a certain machine, a belt of negligible thickness is wrapped around two identical wheels each having a radius of 9, as shown in the accompanying figure. The belt criss-crosses at a point P between the two circles so that the measure of angle NPJ is 60. $\overline{NPS}$ and $\overline{JPK}$ may be considered to be tangent segments.
 (a) Find the degree measure of *minor* arc NJ.
 (b) Find the length of *major* arc NJ.
 (c) Find the length of the belt.

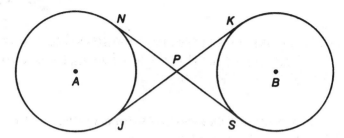

Use the accompanying figure for Exercises 26 and 27.

26. GIVEN: In $\odot O$, $\overline{MR}$ is a diameter, $\overline{OA} \parallel \overline{MS}$, $\overline{OB} \parallel \overline{MT}$, $\overline{AR} \cong \overline{BR}$.
 PROVE: $\overline{SR} \cong \overline{TR}$.

27. GIVEN: In $\odot O$, $\overline{SR} \cong \overline{TR}$, $\overline{OA} \parallel \overline{MS}$, $\overline{OB} \parallel \overline{MT}$.
 PROVE: $\angle AOM \cong \angle BOM$.

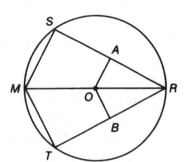

28. GIVEN: In $\odot O$, quadrilateral $OXEY$ is a square.
 PROVE: $\overarc{QP} \cong \overarc{JT}$.

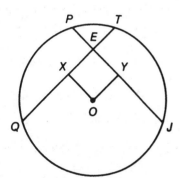

29. GIVEN: $\triangle HBW$ is inscribed in $\odot O$, tangent segment $\overline{AB}$ is tangent at point B, $ABLM$ is a parallelogram.
 PROVE: $\dfrac{BL}{BW} = \dfrac{BM}{BH}$.

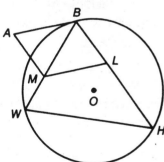

30. GIVEN: In $\odot P$, $\overline{AB}$ is a diameter, $\overline{DB}$ is tangent to
$\odot P$ at B, $\overline{CD} \perp \overline{DB}$.

PROVE: BC is the mean proportional between
CD and AB.

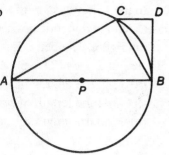

Use the accompanying figure for Exercises 31 and 32.

31. GIVEN: In $\odot O$, $\overline{KJ}$ is a diameter, $\overline{MJ}$ is tangent at
point J, N is the midpoint of $\overparen{LNJ}$.

PROVE: $KL : KJ = KP : KM$.

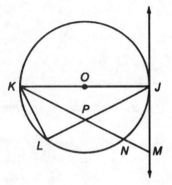

32. GIVEN: In $\odot O$, $\overline{KJ}$ is a diameter, $\overline{MJ}$ is tangent
at point J, $\overline{JP} \cong \overline{JM}$.

PROVE: $KL \cdot KM = JK \cdot LP$.

33. GIVEN: $\overleftrightarrow{TK}$ bisects $\angle NTW$, $\overline{WK} \cong \overline{WT}$.

PROVE:

(a) $\overleftrightarrow{NTP} \parallel \overleftrightarrow{KW}$.
(b) $(TW)^2 = JT \cdot TK$.

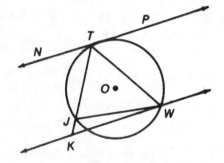

34. Prove that, if two circles are tangent externally, then the common internal tangent
bisects a common external tangent.

35. Prove that, if two circles do not intersect, with one circle lying in the exterior of
the other, then their common external tangent segments are congruent. (HINT:
Consider as separate cases the situations in which the two circles are congruent
and not congruent.)

36. Lines ℓ and m are parallel and each is tangent to circle P. Prove that, if line k is
also tangent to circle P, then the segments determined by the points at which line k
intersects lines ℓ and m and point P intersect at right angles at P.

AREA AND VOLUME

WHAT YOU WILL LEARN

There are convenient formulas for finding the areas and volumes of certain types of plane and solid figures. In this chapter you will learn:

- how to find the areas of triangles, trapezoids, and different types of parallelograms;
- how to compare the areas of similar figures;
- how to find the area of a regular polygon;
- how to find the areas of a circle and a sector;
- how to find the area of a region bounded by a chord and its arc;
- how to find the volumes of familiar types of solid figures.

SECTIONS IN THIS CHAPTER
Areas of a Rectangle, Square, and Parallelogram
Areas of a Triangle and Trapezoid
Comparing Areas
Area of a Regular Polygon
Areas of a Circle, Sector, and Segment
Geometric Solids

Areas of a Rectangle, Square, and Parallelogram

The term *area of a figure* refers to the number of square boxes that the figure can enclose. If a figure can enclose a total of 30 square boxes, then its area is said to be 30 square *units*. If the length of a side of each square box is 1 centimeter (cm), then the area of the figure is 30 square centimeters (cm²). If the length of a side of each square box is 1 inch, then the area of the figure under consideration is expressed as 30 square inches.

Figure 14.1 shows a rectangle that encloses a total of 36 square boxes. The area of this rectangle is 36 square units.

FIGURE 14.1

The value of 36 can be arrived at by multiplying the number of boxes along the length by the number of boxes along the width: Area = 9 × 4 = 36. One of our first goals in this chapter is to develop formulas that will enable us to calculate the areas of figures that we have previously investigated: rectangle, square, parallelogram, triangle, rhombus, trapezoid, regular polygon, and circle. We begin by summarizing some fundamental area postulates.

AREA POSTULATES

POSTULATE 14.1 For any given closed region and unit of measurement, there is a positive number that represents the area of the region.

POSTULATE 14.2 If two figures are congruent, then they have equal areas.

POSTULATE 14.3 The area A of a rectangle is equal to the product of its length (ℓ) and width (w).

Sometimes one side of a rectangle is referred to as the *base* and an adjacent side (which is perpendicular to the base) is called the *altitude* or *height*. The formula for the area of a rectangle may be expressed as $A = bh$, where b represents the length of the base and h represents the height. The terms *base* and *height* will recur in our investigations of the areas of other figures. In each instance, the height or altitude will always be a segment that is perpendicular to the side that is specified to be the base. Also, when the terms *base* and *altitude* are used in connection with area, we will always understand these terms to mean the *length* of the base and the *length* of the altitude.

Find the area of a rectangle if its base is 12 centimeters and its diagonal has a length of 13 centimeters.

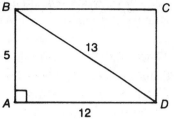

SOLUTION

Triangle BAD is a 5-12-13 right triangle where 12 is the base and 5 is the height. Hence,

$$A = bh$$
$$= 12 \times 5$$
$$= 60 \text{ cm}^2$$

A *square* is a rectangle so that the area formula for to a rectangle also applies to a square. See Figure 14.2. If the length of a side of a square is represented by s, then the area of a square is given by the relationship $A = s \times s$ or $A = s^2$.

FIGURE 14.2

A certain rectangle and square are *equivalent* (have the same area). The base of the rectangle exceeds three times its altitude by 4. If the length of a side of the square is 8, find the dimensions of the rectangle.

SOLUTION

Let x = altitude of rectangle
Then $3x + 4$ = base of rectangle.

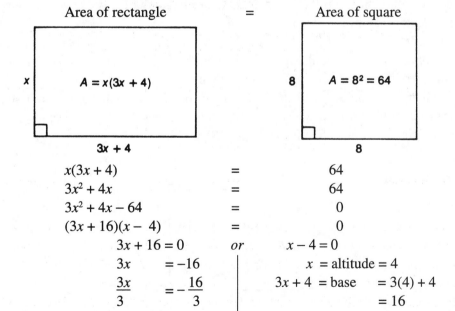

$$x(3x + 4) = 64$$
$$3x^2 + 4x = 64$$
$$3x^2 + 4x - 64 = 0$$
$$(3x + 16)(x - 4) = 0$$

$3x + 16 = 0$ *or* $x - 4 = 0$

$3x = -16$ x = altitude = 4

$\dfrac{3x}{3} = -\dfrac{16}{3}$ $3x + 4$ = base = $3(4) + 4$

 = 16

Reject this solution since x (the length of a side) cannot be negative.

It will sometimes be convenient to subdivide a region into component regions. For example, in Figure 14.3, diagonals $\overline{AC}$ and $\overline{AD}$ separate pentagon $ABCDE$ into three triangular regions such that: Area pentagon $ABCDE$ = area $\triangle ABC$ + area $\triangle ACD$ + area $\triangle ADE$.

The generalization of this notion is presented as Postulate 14.4.

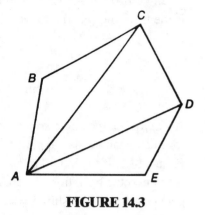

FIGURE 14.3

POSTULATE 14.4 AREA ADDITION

The area of a closed region is equal to the sum of the areas of any non-overlapping divisions of that region.

A rectangle is a special type of parallelogram. It therefore seems reasonable that there may be some relationship between the area formulas for these two figures. We will seek to establish that a parallelogram has the same area as a rectangle having the same base and altitude as the parallelogram. We begin our analysis by relating the concept of base and altitude to a parallelogram.

Any side of a parallelogram may be identified as the *base* of the parallelogram. The *height* or *altitude* of a parallelogram is the length of a segment drawn perpendicular to the base from any point on the side opposite the base.

In Figure 14.4, $\overline{SJ}$, $\overline{AK}$, and $\overline{BL}$ are examples of altitudes, each drawn to base $\overline{RW}$. Since parallel lines are everywhere equidistant, altitudes drawn to a given base of a parallelogram are equal, that is, $SJ = AK = BL = \ldots$.

FIGURE 14.4

Next consider $\square ABCD$ with altitude $\overline{BH}$ drawn to base $\overline{AD}$ (Figure 14.5). If right triangle AHB were cut off the figure and then slid over to the right so that $\overline{AB}$ and $\overline{CD}$ were made to coincide, then the resulting figure would be a rectangle. This is illustrated in Figure 14.6, where right angle DKC corresponds to right angle AHB.

FIGURE 14.5 **FIGURE 14.6**

The area of rectangle $BHKC$ (see Figure 14.6) is given by the relationship $A = HK \times BH$, where $\overline{BH}$ is the altitude and $\overline{HK}$ is the base of the rectangle. Since we have not thrown away or created any additional area, the area of the original parallelogram equals the area of the newly formed rectangle. Hence, the area of $\square ABCD$ also equals $HK \times BH$. Since $\triangle AHB \cong \triangle DKC$, $\overline{AH} \cong \overline{DK}$. This implies that $AD = HK$. Substituting AD for HK in the area relationship we obtain the following:

$$\text{Area of } \square ABCD = HK \times BH$$
$$= AD \times HK$$

Note that $\overline{AD}$ is the base of the parallelogram and $\overline{BH}$ is an altitude drawn to that base. This relationship is stated formally in Theorem 14.1.

THEOREM 14.1 AREA OF A PARALLELOGRAM

The area of a parallelogram is equal to the product of the lengths of the base and the altitude drawn to that base.

Thus, $A = bh$.

EXAMPLE 14.3

A pair of adjacent sides of a parallelogram are 6 and 10 centimeters in length. If the measure of their included angle is 30, find the area of the parallelogram.

SOLUTION

In $\square ABCD$, altitude $BH = 3$ since the length of the side opposite a $30°$ angle in a 30-60 right triangle is one-half the length of the hypotenuse (side $\overline{AB}$).

$$\begin{aligned}
\text{Area of } \square ABCD &= bh \\
&= AD \times BH \\
&= 10 \times 3 \\
&= 30 \text{ cm}^2
\end{aligned}$$

Areas of a Triangle and Trapezoid

The formula to find the area of a triangle can be derived from the area of a parallelogram relationship. Then, once we know how to find the area of a triangle, we can develop a formula to find the area of a trapezoid.

Recall that a diagonal separates a parallelogram into two congruent triangles. Suppose that the area of a parallelogram is 60 square units. If a diagonal of the parallelogram is drawn (Figure 14.7), what will be the area of each triangle? Since congruent triangles have equal areas, each triangle will have an area of 30 square units. Thus, the area of each triangle is one-half the area of the parallelogram.

$$\begin{aligned}
\text{Area of } \triangle ABD &= \frac{1}{2} \text{ area } \square ABCD \\
&= \frac{1}{2} \text{ base } \times \text{ height} \\
&= \frac{1}{2} AD \times BH
\end{aligned}$$

This analysis suggests Theorem 14.2.

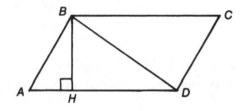

FIGURE 14.7

THEOREM 14.2 AREA OF A TRIANGLE

The area of a triangle is equal to one-half the product of the lengths of the base and the altitude drawn to that base.

Thus, $A = \frac{1}{2}bh$.

EXAMPLE 14.4

Find the area of each of the following triangles.

a.

b.

c.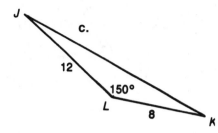

SOLUTION

a. In a right triangle either leg may be considered the base while the other leg then becomes the altitude. The area of a *right* triangle is therefore equal to one-half the product of the lengths of the legs:

$$A = \frac{1}{2}(\text{leg}_1)(\text{leg}_2)$$
$$= \frac{1}{2}(6)(4)$$
$$= 12 \text{ square units}$$

b. From vertex T, drop an altitude to base $\overline{RS}$. Since $\overline{TH}$ is the leg opposite a 45° angle in a 45-45 right triangle, the length of $\overline{TH}$ is equal to one-half the length of the hypotenuse ($\overline{RT}$) multiplied by $\sqrt{2}$:

$$TH = \frac{1}{2}(8)\sqrt{2} = 4\sqrt{2}.$$

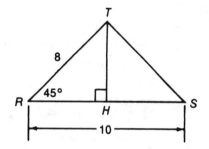

$$A = \frac{1}{2}(RS)(TH)$$
$$= \frac{1}{2}(10)(4\sqrt{2})$$
$$= 5(4\sqrt{2})$$
$$= 20\sqrt{2} \text{ square units}$$

c. From vertex J, drop an altitude to base $\overline{LK}$, extended.

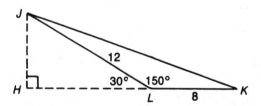

Angle JLH has measure 30 so that the length of altitude $\overline{JH}$ (the leg opposite the 30° angle in a 30-60 right triangle) is equal to one-half the length of the hypotenuse ($\overline{JL}$).

Hence, $JH = \frac{1}{2}(12) = 6$.

$$A = \frac{1}{2}(LK)(JH)$$
$$= \frac{1}{2}(8)(6)$$
$$= 24 \text{ square units}$$

To find the area of a *trapezoid*, we draw a diagonal so that the trapezoid is divided into two triangles. The area of the trapezoid is given by the sum of the areas of the two triangles. In Figure 14.8, the lengths of perpendicular segments $\overline{BX}$ and $\overline{DY}$ both represent the distance between bases $\overline{AD}$ and $\overline{BC}$. Since parallel lines are everywhere equidistant, $DY = BX$. For convenience, we will refer to the length of the altitude by the letter h so that $h = DY = BX$. We may then write

$$\text{Area trapezoid } ABCD = \text{area } \triangle ABD + \text{area } \triangle BCD$$

$$= \frac{1}{2}(AD)(h) + \frac{1}{2}(BC)(h)$$

$$= \frac{1}{2}h(AD + BC)$$

FIGURE 14.8

In this relationship, the terms inside the parentheses represent the lengths of the bases of the trapezoid and the letter h represents the length of an altitude of the trapezoid. This result is stated formally in Theorem 14.3.

THEOREM 14.3 AREA OF A TRAPEZOID

The area of a trapezoid is equal to one-half the product of the length of an altitude and the sum of the lengths of the bases.

Thus, $A = \frac{1}{2}h(b_1 + b_2)$.

EXAMPLE 14.5 Find the area of an isosceles trapezoid whose bases are 8 and 20 and whose lower base angle has a measure of 45.

SOLUTION

Since $\overline{BH}$ is the leg opposite the 45 degree angle in a 45-45 right triangle, $BH = AH = 6$.

$$\text{Area } ABCD = \frac{1}{2}BH(AD + BC)$$

$$= \frac{1}{2}(6)(20 + 8)$$

$$= 3(28)$$

$$= 84 \text{ square units}$$

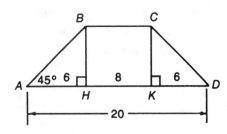

The area of a triangle formula may also be used to derive convenient formulas for finding the areas of equilateral triangles and rhombuses. The formula for the area of an equilateral triangle may be expressed exclusively in terms of the length of a side s of the triangle by representing the length of the altitude in terms of s and then applying the area of a triangle relationship.

Let's illustrate by considering equilateral triangle ABC and drawing altitude $\overline{BD}$ to side $\overline{AC}$ (Figure 14.9).

FIGURE 14.9

An equilateral triangle is also equiangular so that altitude $\overline{BD}$ divides the triangle into two 30-60 right triangles. In $\triangle ADB$, the length of $\overline{AD}$ (the side opposite the 30 degree angle) is one-half the length of the hypotenuse ($\overline{AB}$): $AD = \frac{1}{2} AB = \frac{1}{2} s$. The length of altitude $\overline{BD}$ (the leg opposite the 60 degree angle) is equal to one-half the length of the hypotenuse multiplied by $\sqrt{3}$: $BD = \frac{1}{2} AB \sqrt{3} = \frac{1}{2} s \sqrt{3}$.

$$\text{Area } \triangle ABC = \frac{1}{2}(AC)(BD)$$
$$= \frac{1}{2}(s)\left(\frac{1}{2} s\sqrt{3}\right)$$
$$= \frac{1}{4} s^2 \sqrt{3}$$

THEOREM 14.4 AREA OF AN EQUILATERAL TRIANGLE

The area of an equilateral triangle is equal to one-fourth of the square of the length of a side multiplied by $\sqrt{3}$.

Thus, $A = \dfrac{s^2}{4}\sqrt{3}.$

EXAMPLE 14.6 Find the area of an equilateral triangle whose perimeter is 24.

SOLUTION

If the perimeter of the triangle is 24, then the length of each side is 8.

$$A = \frac{s^2}{4}\sqrt{3}$$
$$= \frac{8^2}{4}\sqrt{3}$$
$$= \frac{64}{4}\sqrt{3}$$
$$= 16\sqrt{3} \text{ square units}$$

EXAMPLE 14.7 Find the length of a side of an equilateral triangle that has an area of $25\sqrt{3}$ square centimeters.

SOLUTION

$$A = \frac{s^2}{4}\sqrt{3}$$
$$25\sqrt{3} = \frac{s^2}{4}\sqrt{3}$$
$$25 = \frac{s^2}{4}$$
$$s^2 = 100$$
$$s = \sqrt{100} = 10 \text{ cm}$$

The diagonals of a rhombus divide the rhombus into four congruent triangles. The area of the rhombus can be obtained by determining the area of one of these triangles and then multiplying it by 4. For convenience, we will refer to the lengths of the two diagonals of a rhombus as d_1 and d_2. It can be shown that the area of any one of the four triangles formed by the diagonals is $\frac{1}{8}(d_1 \cdot d_2)$. To find the area of the rhombus, we multiply $\frac{1}{8}(d_1 \cdot d_2)$ by 4, which gives us $\frac{1}{2}(d_1 \cdot d_2)$.

THEOREM 14.5 AREA OF A RHOMBUS

The area of a rhombus is equal to one-half the product of the lengths of the diagonals.

Thus, $A = \frac{1}{2}(d_1 \cdot d_2)$.

In a square the diagonals are congruent, so

Area of square $= \frac{1}{2}d^2$

where d represents the length of each diagonal.

EXAMPLE 14.8 Find the area of a rhombus if the lengths of its diagonals are 10 and 14.

SOLUTION

$$A = \frac{1}{2}(d_1 \cdot d_2)$$
$$= \frac{1}{2}(10 \cdot 14)$$
$$= \frac{1}{2}(140)$$
$$= 70 \text{ square units}$$

EXAMPLE 14.9 The length of one diagonal of a rhombus is three times the length of the other diagonal. If the area of the rhombus is 54 square units, find the length of each diagonal.

SOLUTION

Let x = length of shorter diagonal.
Then $3x$ = length of longer diagonal.

$$A = \frac{1}{2}(d_1 \cdot d_2)$$
$$54 = \frac{1}{2}(3x \cdot x)$$
$$108 = 3x^2$$
$$x^2 = \frac{108}{3} = 36$$
$$x = \sqrt{36}$$
$$= \text{length of shorter diagonal} = 6$$
$$3x = \text{length of longer diagonal} = 18$$

Comparing Areas

COMPARING AREAS OF EQUIVALENT TRIANGLES

Two triangles that are congruent must have the same area. Can two triangles that are *not* congruent have the same area? In the accompanying diagram, $ABCD$ is a rectangle. Triangles AXD and AYD are not necessarily congruent. How do the areas of triangles AXD and AYD compare? Notice that each of these triangles has the same base ($\overline{AD}$) and a height that is numerically equal to the width of the rectangle. Therefore,

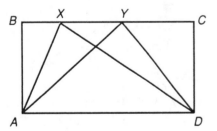

$$\text{Area } \triangle AXD = \text{Area } \triangle AYD = \frac{1}{2}(AD \times AB)$$

THEOREM 14.6 EQUIVALENT TRIANGLES

If two triangles have equal bases and equal altitudes, then their areas are equal.

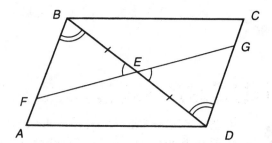

Since area can be represented by a number, areas can be added or subtracted. For example, in the accompanying diagram $ABCD$ is a parallelogram and E is the midpoint of diagonal $\overline{BD}$.

- How do the areas of triangles ABD and CBD compare? Since a diagonal divides a parallelogram into two congruent triangles, the area of triangle ABD is equal to the area of triangle CBD.
- How do the areas of triangles FEB and GED compare? Since these triangles are congruent by the ASA postulate, they have the same area.
- How do the areas of quadrilaterals $AFED$ and $CGEB$ compare? By subtracting equal areas ($\triangle FEB$ and $\triangle GED$) from equal areas ($\triangle ABD$ and $\triangle CBD$, respectively), we see that the areas of the two resulting quadrilaterals must also be equal:

$$\begin{array}{r} \text{Area } \triangle ABD = \text{area } \triangle CBD \\ -\text{Area } \triangle FEB = \text{area } \triangle GED \\ \hline \text{Area } AFED = \text{area } CGEB \end{array}$$

COMPARING AREAS OF SIMILAR POLYGONS

Recall that if two triangles are similar, then the ratio of the lengths of any pair of corresponding sides is the same as the ratio of the lengths of any pair of corresponding altitudes (or corresponding medians and angle bisectors). Suppose $\triangle ABC \sim \triangle RST$ and that their ratio of similitude is $3:1$. For example, an interesting relationship arises when we take the ratio of the areas of the two similar triangles in Figure 14.10:

$$\frac{\text{Area } \triangle ABC}{\text{Area } \triangle RST} = \frac{\frac{1}{2}(AC)(BX)}{\frac{1}{2}(RT)(SY)} = \frac{\frac{1}{2}(15)(3)}{\frac{1}{2}(5)(1)} = \frac{9}{1}$$

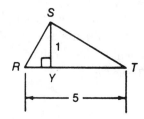

FIGURE 14.10

How does the ratio of the areas of the triangles compare with the ratio of the lengths of a pair of corresponding sides? Since the ratio of similitude is 3:1,

$$\frac{\text{Area } \triangle ABC}{\text{Area } \triangle RST} = \frac{9}{1} = \left(\frac{3}{1}\right)^2 = \left(\frac{\text{length of a side in } \triangle ABC}{\text{length of the corresponding side in } \triangle RST}\right)^2$$

THEOREM 14.7 COMPARING AREAS OF SIMILAR TRIANGLES

If two triangles are similar, then the ratio of their areas is equal to the square of the ratio of the lengths of any pair of corresponding sides:

$$\frac{\text{Area } \triangle I}{\text{Area } \triangle II} = \left(\frac{\text{side}_I}{\text{side}_{II}}\right)^2$$

where $\triangle I$ and $\triangle II$ refer to a pair of similar triangles and side_I and side_{II} represent the lengths of a pair of corresponding sides in triangles I and II, respectively.

Theorem 14.7 may be generalized to apply to any pair of similar *polygons*. Furthermore:

- Since the lengths of corresponding sides, altitudes, medians, and angle bisectors are in proportion, the ratio of the areas of a pair of similar triangles is equal to the square of the ratio of any pair of these corresponding segments.
- Since the perimeters of similar polygons have the same ratio as the lengths of any pair of corresponding sides, the ratio of the areas of similar polygons is equal to the square of the ratio of their perimeters.

 EXAMPLE 14.10 The ratio of similitude of two similar triangles is 2:3. If the area of the smaller triangle is 12 square units, find the area of the larger triangle.

SOLUTION

$$\frac{\text{Area of smaller } \triangle}{\text{Area of larger } \triangle} = \left(\frac{2}{3}\right)^2$$

Let x = area of larger triangle. Then:

$$\frac{12}{x} = \frac{4}{9}$$
$$4x = 108$$
$$x = \frac{108}{4} = 27$$

EXAMPLE 14.11

The areas of two similar polygons are 25 and 81 square units. If the length of a side of the larger polygon is 72, find the length of the corresponding side of the smaller polygon.

SOLUTION

Let x = length of the corresponding side of the smaller polygon.

$$\frac{\text{Area of smaller polygon}}{\text{Area of larger polygon}} = \left(\frac{x}{72}\right)^2$$

$$\frac{25}{81} = \left(\frac{x}{72}\right)^2$$

Solve for x by first taking the square root of both sides of the proportion:

$$\sqrt{\frac{25}{81}} = \frac{x}{72}$$

$$\frac{5}{9} = \frac{x}{72}$$

$$9x = 360$$

$$x = \frac{360}{9} = 40$$

Area of a Regular Polygon

We have intentionally restricted our attention to finding the areas of familiar three-sided and four-sided polygons. As a general rule, it is difficult to develop convenient formulas for other types of polygons. There is, however, one notable exception. A formula for the area of a *regular* polygon can be derived by subdividing the regular polygon into a set of congruent triangles and then summing the areas of these triangles. Throughout our analysis we shall assume that *circles having the same center can be inscribed and circumscribed about a regular polygon*. Figures 14.11 and 14.12 identify some important terms relating to regular polygons and their circumscribed and inscribed circles. This information will be needed when we eventually turn our attention to finding the area of a regular polygon.

In Figure 14.11:

* Point O is the center of the circle inscribed and the circle circumscribed about regular pentagon *ABCDE*. Point O is referred to as the *center* of a regular polygon.

FIGURE 14.11

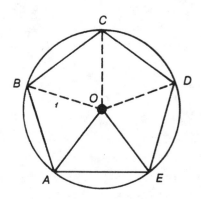

FIGURE 14.12

- $\overline{OX}$ is a segment drawn from the center of the regular pentagon to the point at which the inscribed circle is tangent to side $\overline{AE}$. $\overline{OX}$ is called an *apothem*. Apothems may be drawn to each of the other sides.

In Figure 14.12:

- $\overline{OA}$ and $\overline{OE}$ are *radii* of regular pentagon *ABCDE*. A *radius* of a regular polygon is a segment drawn from the center to any vertex of the polygon.
- Angle *AOE* is a *central angle* of regular pentagon *ABCDE*. A *central angle* of a regular polygon is formed by drawing two radii to consecutive vertices of the polygon.

DEFINITIONS RELATING TO REGULAR POLYGONS

- The *center* of a regular polygon is the common center of its inscribed and circumscribed circles.
- An *apothem* of a regular polygon is a segment whose end points are the center of the polygon and a point at which the inscribed circle is tangent to a side. An apothem of a regular polygon is also a radius of the *inscribed* circle.
- A *radius* of a regular polygon is a segment whose end points are the center of the polygon and a vertex of the polygon. A radius of a regular polygon is also a radius of the *circumscribed* circle.
- A *central angle* of a regular polygon is an angle whose vertex is the center of the polygon and whose sides are radii drawn to the end points of the same side of the polygon. An *n*-sided regular polygon will have *n* central angles.

In Figure 14.12, the sides of the regular pentagon divide the circle into five congruent arcs. It follows that each central angle must have the same measure. Since there are five central angles, the measure of each central angle may be found by dividing 360 by 5, obtaining a measure of 72 for each of the central angles. In general, the radii of a regular polygon divide the polygon into triangles which may be proven congruent to one another by SAS, where the congruent sides are the radii and the congruent included angles are the central angles. Using the CPCTC principle, we may further conclude that each radius bisects the angle located at the vertex to which it is drawn.

THEOREM 14.8 ANGLES OF A REGULAR POLYGON

- The central angles of a regular polygon are congruent.
- The measure of a central angle of a regular polygon is equal to 360 divided by the number of sides of the polygon.
- The radii of a regular polygon bisect the interior angles of the regular polygon.

EXAMPLE
14.12

The length of a side of a regular hexagon is 14. Find each of the following:
a. The measure of a central angle.
b. The length of a radius of the hexagon.
c. The length of an apothem.

SOLUTION

a.

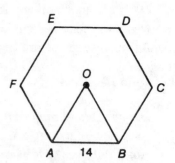

$$m \angle AOB = \frac{360}{n}$$

$$= \frac{360}{6}$$

$$= 60$$

b. Since the central angle has measure 60, and $\overline{OA} \cong \overline{OB}$, angles OAB and OBA must be congruent and be equal to 60. Since $\triangle AOB$ is equiangular, it is also equilateral. Hence, the radius of the hexagon must have the same length as $\overline{AB}$ and is therefore 14.

c. Let us focus on triangle AOB and draw the apothem to side $\overline{AB}$.

Since an apothem is drawn to a point at which the inscribed circle is tangent to a side of the polygon, the apothem must be perpendicular to the side to which it is drawn. Furthermore, it bisects the side to which it is drawn. As an illustration observe that apothem $\overline{OX}$ divides triangle AOB into two congruent right triangles. $\triangle AOX \cong \triangle BOX$ by Hy-leg since

$$\overline{OA} \cong \overline{OB} \quad \text{(Hy)}$$
$$\overline{OX} \cong \overline{OX} \quad \text{(Leg)}$$

By CPCTC, $\overline{AX} \cong \overline{BX}$. In addition, $\angle AOX \cong \angle BOX$.

For the problem at hand, the apothem $\overline{OX}$ is the leg opposite the 60° angle in a 30-60 right triangle (that is, in $\triangle AOX$):

$$OX = \frac{1}{2} OA\sqrt{3}$$
$$= \frac{1}{2}(14)\sqrt{3}$$
$$= 7\sqrt{3}$$

Theorem 14.9 summarizes the properties of an apothem.

THEOREM 14.9 PROPERTIES OF AN APOTHEM

- An apothem of a regular polygon is the perpendicular bisector of the side to which it is drawn.
- An apothem bisects the central angle determined by the side to which it is drawn.

We are now in a position to develop a convenient formula for finding the area of a *regular* polygon. Since an *n*-sided regular polygon is composed of *n* identical triangles, the formula for the area of a regular polygon is a generalization of the formula for the area of a triangle. The apothem of a regular polygon corresponds to the altitude of a triangle, and the perimeter of a regular polygon corresponds to the base of a triangle. Therefore, the area of a regular polygon may be found by multiplying one-half the length of the apothem by the perimeter of the regular polygon.

THEOREM 14.10 AREA OF A REGULAR POLYGON

The area (*A*) of a regular polygon is equal to one-half the product of the length of an apothem (*a*) and its perimeter (*p*).

Thus, $A = \dfrac{1}{2} ap$.

OUTLINE OF PROOF

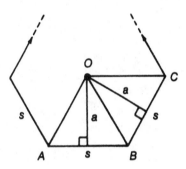

GIVEN: $ABC \ldots$ is a regular polygon having *n* sides. Let:
 a = length of an apothem,
 s = length of each side,
 p = perimeter of $ABC.\ldots$

PROVE: Area $ABC \ldots = \dfrac{1}{2} ap$.

• Draw the radii of the polygon.
• Find the sum of the areas of all the triangles thus formed:

$$\text{Area } ABC\ldots = \text{area } \triangle AOB + \text{area } \triangle BOC + \cdots$$
$$= \tfrac{1}{2} as \quad\quad + \tfrac{1}{2} as \quad\quad + \cdots + \tfrac{1}{2} as$$
$$= \tfrac{1}{2} a(s + s + \cdots + s)$$
$$= \tfrac{1}{2} ap$$

EXAMPLE 14.13

The length of a side of a regular pentagon is 20. Find each of the following:
a. The length of the apothem correct to the *nearest hundredth*.
b. The area correct to the *nearest whole number*.

SOLUTION

First draw a representative triangle of the polygon.

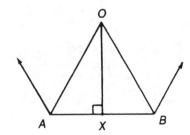

a. Since an apothem is the perpendicular bisector of the side to which it is drawn, $\triangle OAX$ is a right triangle and $AX = \frac{1}{2}(20) = 10$. The measure of central angle $AOB = \frac{360}{5} = 72$. Apothem $\overline{AX}$ bisects central angle AOB, so $m\measuredangle AOX = \frac{1}{2}(72) = 36$. We must find OX by using an appropriate trigonometric ratio. The arithmetic is simplest if we find the tangent of angle OAX. Since $m\measuredangle OAX = 90 - 36 = 54$, we may write the following:

$$\tan 54° = \frac{\text{side opposite } \measuredangle}{\text{side adjacent } \measuredangle}$$
$$= \frac{OX}{AX}$$
$$1.3764 = \frac{OX}{10}$$
$$OX = 10(1.3764)$$
$$= 13.76 \text{ correct to the } \textit{nearest hundredth}$$

b. If the length of a side of a regular pentagon is 20, then its perimeter is 5 times 20 or 100.

$$A = \frac{1}{2}ap$$
$$= \frac{1}{2}(13.76)(100)$$
$$= \frac{1}{2}(1376)$$
$$= 688 \text{ square units correct to the } \textit{nearest whole number}$$

Areas of a Circle, Sector, and Segment

Consider what happens if the number of sides, n, of an inscribed regular polygon gets larger, as shown in Figure 14.13.

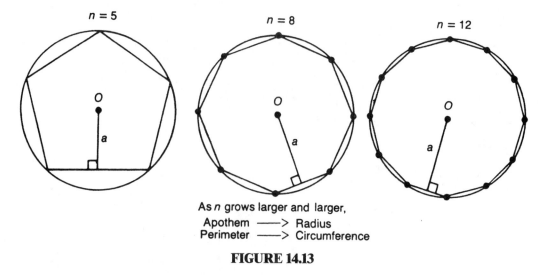

As n grows larger and larger,
Apothem ——> Radius
Perimeter ——> Circumference

FIGURE 14.13

As the number of sides increases indefinitely without being constrained by an upper bound, the area of the inscribed regular polygon and the area of the circle become indistinguishable. Under these circumstances, we may make the following substitutions in the formula for the area of a regular polygon where r represents the radius of the circle and C its circumference:

As the number of sides becomes infinitely large

$$\text{Area of regular polygon} = \frac{1}{2}\ a\ \ p$$

$$\text{Area of circumscribed circle} = \frac{1}{2}\ r\ \ C$$

or

$$= \frac{1}{2}(r)(2\pi r)$$

$$\text{Area of circle} = \pi r^2$$

THEOREM 14.11 AREA OF A CIRCLE

The area (A) of a circle is equal to the product of the constant π and the square of the length of the radius (r) of the circle.

Thus, $A = \pi r^2$.

a. Find the area of circle O.

b. Find the area of the shaded region.

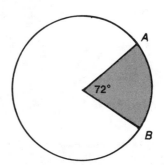

SOLUTION

a. $A = \pi r^2$

$= \pi (10)^2$

$= 100\pi$ square units

b. The shaded region is called a *sector* of the circle. In a manner analogous to determining the length of an arc of a circle, we form the following proportion:

$$\frac{\text{Area of sector}}{\text{Area of circle}} = \frac{\text{degree measure of sector arc}}{360°}$$

Let x = area of the sector.

$$\frac{x}{100\pi} = \frac{72°}{360°}$$

$$\pi = \frac{1}{5}$$

$$5x = 100\pi$$

$$x = 20\pi \text{ square units}$$

Suppose that in part **b** of Example 14.14 chord $\overline{AB}$ was drawn. Then,

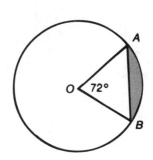

if it was known that the area of triangle *AOB* was 50 square units, what would be the area of the region bounded by chord $\overline{AB}$ and the minor arc that it cuts off? The region described is called a *segment* of the circle. To find the area of this segment, all we need do is to subtract the area of triangle *AOB* from the area of sector *AOB*.

$$
\begin{aligned}
\text{Area segment } \overline{AB} &= \text{area sector } AOB - \text{area } \triangle AOB \\
&= 20\pi - 50 \text{ square units}
\end{aligned}
$$

Unless otherwise directed, we generally leave the answer in this form.

Let's now summarize the concepts introduced in Example 14.14.

DEFINITIONS OF SECTOR AND SEGMENT OF A CIRCLE

- A *sector of a circle* is a region of a circle bounded by two radii and the minor arc they determine.
- A *segment of a circle* is a region of a circle bounded by a chord and the minor arc that it cuts off.

THEOREM 14.12 AREA OF A SECTOR

The area of a sector of a circle is determined by the following proportion where *n* represents the degree measure of its minor arc:

$$
\frac{\text{Area of sector}}{\text{Area of circle}} = \frac{n}{360}
$$

$$
\frac{\text{Area sector } AOB}{\pi r^2} = \frac{n}{360}
$$

Geometric Solids

The amount of space that a solid encloses is represented by its volume. The *volume* of a solid (Figure 14.14), expressed in cubic units, is the number of cubes having an edge length of 1 unit that the solid can enclose.

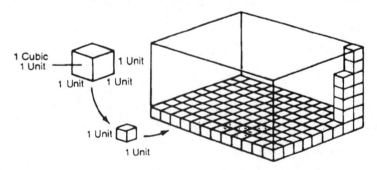

FIGURE 14.14 Volume of a solid

SPECIAL PRISMS

A polygon is "flat" since all its sides lie in the same plane. The figure that is the counterpart of a polygon in space is called a *polyhedron*.

A *prism* (Figure 14.15) is a special type of polyhedron. The sides of a prism are called *faces*. All prisms have these two properties:

- Two of the faces, called *bases*, are congruent polygons lying in parallel planes.
- The faces that are not bases, called *lateral faces*, are parallelograms.

A lateral face ⎯

2 Bases
5 Lateral faces

FIGURE 14.15 Prism

A *rectangular solid* (Figure 14.16) is a prism whose two bases and four lateral faces are rectangles. If the bases and lateral faces are squares, the rectangular solid is called a *cube*.

FIGURE 14.16 Rectangular solid

VOLUME OF A RECTANGULAR SOLID

The volume (V) of *any* prism is equal to the product of the area of its base (B) and its height (h):

$$\text{Volume of a prism: } V = Bh.$$

By using this relationship, special formulas can be derived for determining the volumes of a cube (Figure 14.17) and a rectangular solid (Figure 14.18):

FIGURE 14.17 Cube

Volume of a cube:
$$V = Bh = \left(e^2\right)e$$
$$= e^3$$

FIGURE 14.18 Rectangular solid

Volume of rectangular solid:
$$V = Bh = (\ell \times w)h$$
$$= \ell wh$$

 EXAMPLE 14.15 What is the volume of a rectangular solid having a length, width, and height of 5 centimeters, 4 centimeters, and 3 centimeters, respectively?

SOLUTION

$$V = \ell wh = (5)(4)(3) = 60 \text{ cm}^3$$

EXAMPLE 14.16 A rectangular solid has the same volume as a cube whose edge length is 6. What is the height of the rectangular solid if its length is 8 and its width is 3?

SOLUTION

Volume of rectangular solid = volume of cube

$$\ell wh = e^3$$
$$(8)(3)h = 6^3$$
$$24h = 216$$
$$h = \frac{216}{24} = 9$$

EXAMPLE 14.17 What is the change in volume of a cube whose edge length is doubled?

SOLUTION

In the formula $V = e^3$, replace e with $2e$. The new volume is

$$V = (2e)^3 = 8e^3.$$

The new volume is 8 times as great as the original volume.

VOLUMES OF A CYLINDER AND CONE

A *cylinder* (Figure 14.19) has *two* bases that are congruent circles lying in parallel planes. Like the volume of a prism, a solid that also has two congruent bases, the volume of a cylinder is given by the formula $V = Bh$. In this formula, B represents the area of one of the two identical bases. The area of a cylinder's circular base is πr^2.

FIGURE 14.19 Cylinder

Volume of a cylinder:
$$V = \pi r^2 h$$

FIGURE 14.20 Cone

Volume of a cone:
$$V = \frac{1}{3} \pi r^2 h$$

A *cone* (Figure 14.20), has one base and a volume given by the formula $V = \frac{1}{3} Bh$. Here B represents the area, πr^2, of the circular base of the cone.

EXAMPLE 14.18 The radius of a base of a cylinder is 5 and its height is 6. What is its volume?

SOLUTION

$$V = \pi r^2 h$$
$$= \pi (5^2)6 = \pi (25)(6) = 150 \pi$$

The *slant height*, ℓ, of a cone is a segment that connects the vertex to a point on the circular base. What is the volume of a cone having a radius of 6 and a slant height of 10?

SOLUTION

The slant height is the hypotenuse of a right triangle whose legs are a radius and the altitude of the cone. The lengths of the sides of this triangle form a 6-8-10 right triangle, where 8 represents the height of the cone. Therefore:

$$V = \frac{1}{3}\pi r^2 h$$
$$= \frac{1}{3}\pi(6)^2 \cdot 8$$
$$= \frac{1}{\cancel{3}}\pi(\cancel{36})\cdot\overset{12}{}$$
$$= \frac{1}{96\pi}$$

The circumference of the base of a cone is 8π centimeters. If the volume of the cone is 16π cubic centimeters, what is the height?

SOLUTION

Since $C = \pi D = 8\pi$, the diameter is 8 cm; therefore, the radius of the base is 4 cm.

$$V = \frac{1}{3}\pi r^2 h$$
$$16\pi = \frac{1}{3}\pi(4^2)h$$
$$48\pi = 16\pi h$$
$$\frac{48\pi}{16\pi} = h$$
$$3 = h$$

The height of the cone is 3 cm.

VOLUME OF A SPHERE

A sphere (Figure 14.21) is a solid that represents the set of all points in space that are at a given distance from a fixed point.

FIGURE 14.21 Sphere

Volume of a sphere:
$$V = \frac{4}{3}\pi r^3$$

EXAMPLE 14.21 Find the volume of a sphere whose diameter is 6 centimeters.

SOLUTION

$$\text{Radius} = \frac{1}{2}(6) = 3$$

$$V = \frac{4}{3}\pi(3)^3$$

$$= \frac{4}{3}\pi(27)$$

$$= 4\pi(9)$$

$$= 36\pi \text{ cm}^3$$

SUMMARY OF VOLUME FORMULAS

Figure	Volume Formula
Rectangular box	$V = \text{length} \times \text{width} \times \text{height}$ $= \ell w h$
Cube	$V = \text{edge} \times \text{edge} \times \text{edge}$ $= e^3$
Cylinder	$V = \pi \times (\text{radius})^2 \times \text{height}$ $= \pi r^2 h$
Cone	$V = \frac{1}{3} \times \pi \times (\text{radius})^2 \times \text{height}$ $= \frac{1}{3}\pi r^2 h$
Sphere	$V = \frac{4}{3} \times \pi \times (\text{radius})^3$ $= \frac{4}{3}\pi r^3$

REVIEW EXERCISES FOR CHAPTER 14

1. Find the area of each of the following:
 (a) A rectangle whose base measures 6 and whose diagonal measures 10.
 (b) A square whose diagonal measures 8.
 (c) A parallelogram having two adjacent sides of 12 and 15 centimeters and an included angle of measure 60.

2. Find the dimensions of each of the following:
 (a) A square whose area is 144 square centimeters.
 (b) A rectangle whose area is 75 square units and whose base and altitude are in the ratio of $3:1$.
 (c) A rectangle having an area of 135 square units, a base represented by $x + 2$, and an altitude represented as $2x + 1$.

For Exercises 3 to 11, find the area of each of the figures. Whenever appropriate, answers may be left in radical form.

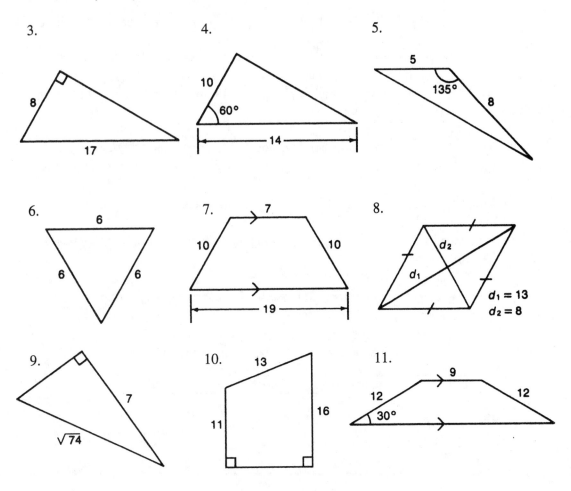

3.

4.

5.

6.

7.

8.

9.

10.

11.

12. In each case, the number of square units in the area of an equilateral triangle is given. Find the length of a side of the triangle.
 (a) $16\sqrt{3}$ **(b)** $11\sqrt{3}$ **(c)** 4

13. Find the length of the shorter diagonal of a rhombus if:
 (a) The length of the longer diagonal is 15 and the area is 90.
 (b) The lengths of the diagonals are in the ratio of $2:3$ and the area of the rhombus is 147.

14. Find the length of an altitude of a trapezoid if:
 (a) Its area is 72 and the sum of the lengths of the bases is 36.
 (b) Its area is 80 and its median is 16.
 (c) The sum of the lengths of the bases is numerically equal to one-third of the area of the trapezoid.

15. Find the area of a rhombus if its perimeter is 68 and the length of one of its diagonals is 16.

16. Find the area of a triangle if the lengths of a pair of adjacent sides are 6 and 14 and the measure of the included angle is:
 (a) 90 **(b)** 30 **(c)** 120 (Leave answer in radical form.)

17. $\triangle JKL \sim \triangle RST$. If $JL = 20$ and $RT = 15$, find the ratio of the areas of the triangles.

18. The areas of two similar polygons are 81 square units and 121 square units.
 (a) Find their ratio of similitude.
 (b) If the perimeter of the smaller polygon is 45, find the perimeter of the larger polygon.

19. The lengths of a pair of corresponding sides of a pair of similar triangles are in the ratio of $5:8$. If the area of the smaller triangle is 75, find the area of the larger triangle.

20. Find the measure of a central angle for a regular octagon.

21. If the measure of an interior angle of a regular polygon is 150, find the measure of a central angle.

22. Find the area of a regular hexagon inscribed in a circle having a diameter of 20 centimeters.

23. Fill in the following table for a regular polygon having four sides:

	Side	Radius	Apothem	Area
(a)	6 cm	?	?	?
(b)	?	?	?	49 cm²
(c)	?	?	5 cm	?
(d)	?	8 cm	?	?

24. A circle having a radius of 6 centimeters is inscribed in a regular hexagon. Another circle whose radius is 6 centimeters is circumscribed about another regular hexagon. Find the ratio of the area of the smaller hexagon to the area of the larger hexagon.

25. The length of a side of a regular decagon is 20 centimeters.
 (a) Find the length of the apothem correct to the *nearest tenth* of a centimeter.
 (b) Using the answer obtained in part **a**, find the area of the decagon.

Unless otherwise indicated, answers for Exercises 26 to 29 may be left in terms of π.

26. Fill in the following table:

	Radius	Diameter	Circumference	Area
(a)	5	?	?	
(b)	?	9	?	?
(c)	?	?	?	49π
(d)	?	?	18π	?
(e)	?	?	?	$\dfrac{64}{25}\pi$

27. The ratio of the areas of two circles is 1:9. If the radius of the smaller circle is 5 centimeters, find the length of the radius of the larger circle.

28. The ratio of the lengths of the radii of two circles is 4:25. If the area of the smaller circle is 8 square units, find the area of the larger circle.

29. The diameters of two concentric circles are 8 and 12. Find the area of the ring-shaped region (called an *annulus*) bounded by the two circles.

30. Fill in the following table:

	Radius of Circle	Area of Circle	Degree Measure of Sector Arc	Area of Sector
(a)	12	?	45°	?
(b)	?	36π	?	12π
(c)	?	?	72°	20π
(d)	9	?	160°	?

31. In circle O, radii OA and OB are drawn so that OA is equal to the length of chord $\overline{AB}$. If $OA = 12$, find the area of:
 (a) Sector AOB.
 (b) Segment AB.

32. A square whose side is 8 centimeters in length is inscribed in a circle. Find the area of the segment formed by a side of the square and its intercepted arc.

33. What is the ratio of the area of the inscribed and circumscribed circles of a square having a side of 6 centimeters?

For Exercises 34 and 35, find the area of the shaded region in each figure.

34.

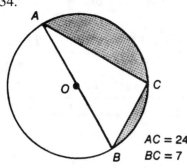

$AC = 24$
$BC = 7$

35.

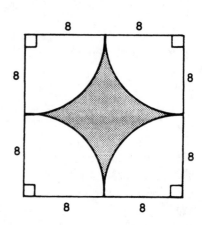

36. A regular pentagon is inscribed in a circle having a radius of 20 centimeters. Find the area of the sector formed by drawing radii to a pair of consecutive vertices of the pentagon.

37. An isosceles right triangle whose hypotenuse has a length of 16 is inscribed in a circle. Find the area of the segment formed by a leg of the right triangle and its intercepted arc.

38. A regular hexagon is inscribed in a circle. The length of the apothem is $4\sqrt{3}$.
 (a) Find the area of the circle.
 (b) Find the area of the hexagon.
 (c) Find the area of the segment cut off by a side of the hexagon.

39. GIVEN: G is the midpoint of $\overline{CV}$.
 PROVE: (a) Area $\triangle CLG$ = area $\triangle VLG$.
 (b) Area $\triangle BLC$ = area $\triangle BLV$.

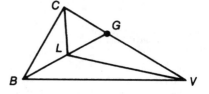

40. In a regular polygon of nine sides, the length of each side is 8.
 (a) Find the length of the apothem to the *nearest integer*.
 (b) Using the answer obtained in part **a**, find the area of the polygon.

41. GIVEN: $\triangle BCD$, $\overline{BC} \cong \overline{BD}$; points A
 and E are exterior to $\triangle BCD$.
 $\overline{AB}$ and $\overline{BE}$ are drawn so that
 $\overline{AB} \cong \overline{BE}$; $\overline{AC}$ and $\overline{DE}$ are
 drawn; $\angle ABD \cong \angle EBC$.

 PROVE: (a) $\triangle ABC \cong \triangle EBD$.
 (b) Area polygon $ABDC$ =
 area polygon $BEDC$.

42. GIVEN: $\square ABCD, \overline{BE} \cong \overline{FD}.$
 PROVE: Area of $\triangle FAC$ = area $\square ABCD$.

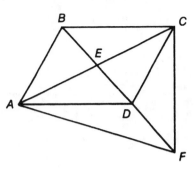

43. Prove, using algebraic methods, that for an equilateral triangle:
 (a) The length of the radius of the inscribed circle is one-third the length of the altitude of the triangle.
 (b) The areas of the inscribed and circumscribed circles are in the ratio of $1:4$.

44. If a cube and a 3-inch-by-8-inch-by-9 inch rectangular box have the same volume, find:
 (a) The perimeter of the cube. **(b)** The area of a side of the cube.

45. If the radius of a sphere is tripled, by what number is its volume multiplied?

46. The circumference of a cylinder is 10π centimeters and its height is 3.5 centimeters. What is the volume of the cylinder?

47. The surface area of a rectangular solid is the sum of the areas of its six sides.
 (a) What is the surface area of a cube whose volume is 125 cubic inches?
 (b) What is the surface area of a 3-by-4-by-5 rectangular box?
 (c) What is the volume of a cube whose surface area is 96 square units?

48. A right triangle whose vertical leg measures 8 centimeters and and whose horizontal leg measures 15 centimeters is revolved in space about the shorter leg. What is the volume of the resulting cone?

49. The length, width, and height of a rectangular solid are in the ratio of $3:2:1$. If the volume of the box is 48 cubic units, what is the total surface area of the box?

50. If the edge length of a cube is 4, what is the distance from one corner of the cube to the furthest corner on the opposite face of the cube?

COORDINATE GEOMETRY

WHAT YOU WILL LEARN

The French mathematician Rene Descartes (1596–1650) developed a method for representing points and equations graphically by drawing a rectangular grid, called a *coordinate plane*, in which horizontal and vertical number lines intersect at their zero points. In this chapter you will learn:

- the way to plot points and equations in the coordinate plane;
- the way to find areas of figures in the coordinate plane;
- the use of formulas to find the midpoint, length, and slope of a line segment;
- the general form of an equation of a line;
- the general form of an equation of a circle;
- the way to prove geometric relationships using coordinates.

The Coordinate Plane

You can create a *coordinate plane* by drawing a horizontal number line called the *x-axis* and a vertical number line called the *y-axis*. The x-axis and the y-axis are referred to as the *coordinate axes*, and their point of intersection is called the *origin*. The location of a point in the coordinate plane is determined by two numbers: the *x-coordinate* and the *y-coordinate* of the point. The coordinates of a point are always written as an ordered pair of numbers having the form (x, y). For example, $(2, 3)$ represents the point whose x-coordinate is 2 and whose y-coordinate is 3. This point is named point A in Figure 15.1.

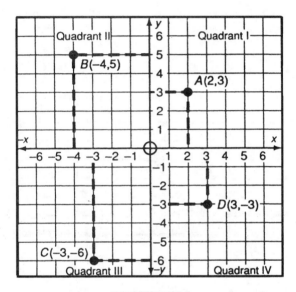

FIGURE 15.1

The x-coordinate of a point is sometimes called the *abscissa*, and the y-coordinate is known as the *ordinate*. Notice in Figure 15.1 that:

- The coordinates of the origin are $(0, 0)$.
- A point is named by writing a single capital letter followed by the coordinates of the point.
- Point A is located by beginning at the origin and moving along the x-axis 2 units to the *right* (since 2 is *positive*), stopping at $x = 2$, and then moving 3 units vertically *up* (since 3 is *positive*).
- The sign of the x-coordinate (positive or negative) tells whether the point is located horizontally to the right of the origin ($x > 0$) or to the left of the origin ($x < 0$). Similarly, the sign of the y-coordinate (positive or negative) tells whether the point is located vertically above the origin ($y > 0$) or below the origin ($y < 0$).
- The coordinate axes divide the coordinate plane into four *quadrants*. Points A, B, C, and D lie in different quadrants. The signs of the x- and y-coordinates of a point determine the quadrant in which the point is located.

Example	Signs of Coordinates	Location of Point
$A(2, 3)$	$(+, +)$	Quadrant I
$B(-4, 5)$	$(-, +)$	Quadrant II
$C(-3, -6)$	$(-, -)$	Quadrant III
$D(3, -3)$	$(+, -)$	Quadrant IV

Finding Area Using Coordinates

DRAWING AN ALTITUDE TO FIND AREA

If one side of a triangle or quadrilateral is parallel to a coordinate axis, then the area of the figure can be determined by drawing an altitude to this side and using the appropriate area formula.

 EXAMPLE 15.1 Graph a parallelogram whose vertices are $A(2, 2)$, $B(5, 6)$, $C(13, 6)$, and $D(10, 2)$, and then find its area.

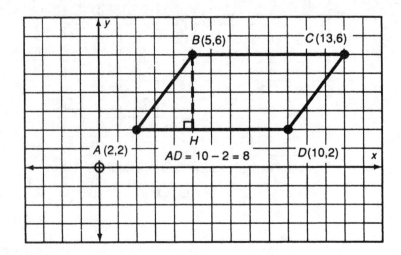

SOLUTION

In the accompanying graph, altitude $\overline{BH}$ has been drawn to $\overline{AD}$. By counting boxes we see that $AD = 8$ and $BH = 4$.

Therefore,

$$
\begin{aligned}
\text{Area } \square\, ABCD &= bh \\
&= AD \times BH \\
&= 8 \times 4 \\
&= 32 \text{ square units}
\end{aligned}
$$

 EXAMPLE 15.2 Graph the trapezoid whose vertices are $A(-4, 0)$, $B(-4, 3)$, $C(0, 6)$, and $D(0, 0)$, and then find its area.

SOLUTION

In the accompanying graph,
$\overline{AB}$ and $\overline{CD}$ are the bases and $\overline{AD}$
an altitude of the trapezoid.

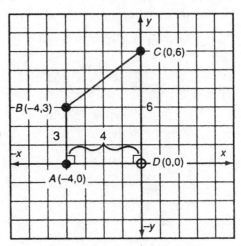

$$\text{Area trapezoid } ABCD = \frac{1}{2}h(b_1 + b_2)$$

$$= \frac{1}{2}(AD)(AB + CD)$$

$$= \frac{1}{2}(4)(3 + 6)$$

$$= 2(9)$$

$$= 18 \text{ square units}$$

USING SUBTRACTION TO FIND AREA

Sometimes we need to find the area of
a quadrilateral (or triangle) that does not have
a vertical or horizontal side. For example,
in Figure 15.2 the area of quadrilateral *ABCD*
can be obtained *indirectly* by subtracting
the sum of the areas of right triangles I, II, III,
and IV from the area of rectangle *WXYZ*.

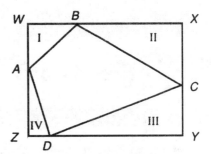

FIGURE 15.2

EXAMPLE 15.3 Find the area of the quadrilateral whose vertices are $A(-2, 2)$, $B(2, 5)$, $C(8, 1)$, and $D(-1, -2)$.

SOLUTION

The area of quadrilateral *ABCD* is calculated *indirectly* as follows:

- Circumscribe a rectangle about quadrilateral *ABCD* by drawing intersecting horizontal and vertical segments through the vertices of the quadrilateral as shown in the accompanying diagram.
- Find the area of the rectangle:
$$\text{Area rectangle } WXYZ = (ZY)(YX) = (10)(7) = 70$$
- Find the sum of the areas of the right triangles in the four corners of the rectangle. Keep in mind that the area of a right triangle is equal to one-half the product of the lengths of the legs of the triangle.

$$\text{Sum} = \text{Area right } \triangle BWA = \frac{1}{2}(BW)(WA) = \frac{1}{2}(4)(3) = 6$$

$$+ \text{ Area right } \triangle BXC = \frac{1}{2}(BX)(XC) = \frac{1}{2}(6)(4) = 12$$

$$+ \text{ Area right } \triangle DYC = \frac{1}{2}(DY)(YC) = \frac{1}{2}(9)(3) = 13.5$$

$$+ \text{ Area right } \triangle DZA = \frac{1}{2}(DZ)(ZA) = \frac{1}{2}(1)(4) = 2$$

$$\text{Sum of } \triangle \text{ areas} = 33.5$$

- Subtract the sum of the areas of the right triangles from the area of the rectangle.

 Area quad $ABCD$ = Area rect $WXYZ$ − Sum of areas of right triangles

 = 70 − 33.5

 = 36.5

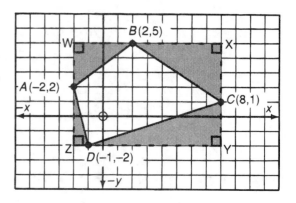

The Midpoint and Distance Formulas

FINDING THE MIDPOINT

In Figure 15.3, what are the coordinates of the *midpoint* of $\overline{AB}$? In general, the coordinates of the midpoint of a segment joining two points are found by taking the average of their x-coordinates and then finding the average of their y-coordinates. In this case:

$$\frac{2+10}{2} = 6 \quad \text{and} \quad \frac{5+5}{2} = 5$$

Hence, the midpoint of $\overline{AB}$ will be located at $M(6, 5)$.

FIGURE 15.3

THEOREM 15.1 THE MIDPOINT FORMULA

The coordinates of the midpoint $M(x_m, y_m)$ of a segment whose end points are $A(x_1, y_1)$ and $B(x_2, y_2)$ may be found using the formulas:

and
$$x_m = \frac{x_1 + x_2}{2}$$
$$y_m = \frac{y_1 + y_2}{2}$$

EXAMPLE 15.4

Find the coordinates of the midpoint of a segment whose end points are $H(4, 9)$ and $K(-10, 1)$.

SOLUTION

$x_1 \rightarrow H(4, 9)$ $x_2 \rightarrow K(-10, 1)$
$\uparrow y_1$ $\uparrow y_2$

$$x_m = \frac{x_1 + x_2}{2} \qquad\qquad y_m = \frac{y_1 + y_2}{2}$$

$$= \frac{4 + (-10)}{2} \qquad\qquad = \frac{9 + 1}{2}$$

$$= \frac{-6}{2} \qquad\qquad\qquad = \frac{10}{2}$$

$$= -3 \qquad\qquad\qquad = 5$$

The coordinates of the midpoint of $\overline{HK}$ are $(-3, 5)$.

EXAMPLE 15.5

$M(7, -1)$ is the midpoint of $\overline{WL}$. If the coordinates of end point W are $(5, 4)$, find the coordinates of end point L.

SOLUTION

$$x_m = \frac{5 + x}{2}$$

$W(5, 4)$ $L(x, y)$

$$y_m = \frac{4 + y}{2}$$

$$x_m = 7 = \frac{5 + x}{2} \qquad\qquad y_m = -1 = \frac{4 + y}{2}$$

Multiply each side of the equation by 2:

$$2(7) = 2\left(\frac{5 + x}{2}\right) \qquad\qquad 2(-1) = 2\left(\frac{4 + y}{2}\right)$$

$$14 = 5 + x \qquad\qquad\qquad -2 = 4 + y$$

$$9 = x \qquad\qquad\qquad\qquad -6 = y$$

Hence, the coordinates of point L are $(9, -6)$.

EXAMPLE 15.6

The coordinates of quadrilateral $ABCD$ are $A(-3, 0)$, $B(4, 7)$, $C(9, 2)$, and $D(2, -5)$. Prove that $ABCD$ is a parallelogram.

SOLUTION

If the diagonals of a quadrilateral bisect each other, then the quadrilateral is a parallelogram. In the diagram, diagonals $\overline{AC}$ and $\overline{BD}$ intersect at point E. If the coordinates of point E are the coordinates of the midpoints of $\overline{AC}$ and $\overline{BD}$, then the diagonals bisect each other, and the quadrilateral is a parallelogram.

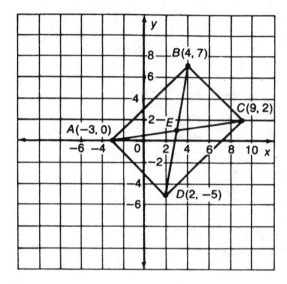

To find the midpoint of $\overline{AC}$, let $(x_1, y_1) = (-3, 0)$ and $(x_2, y_2) = (9, 2)$:

$$x_m = \frac{-3 + 9}{2} = \frac{6}{2} = 3$$

$$y_m = \frac{0 + 2}{2} = \frac{2}{2} = 1$$

Hence, the coordinates of the midpoint of $\overline{AC}$ are $(3, 1)$.

To find the midpoint of $\overline{BD}$, let $(x_1, y_1) = (4, 7)$ and $(x_2, y_2) = (2, -5)$:

$$x_m = \frac{4 + 2}{2} = \frac{6}{2} = 3$$

$$y_m = \frac{7 + (-5)}{2} = \frac{2}{2} = 1$$

The coordinates of the midpoint of $\overline{BD}$ are $(3, 1)$.

Since the diagonals have the same midpoint, they bisect each other, thereby establishing that the quadrilateral is a parallelogram.

FINDING THE DISTANCE BETWEEN TWO POINTS

The distance between two points in the coordinate plane is the length of the segment that joins the two points. Figures 15.4 and 15.5 illustrate how to find the distance between two points that determine a horizontal or a vertical line. In each case, the distance between the two given points is found by subtracting a single pair of coordi-

nates. The length of a *horizontal* segment is obtained by calculating the difference in the *x*-coordinates of the two points. The length of a *vertical* segment (Figure 15.5) is found by subtracting the *y*-coordinates of the two points.

FIGURE 15.4 **FIGURE 15.5**

The Greek letter Δ (delta) is sometimes used to represent the difference or change in a pair of values. In general, if the coordinates of two points are $P(x_1, y_1)$ and $Q(x_2, y_2)$, then $\Delta x = x_2 - x_1$ and $\Delta y = y_2 - y_1$. For example, if the coordinates of P and Q are $P(1, 7)$ and $Q(5, 9)$, then $\Delta x = 5 - 1 = 4$ and $\Delta y = 9 - 7 = 2$.

THEOREM 15.2 LENGTHS OF HORIZONTAL AND VERTICAL SEGMENTS

$$PQ = \Delta x = x_2 - x_1 \ (x_2 > x_1)$$ $$RW = \Delta y = y_2 - y_1 \ (y_2 > y_1)$$

Finding the distance between two points that determine a slanted line is a bit more complicated. Figure 15.6 demonstrates that the distance between points A and B can be found indirectly by first forming a right triangle in which $\overline{AB}$ is the hypotenuse, and horizontal and vertical segments are legs of the triangle. After the lengths of the legs of this right triangle have been determined, a straightforward application of the Pythagorean Theorem leads to the length of $\overline{AB}$, which represents the distance between points A and B. The general formula for determining the distance between any two given points is stated in Theorem 15.3.

FIGURE 15.6

THEOREM 15.3 THE DISTANCE FORMULA

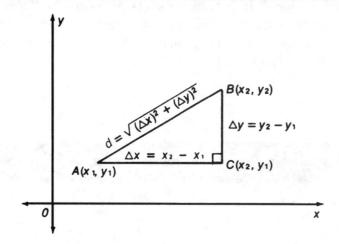

The distance, d, between points $A(x_1, y_1)$ and $B(x_2, y_2)$ may be found using the formula:

$$d = \sqrt{(\Delta x)^2 + (\Delta y)^2}$$

where

$$\Delta x = x_2 - x_1 \quad \text{and} \quad \Delta y = y_2 - y_1$$

Prove that the parallelogram whose coordinates are $A(-3, 0)$, $B(4, 7)$, $C(9, 2)$, and $D(2, -5)$ is a rectangle. (See the diagram of Example 15.6.)

SOLUTION

A parallelogram is a rectangle if its diagonals are equal in length. Let us find and then compare the lengths of diagonals $\overline{AC}$ and $\overline{BD}$.

To find the length of $\overline{AC}$, let $(x_1, y_1) = A(-3, 0)$ and $(x_2, y_2) = C(9, 2)$. Therefore:

$$\Delta x = x_2 - x_1 \qquad\qquad \Delta y = y_2 - y_1$$
$$= 9 - (-3) = 9 + 3 \qquad\qquad = 2 - 0$$
$$= 12 \qquad\qquad\qquad = 2$$

$$AC = \sqrt{(\Delta x)^2 + (\Delta y)^2}$$
$$= \sqrt{(12)^2 + (2)^2}$$
$$= \sqrt{144 + 4}$$
$$= \sqrt{148}$$

To find the length of $\overline{BD}$, let $(x_1, y_1) = B(4, 7)$ and $(x_2, y_2) = D(2, -5)$. Therefore:

$$\Delta x = x_2 - x_1 \qquad\qquad \Delta y = y_2 - y_1$$
$$= 2 - 4 \qquad\qquad\qquad = -5 - 7$$
$$= -2 \qquad\qquad\qquad = -12$$

$$BD = \sqrt{(\Delta x)^2 + (\Delta y)^2}$$
$$= \sqrt{(-2)^2 + (-12)^2}$$
$$= \sqrt{4 + 144}$$
$$= \sqrt{148}$$

Since AC and BD are each equal to $\sqrt{148}$, they must be equal to each other. Hence, $AC = BD$, and therefore parallelogram $ABCD$ is a rectangle.

Slope of a Line

If you imagine that a line represents a hill, it is obvious that some lines will be more difficult to walk up than other lines.

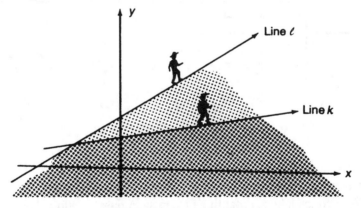

Line ℓ would be more difficult to climb than line k since line ℓ is *steeper* than line k. Another name for steepness is *slope*. The slope of a line may be expressed as a number. To do this, select any two different points on the line. In traveling from one point to the other, compare the change in the vertical distance (Δy) to the change in the horizontal distance (Δx). The ratio of these quantities $\left(\dfrac{\Delta y}{\Delta x}\right)$ represents the slope of the line.

DEFINITION OF SLOPE

The *slope, m,* of a nonvertical line that passes through points $A(x_1, y_1)$ and $B(x_2, y_2)$ is given by the ratio of the change in the values of their y-coordinates to the change in the value of their x-coordinates:

$$\text{Slope} = m = \frac{\Delta y}{\Delta x} = \frac{y_2 - y_1}{x_2 - x_1}$$

(a)

(b)

(c)

(d)

FIGURE 15.7

As the series of graphs in Figure 15.7 illustrates, the steeper the line, the larger the value of its slope. Since a horizontal line has no steepness, its slope has a numerical value of 0. A vertical line is impossible to "walk up," so its slope is not defined. Mathematically speaking:

- The y-coordinate of every point on a *horizontal* line is the same, so $\Delta y = 0$. For any two points on a horizontal line, the slope formula may be expressed as

$$m = \frac{\Delta y}{\Delta x} = \frac{0}{\Delta x}$$

Since 0 divided by any number is 0, the slope of a horizontal line is always 0.

- The x-coordinate of every point on a *vertical* line is the same, so $\Delta x = 0$. For any two points on a vertical line, the slope formula may be expressed as

$$m = \frac{\Delta y}{\Delta x} = \frac{\Delta y}{0}$$

Since division by 0 is not defined, the slope of a vertical line is not defined.

For each of the following pairs of values, plot the points and draw a line through the points. Use the slope formula to determine the slope of each line.

a. $A(4, 2)$ and $B(6, 5)$
b. $P(0, 3)$ and $Q(5, -1)$
c. $J(4, 3)$ and $K(8, 3)$
d. $W(2, 1)$ and $C(2, 7)$

SOLUTION

a. See diagram **a**. To find the slope, observe that:

$$\Delta y = 5 - 2 = 3$$

$$A(4, 2) \qquad\qquad\qquad\qquad B(6, 5)$$

$$\Delta x = 6 - 4 = 2$$

$$\text{Slope of } \overleftrightarrow{AB} = m = \frac{\Delta y}{\Delta x} = \frac{y_2 - y_1}{x_2 - x_1} = \frac{5 - 2}{6 - 4} = \frac{3}{2}$$

The slope of $\overleftrightarrow{AB}$ is $\frac{3}{2}$. The slope of $\overleftrightarrow{AB}$ was calculated assuming that point B was the second point. When using the slope formula either of the two points may be considered the second point. For example, let's repeat the slope calculation considering point A to be the second point (that is, $x_2 = 4$ and $y_2 = 2$):

$$m = \frac{\Delta y}{\Delta x} = \frac{y_2 - y_1}{x_2 - x_1} = \frac{2 - 5}{4 - 6} = \frac{-3}{-2} = \frac{3}{2}$$

The same value, $\frac{3}{2}$, is obtained for the slope of $\overleftrightarrow{AB}$, regardless of whether A is taken as the first or second point.

b. See diagram **b**. Let $(x_1, y_1) = (0, 3)$ and $(x_2, y_2) = (5, -1)$:

$$m = \frac{\Delta y}{\Delta x} = \frac{y_2 - y_1}{x_2 - x_1} = \frac{-1 - 3}{5 - 0} = \frac{-4}{5}$$

The slope of $\overleftrightarrow{PQ}$ is $\frac{-4}{5}$. The slope of $\overleftrightarrow{AB}$ in part **a** was *positive* in value, but the slope of $\overleftrightarrow{PQ}$ has a *negative* value. Compare the directions of these lines.

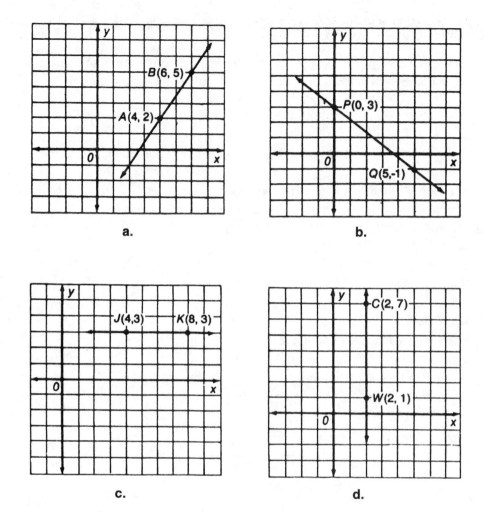

a.

b.

c.

d.

Notice that, if a line has a positive slope, it climbs up *and* to the right; if a line has a negative slope, it falls down *and* to the right. These observations may be restated as follows: If, as you move along a line from left to right, the values of the *x*-coordinates increase and the values of the *y*-coordinates also increase, then the line has a positive slope; but if, as the values of the *x*-coordinates increase, the values of the *y*-coordinates decrease, then the line has a negative slope.

c. See diagram **c**. Since the line is horizontal, there is no change in y, so $\Delta y = 0$.

$$m = \frac{\Delta y}{\Delta x} = \frac{0}{\Delta x} = 0$$

d. See diagram **d**. Since the line is vertical, there is no change in x, so $\Delta x = 0$. The slope is undefined.

A special relationship exists between the slopes of parallel lines. Since two nonvertical lines that are parallel have the same steepness, their slopes are equal. Conversely, if it is known that two nonvertical lines have the same slope, then the lines must be parallel.

THEOREM 15.4 SLOPES OF PARALLEL LINES

- If two nonvertical lines are parallel, then their slopes are equal.
- If two nonvertical lines have the same slope, then they are parallel.

Slope = m Slope = m

EXAMPLE 15.9

Show that the quadrilateral whose vertices are $A(-3, -8)$, $B(-2, 1)$, $C(2, 5)$, and $D(7, 2)$ is a trapezoid.

SOLUTION

Since a trapezoid is a quadrilateral that has *exactly* one pair of parallel sides, we must show that one pair of sides have the same slope and one pair of sides have different slopes.

- Slope of $\overline{AB}$ = $\dfrac{\Delta y}{\Delta x} = \dfrac{1-(-8)}{-2-(-3)} = \dfrac{1+8}{-2+3} = \dfrac{9}{1} = 9$

- Slope of $\overline{BC}$ = $\dfrac{\Delta y}{\Delta x} = \dfrac{5-1}{2-(-2)} = \dfrac{4}{2+2} = \dfrac{4}{4} = 1$

- Slope of $\overline{CD}$ = $\dfrac{\Delta y}{\Delta x} = \dfrac{2-5}{7-2} = \dfrac{-3}{5} = -\dfrac{3}{5}$

- Slope of $\overline{AD}$ = $\dfrac{\Delta y}{\Delta x} = \dfrac{2-(-8)}{7-(-3)} = \dfrac{2+8}{7+3} = \dfrac{10}{10} = 1$

Since the slope of $\overline{BC}$ = the slope of $\overline{AD}$, $\overline{BC}$, $\parallel \overline{AD}$ and $\overline{BC}$ and $\overline{AD}$ are the bases of the trapezoid.

Since the slope of $\overline{AB} \neq$ slope of $\overline{CD}$, $\overline{AB}$ and $\overline{CD}$ are the legs or nonparallel sides of the trapezoid.

A less obvious relationship exists between the slopes of *perpendicular* lines. If two lines are perpendicular, the slope of one line will be the negative reciprocal of the slope of the other line. For example, suppose line ℓ is perpendicular to line k. If the slope of line ℓ is $\dfrac{2}{3}$, then the slope of line k is $-\dfrac{3}{2}$. Conversely, if it is known that the slopes of two lines are negative reciprocals of one another (or, equivalently, that the product of their slopes is negative), then the lines must be perpendicular.

THEOREM 15.5 SLOPES OF PERPENDICULAR LINES

- If two lines are perpendicular, then their slopes are negative reciprocals of one another.
- If the slopes of two lines are negative reciprocals of one another, then the lines are perpendicular.

EXAMPLE 15.10

Show by means of slope that the triangle whose vertices are $A(2, 0)$, $B(11, 8)$, and $C(6, 10)$ is a right triangle.

SOLUTION

A triangle is a right triangle if two sides are perpendicular. Let us find and then compare the slopes of the three sides of the triangle.

- Slope of $\overline{AB}$ = $\dfrac{\Delta y}{\Delta x} = \dfrac{8 - 0}{11 - 2} = \dfrac{8}{9}$

- Slope of $\overline{AC}$ = $\dfrac{\Delta y}{\Delta x} = \dfrac{10 - 0}{6 - 2} = \dfrac{10}{4} = \dfrac{5}{2}$

- Slope of $\overline{BC}$ = $\dfrac{\Delta y}{\Delta x} = \dfrac{10 - 8}{6 - 11} = \dfrac{2}{-5} = -\dfrac{2}{5}$

Notice that the slopes of $\overline{AC}$ and $\overline{BC}$ are negative reciprocals. Hence, $\overline{AC} \perp \overline{BC}$, and angle C of $\triangle ABC$ is a right angle.

Equation of a Line

For each point that lies on a given line, the same relationship between the values of the x- and y-coordinates of these points must hold. For example, the coordinates of three representative points on line ℓ in Figure 15.8 are $(0, 1)$, $(1, 3)$, and $(2, 5)$. In comparing the x- and y-coordinates of these points, do you see a pattern? In each case the y-coordinate is arrived at by multiplying the corresponding x-coordinate by 2 and then adding 1.

A general rule that expresses this relationship is the equation $y = 2x + 1$, which is referred to as *an equation of line ℓ*. The x- and y-coordinates for *every* point on line ℓ in Figure 15.8 must satisfy this equation. Without looking at the figure, can you predict the y-coordinate of a point that lies of line ℓ if its x-coordinate is -2? All you need do is substitute -2 for x in the equation $y = 2x + 1$:

$$y = 2x + 1$$
$$= 2(-2) + 1$$
$$= -4 + 1$$
$$= -3$$

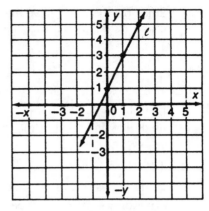

FIGURE 15.8

Now check Figure 15.8. Notice that the point whose coordinates are (–2, –3) lies on line ℓ.

In the equation $y = 2x + 1$, let's see if any special significance can be attached to the numbers 2 and 1. Referring to Figure 15.8, notice that the line crosses the y-axis at the y-value of 1. We call this point the y-*intercept* of a line. The slope of the line can be found by considering any two points on the line, say (1, 3) and (2, 5). For these points, if we let $x_1 = 1$, $y_1 = 3$, $x_2 = 2$, and $y^2 = 5$, we have

$$m = \frac{\Delta y}{\Delta x} = \frac{5-3}{2-1} = \frac{2}{1} = 2$$

The slope of the line is 2. The coefficient of x in the equation $y = 2x + 1$ is also 2. Hence,

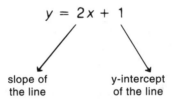

slope of y-intercept
the line of the line

In general, whenever an equation of a nonvertical line is written in the form

$$y = mx + b,$$

the equation is said to be written in *slope-intercept* form since m represents the slope of the line and b is its y-intercept.

Line p is parallel to the line $y + 4x = 3$, and line q is perpendicular to the line $y + 4x = 3$.
 a. What is the slope of line p?
 b. What is the slope of the line q?

SOLUTION
 Write an equation of the given line in $y = mx + b$ form $y = -4x + 3$. The slope of the line is $m = -4$.

a. Since parallel lines have the same slope, the slope of line p is –4.

b. Since perpendicular lines have slopes that are negative reciprocals, the slope of line q is $\frac{1}{4}$.

EXAMPLE 15.12 The line whose equation is $y - 3x = b$ passes through point $A(2, 5)$.

a. Find the slope of the line.

b. Find the y-intercept of the line.

SOLUTION

a. To find the slope of the line, put the equation of the line in $y = mx + b$ form. The equation $y - 3x = b$ may be written as $y = 3x + b$, which means that the slope of the line is 3.

b. In the equation $y = 3x + b$, b represents the y-intercept of the line. Since point A lies on the line, its coordinates must satisfy the equation. To find the value of b, replace x by 2 and y by 5 in the original equation.

$$y - 3x = b$$
$$5 - 3 \cdot 2 = b$$
$$5 - 6 = b$$
$$-1 = b$$

The y-intercept is –1.

WRITING EQUATIONS OF LINES

If a line has a slope of 3 ($m = 3$) and a y-intercept of –4($b = -4$), then an equation of this line may be written by replacing m and b in $y = mx + b$ with their numerical values: $y = 3x - 4$.

EXAMPLE 15.13 Write an equation of the line whose y-intercept is 1 and that is perpendicular to the line $y = \frac{1}{2}x - 4$.

SOLUTION

The slope of the given line is $\frac{1}{2}$. Since the lines are perpendicular, the slope m of the desired line is the negative reciprocal of $\frac{1}{2}$, which is –2. An equation of the desired line is $y = -2x + 1$.

If (1) the slope m of a line and the coordinates of a point $P(a, b)$ on the line, or (2) the coordinates of two points on the line are known, then it is usually more convenient to write an equation of the line using the *point-slope* form:

$$y - b = m(x - a).$$

EXAMPLE 15.14 Write an equation of a line that is parallel to the line $y + 2x = 5$ and passes through point (1, 4).

SOLUTION

Since the lines are parallel, the slope m of the desired line is equal to the slope of the line $y + 2x = 5$. The line $y + 2x = 5$ may be written in the slope-intercept form as $y = -2x + 5$. Therefore, its slope is -2, so $m = -2$.

Method 1 (Slope-Intercept Form)	Method 2 (Point-Slope Form)
Find b by replacing m by -2, x by 1, and y by 4: $y = mx + b$ $4 = -2(1) + b$ $4 = -2 + b$ $6 = b$ Therefore, $y = -2x + 6$	Since $(a, b) = (1, 4)$, replace m by -2, a by 1, and b by 4: $y - b = m(x - a)$ $y - 4 = -2(x - 1)$

Keep in mind that the slope-intercept and point-slope forms of an equation of a line are equivalent. For example, the equation $y - 4 = -2(x - 1)$ may be expressed in $y = mx + b$ form as follows:

$$y - 4 = -2(x - 1)$$
$$= -2x + 2$$
$$y = -2x + 2 + 4$$
$$= -2x + 6$$

 EXAMPLE 15.15 Write an equation of the line that passes through the origin and is perpendicular to the line whose equation is $y = 3x - 6$.

SOLUTION

The slope of the line whose equation is $y = 3x - 6$ is 3, so the slope m of a perpendicular line is $-\dfrac{1}{3}$ (the negative reciprocal of 3). Since the desired line passes through the origin, $(a, b) = (0, 0)$. Therefore,

$$y - b = m(x - a)$$
$$y - 0 = -\frac{1}{3}(x - 0)$$

or

$$y = -\frac{1}{3}x$$

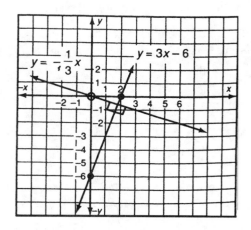

EXAMPLE 15.16 Write an equation of the line that contains points $A(6, 0)$ and $B(2, -6)$.

SOLUTION

Use the point-slope form. First, find the slope m of $\overleftrightarrow{AB}$.

$$m = \frac{-6-0}{2-6} = \frac{-6}{-4} = \frac{3}{2}$$

Next, choose either point and then use its x- and y-coordinates. Consider point A. Then $(a, b) = (6, 0)$.

$$y - b = m(x - a)$$
$$y - 0 = \frac{3}{2}(x - 6)$$
$$y = \frac{3}{2}(x - 6)$$

Multiplying each term inside the parentheses by $\frac{3}{2}$ gives

$$y = \frac{3}{2}x - 9$$

EQUATIONS OF HORIZONTAL AND VERTICAL LINES

Until now we have been considering only *oblique* or *slanted* lines. Oblique lines have nonzero slopes. The slope of a horizontal line is 0, while the slope of a vertical line is undefined. The forms of the equations of horizontal and vertical lines, therefore, require special consideration.

Since each point on a horizontal line has the same y-coordinate, call it b, its equation has the form $y = b$. Note that b is also the y-intercept of the horizontal line, extended, if necessary. Similarly, each point on a vertical line has the same x-coordinate, say a, so that equation of a vertical line takes the form $x = a$. Note that a is the x-intercept of the vertical line, extended, if necessary.

EXAMPLE 15.17

Write an equation of a line that passes through point $A(-1, 2)$ and is parallel to
a. the x-axis
b. the y-axis

SOLUTION

a.

b.

General Form of Equation	Comments
$y = mx + b$	m = slope of line; b = y-intercept
$y - b = m(x - a)$	Form to use when the slope m of the line and the coordinates (a, b) of a point on the line are known. Also use when two points on the line are given, after first using their coordinates to calculate m.
$x = a$	Vertical line (parallel to y-axis) that intersects the x-axis at a.
$y = b$	Horizontal line (parallel to x-axis) that intersects the y-axis at b.

Equation of a Circle

The locus of all points that are 4 units from point $A(2, 3)$ is a circle having A as its center and a radius of 4 units. (See Figure 15.9.) Suppose we wish to find the *equation* that defines this locus. We proceed by choosing a representative point, $P(x, y)$, that satisfies the locus condition. The distance between point A and point P must therefore be 4. A straightforward application of the distance formula, where $\Delta x = x - 2$ and $\Delta y = y - 3$, yields the following result:

$$PA = \sqrt{(\Delta x)^2 + (\Delta y)^2}$$
$$= \sqrt{(x - 2)^2 + (y - 3)^2}$$

Since $PA = 4$, we may write:

$$\sqrt{(x - 2)^2 + (y - 3)^2} = 4$$

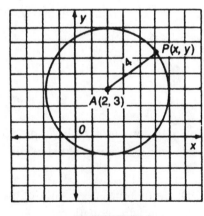

FIGURE 15.9

By squaring both sides of this equation, we obtain

$$(x - 2)^2 + (y - 3)^2 = 16$$

coordinates of the center of the circle

the square of the radius

THEOREM 15.6 EQUATION OF A CIRCLE

A circle whose center is located at $A(p, q)$ and that has a radius r units in length is defined by the equation:

$$(x - p)^2 + (y - q)^2 = r^2$$

If the center of the circle having radius r is the origin, then its equation is

$$x^2 + y^2 = r^2$$

EXAMPLE 15.18

Write an equation for each of the following circles:
a. The center is at $(3, -5)$, and the radius is 7 units in length.
b. The center is at the origin, and the radius is 5 units in length.

SOLUTION
a. Here, $p = 3$, $q = -5$, and $r = 7$.

$$(x - p)^2 + (y - q)^2 = r^2$$
$$(x - 3)^2 + y - (-5)^2 = 7^2$$
$$(x - 3)^2 + (y + 5)^2 = 49$$

b. At the origin, $p = 0$ and $q = 0$. Since $r = 5$,

$$(x - p)^2 + (y - q)^2 = r^2$$
$$(x - 0)^2 + (y - 0)^2 = 5^2$$
$$x^2 + y^2 = 25$$

Notice that in part **b** of Example 15.18 an equation of the form $x^2 + y^2 = r^2$ defines a circle whose center is at the origin and that has a radius r units in length.

EXAMPLE 15.19

Determine the coordinates of the center of a circle and the length of its radius given the following equations:
a. $x^2 + y^2 = 100$
b. $(x + 1)^2 + (y - 4)^2 = 39$

SOLUTION
a. *Center:* origin. *Radius:* $\sqrt{100} = 10$.
b. Rewrite the equation as follows:

$$[x - (-1)]^2 + (y - 4)^2 = 39$$

Center: $(-1, 4)$ *Radius:* $\sqrt{39}$

Proofs Using Coordinates

By placing a polygon in a convenient position in a coordinate plane, and then representing the coordinates of its vertices using letters and zeros, we can easily prove many theorems. We will illustrate this fact by showing how the methods of coordinate geometry can be used to establish that the diagonals of a rectangle are congruent.

Our first concern is to decide where in the coordinate plane a representative rectangle should be placed. Figure 15.10 shows three possible positionings of a rectangle in a coordinate plane. One of these has a clear advantage over the other two placements. We generally look to position a polygon so that:

- The origin is a vertex of the polygon.
- At least one of the sides of the polygon coincides with a coordinate axis.

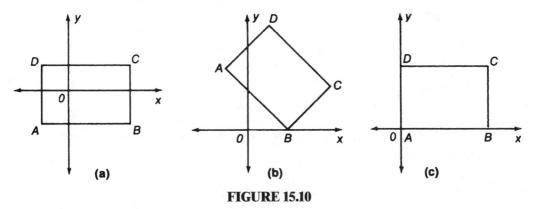

FIGURE 15.10

As we shall see, this positioning tends to simplify matters because it introduces zeros for some of the coordinates of the vertices. Figure 15.10c conforms to the guidelines given above. Since the sides of a rectangle are perpendicular to each other, a pair of adjacent sides of the rectangle can be made to coincide with the coordinate axes so that the origin is a vertex of the rectangle. The actual proof follows.

GIVEN: Rectangle $ABCD$ with diagonals $\overline{AC}$ and $\overline{BD}$.

PROVE: $AC = BD$.

PLAN: Use the distance formula to show $AC = BD$.

PROOF: STEP 1 Find AC.

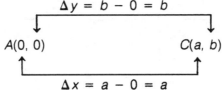

Use the distance formula:

$$AC = \sqrt{(\Delta x)^2 + (\Delta y)^2}$$
$$= \sqrt{(a)^2 + (b)^2}$$
$$= \sqrt{a^2 + b^2}$$

STEP 2 Find *BD*.

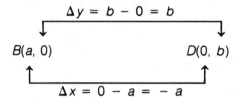

Use the distance formula:

$$BD = \sqrt{(\Delta x)^2 + (\Delta y)^2}$$
$$= \sqrt{(-a)^2 + (b)^2}$$
$$= \sqrt{a^2 + b^2}$$

Since *AC* and *BD* are both equal to $\sqrt{a^2 + b^2}$, they are equal to each other. Since $AC = BD$, $\overline{AC} \cong \overline{BD}$.

 EXAMPLE 15.20 Prove the diagonals of a parallelogram bisect each other.

SOLUTION

Place parallelogram *QRST* in the coordinate plane so that side $\overline{RS}$ coincides with the *x*-axis and one of the end points, point *R*, is the origin, as shown in the diagram. Next, draw $\overline{QT}$ parallel to $\overline{RS}$. Since $\overline{QT}$ is a horizontal line segment, the *y*-coordinates of points *Q* and *T* must be the same, say *b*. Represent the *x*-coordinate of point *Q* by *a*. Since the lengths of $\overline{QT}$ and $\overline{RS}$ must be the same, the *x*-coordinate of point *T* must then be represented by $a + c$.

PLAN: To prove the diagonals of parallelogram *QRST* bisect each other, find and then compare the midpoints of *RT* and *QS*.

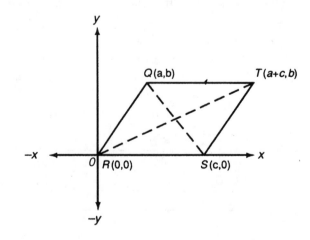

Midpoint of $\overline{RT}$:

$$x_m = \frac{0+(a+c)}{2} = \frac{a+c}{2};$$

$$y_m = \frac{0+b}{2} = \frac{b}{2}$$

The midpoint of diagonal $\overline{RT}$ is $\left(\dfrac{a+c}{2}, \dfrac{b}{2}\right)$.

Midpoint of $\overline{QT}$:

$$x_m = \frac{a+c}{2};$$

$$y_m = \frac{b+0}{2} = \frac{b}{2}$$

The midpoint of diagonal $\overline{QT}$ is $\left(\dfrac{a+c}{2}, \dfrac{b}{2}\right)$.

Since the diagonals have the same midpoint, they bisect each other.

REVIEW EXERCISES FOR CHAPTER 15

In Exercises 1 to 4, draw the triangles whose vertices are given and then find their areas.

1. $A(0, 0)$, $B(5, 0)$, and $C(0, 8)$.

2. $R(-6, 0)$, $S(0, 0)$, and $T(-6, -4)$.

3. $A(-5, 2)$, $B(-3, 6)$, and $C(3, 1)$.

4. $J(2, -7)$, $K(11, 7)$, and $L(8, -3)$.

5. Fill in the following table:

	Point **A**	Point **B**	Midpoint of $\overline{AB}$	Length of $\overline{AB}$
(a)	(3, 8)	(1, 2)	?	?
(b)	(−1, 3)	(−6, −9)	?	?
(c)	(2, 4)	?	(3, −2)	?
(d)	?	(−5, 1)	(−1, −4)	?
(e)	(0, 2a)	(2b, 0)	?	?
(f)	(a, b)	(c, d)	?	?

6. $P(2, -3)$ and $Q(-4, 5)$ are the end points of diameter $\overline{PQ}$ of circle O.
 (a) Find the coordinates of the center of the circle
 (b) Find the length of a radius of the circle.
 (c) Determine whether the point $K(-5, -2)$ lies on the circle.

7. The coordinates of the vertices of quadrilateral *MATH* are $M(-3, 2)$, $A(4, 8)$,

$T(15, 5)$, and $H(8, y)$. If $MATH$ is a parallelogram, find the value of y:
(a) by using midpoint relationships
(b) by using slope relationships

8. The coordinates of the vertices of $\triangle RST$ are $R(4, -4)$, $S(-1, 5)$, and $T(13, 9)$. Find the length of the median drawn to side $\overline{ST}$.

9. The coordinates of the vertices of quadrilateral $JKLM$ are $J(-7, 6)$, $K(-4, 8)$, $L(-2, 5)$, and $M(-5, 3)$. Prove that $JKLM$ is a square. (*Hint:* Show that $JKLM$ is a rectangle having a pair of congruent adjacent sides.)

10. The coordinates of the vertices of a triangle are $J(-10, -6)$, $K(0, 6)$, and $L(6, 1)$.
 (a) By means of slope, show that $\triangle JKL$ is a right triangle.
 (b) Verify your conclusion by showing that the lengths of the sides of $\triangle JKL$ satisfy the converse of the Pythagorean Theorem.
 (c) Find the area of $\triangle JKL$.

11. The coordinates of the vertices of a triangle are $A(2, -3)$, $B(5, 5)$, and $C(11, 3)$.
 (a) Find the slope of a line that is parallel to $\overline{AB}$.
 (b) Find the slope of the altitude drawn to side $\overline{AC}$.
 (c) Find the slope of the median drawn to side $\overline{BC}$.

12. Three points are *collinear* if the slopes of two of the segments determined by selecting any two pairs of points are the same. For example, points A, B, and C are collinear if any one of the following relationships is true: (1) slope of $\overline{AB}$ = slope of $\overline{BC}$; or (2) slope of $\overline{AB}$ = slope of $\overline{AC}$; or (3) slope of $\overline{BC}$ = slope of $\overline{AC}$.
 Determine in each case whether the three points are collinear.
 (a) $A(-4, -5)$, $B(0, -2)$, $C(8, 4)$.
 (b) $R(2, 1)$, $S(10, 7)$, $T(-4, -6)$.
 (c) $J(1, 2)$, $K(5, 8)$, $L(-3, -4)$.

13. Show by means of slope that the quadrilateral whose vertices are $S(-3, 2)$, $T(2, 10)$, $A(10, 5)$, and $R(5, -3)$ is a:
 (a) parallelogram (b) rectangle

14. Write the equation of the line that passes through the point $A(-4, 3)$ and that is parallel to the line $2y - x = 6$.

15. Write the equation of the line that passes through the point $(3, 1)$ and that is perpendicular to the line $3y + 2x = 15$.

16. Write the equation of the line that passes through points A and B with the given coordinates.
 (a) $A(-2, 7)$ and $B(4, 7)$ (b) $A(0, 3)$ and $B(2, 1)$
 (c) $A(6, 8)$ and $B(6, -3)$ (d) $A(1, -3)$ and $B(-1, 5)$

17. Write the equation of the line that contains point $(-5, 2)$ and that is parallel to:
 (**a**) the x-axis (**b**) the y-axis

18. The coordinates of the vertices of a triangle are $A(-1, -4)$, $B(7, 8)$, and $C(9, 6)$. Write the equation of each of the following lines:
 (**a**) The line that passes through point B and is parallel to $\overline{AC}$.
 (**b**) The line that contains the median drawn to side $\overline{BC}$ from vertex A.
 (**c**) The line that contains the altitude drawn to side $\overline{AB}$ from vertex C.

19. Write the equation of the perpendicular bisector of the segment that joins the points $A(3, -7)$ and $B(5, 1)$.

20. The equation of the line that contains the perpendicular bisector of the segment that joins the points $R(-8, t)$ and $S(2, -3)$ is $2y + 5x = k$. Find the values of k and t.

21. Fill in the following table:

	Center	Radius	Equation of Circle
(a)	(0, 0)	4	?
(b)	(3, 4)	6	?
(c)	?	?	$(x - 1)^2 + y^2 = 81$
(d)	?	?	$x^2 + y^2 = 49$
(e)	?	?	$(x - 2)^2 + (y + 5)^2 = 51$

22. The line whose equation is $x = 5$ is tangent to a circle whose center is at the origin. Write the equation of the circle.

23. The center of a circle is located at $O(1, h)$. The line whose equation is $y = kx + 1$ is tangent to circle O at the point $P(3, 6)$. Find the values of k and h.

24. Two circles are tangent externally at point P. The equation of one of the circles is $x^2 + y^2 = 16$. If the other circle has its center on the positive y-axis and has a radius of 5 units, find its equation.

25. The coordinates of the vertices of rectangle $ABCD$ are $A(0, 0)$, $B(0, 4)$, $C(5, 4)$, and $D(5, 0)$. Find the length of a diagonal.

26. Find the area of the quadrilateral whose vertices are given.
 (**a**) $A(-2, 2)$, $B(2, 5)$, $C(8, 1)$, and $D(-1, -2)$.
 (**b**) $T(-4, -4)$, $R(-1, 5)$, $A(6, 5)$, and $P(9, -4)$.
 (**c**) $A(-4, -2)$, $B(0, 5)$, $C(9, 3)$, and $D(7, -4)$.
 (**d**) $M(-3, 5)$, $A(4, 8)$, $T(15, 3)$, and $H(8, -1)$.

27. Given: points $A(1, -1)$, $B(5, 7)$, $C(0, 4)$ and $D(3, k)$.
 (**a**) If $\overline{AB}$ is parallel to $\overline{CD}$, find k.
 (**b**) Write an equation of $\overleftrightarrow{CD}$.
 (**c**) Find the area of quadrilateral $ABDC$.

28. In $\triangle ABC$, the coordinates of A are $(-6, -8)$, of B are $(6, 4)$, and of C are $(-6, 10)$.
 (a) Write an equation of the altitude of $\triangle ABC$ from C to $\overline{AB}$.
 (b) Write an equation of the altitude of $\triangle ABC$ from B to $\overline{AC}$.
 (c) Find the x-coordinate of the point of intersection of the two altitudes in parts **a** and **b**.
 (d) Write an equation of the circle whose center is at the origin and which is tangent to the altitude in part **b**.

29. Quadrilateral *JAKE* has coordinates $J(0, 3a)$, $A(3a, 3a)$, $K(4a, 0)$, $E(-a, 0)$. Prove by coordinate geometry that quadrilateral *JAKE* is an isosceles trapezoid.

In Exercises 30 to 33, prove each theorem using the methods of coordinate geometry.

30. The opposite sides of a parallelogram are congruent.

31. The length of the median drawn to the hypotenuse of a right triangle is equal to one-half the length of the hypotenuse.

32. The segment joining the midpoints of two sides of a triangle is parallel to the third side and one-half of its length.

33. The diagonals of an isosceles trapezoid are congruent.

LOCUS AND CONSTRUCTIONS

WHAT YOU WILL LEARN

Points that satisfy one or more given conditions can be located by drawing a sketch based on geometric principles or by performing a construction using a straightedge to make straight lines and a compass to mark off equal distances. In this chapter you will learn:

- how to locate and describe the set of all points that fit one or more conditions;
- how to use a compass and straightedge to copy angles and segments, to bisect angles and segments, and to construct a line that is perpendicular or parallel to another line.

Describing Points That Fit One Condition

Draw a point on a piece of paper and label it *K*. Take a ruler and locate all points that are 3 inches from point *K*. How many such points can you find? How would you concisely describe the set of *all* points that satisfy the condition that the point be 3 inches from point *K*?

The set of all points that are 3 inches from point *K* forms a *circle* having *K* as its center and a radius of 3 inches. The circle drawn is said to represent the *locus* of all points 3 inches from point *K*.

DEFINITION OF LOCUS

A *locus* (plural: *loci*) is the set of all points, and only those points, that satisfy one or more stated conditions.

Think of a locus as a path consisting of one or more points such that each point along the path conforms to the given condition(s). In our example, the given condition was that the points had to be 3 inches from point *K*. The circular path having point *K* as a center and a radius of 3 inches is the locus.

STEP 1: Draw a diagram that includes all the given information (point *K*) and several representative points that satisfy the given condition (that the point be 3 inches from point *K*).

STEP 2: Keep drawing points until you discover a pattern. Connect the points by using a broken curve or line.

STEP 3: Describe the locus in a complete sentence.

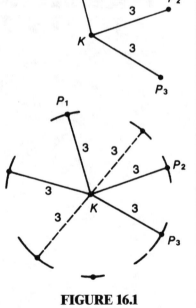

The locus is a circle having point *K* as its center and a radius of 3 inches.

FIGURE 16.1

Figure 16.1 uses our example to summarize the steps to be followed in determining a locus. (Note that the representative points in Figure 16.1 have been labeled P_1, P_2, and P_3. It is a common practice in mathematics to use *subscripts* to name a related set of points. The subscript is the number written one-half line below the letter. The symbol P_1, for example, is read "*P* one" or "*P* sub-one." The symbol P_2 names the point "*P* two" and so on.)

Some additional examples of conditions for determining loci are provided in Table 16.1.

SUMMARY

- The locus of all points at a given distance from a point is a circle having the point as its center and the given distance as the length of its radius.
- The locus of all points equidistant from the sides of an angle is the ray that bisects the angle.
- The loci of all points at a given distance from a line are two lines parallel to the original line, on opposite sides of the original line, and each at the given distance from the line.
- The locus of all points equidistant from two parallel lines is a line parallel to the two lines and halfway between them.
- The locus of all points equidistant from the end points of a line segment is the perpendicular bisector of the segment.
- The loci of all points d units from a circle having a radius of r units are two concentric circles with radii of $r - d$ and $r + d$ units.
- The locus of all points equidistant from two concentric circles is a third concentric circle whose radius is the average of the radii of the given circles.

Describing Points That Fit More Than One Condition

Tree *A* stands 5 meters away from tree *B*. A map indicates that treasure is buried 2 meters from tree *A* (condition 1) *and* 4 meters from tree *B* (condition 2). Where would you dig for the treasure? You can narrow down the possibilities by determining the points at which the loci of the individual conditions intersect. Since the sum of the lengths of the radii of the two circles (6 meters) is greater than the distance between the trees (5 meters), the circles that define the loci for the two conditions will intersect at two different points, as illustrated in Figure 16.2. The treasure is buried at either point *X or* point *Y*.

Suppose the trees were 6 meters apart. Where would you dig? Since the circles that define the loci for the two conditions will be tangent to each other, the desired locus is a single point that corresponds to the point of tangency.

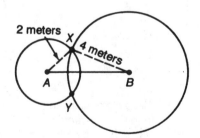

Note: *AB* = 5 meters

FIGURE 16.2

TABLE 16.1

Condition	Diagram	Locus
1. All points that are equidistant from the sides of an angle.		The ray that bisects the angle.
2. All points 4 cm from line ℓ.		Two lines parallel to line ℓ, on opposite sides of ℓ, and a distance of 4 cm from line ℓ.
3. All points that are equidistant from two parallel lines.		A line parallel to the two lines and halfway between them.
4. All points equidistant from the end points of a line segment.		The perpendicular bisector of the segment ($ZA = YA$, for example, since $\triangle ZAM \cong \triangle YAM$ by SAS.)
5. All points d units from a circle having a radius of r units.		Two concentric circles having the same center as the original circle, the smaller circle having a radius of $r - d$ units and the larger circle having a radius of $r + d$ units.
6. All points equidistant from two concentric circles having radii of p and q units.		A circle having the same center as the given circles and a radius of $\dfrac{p + q}{2}$ units.

In general, the loci that satisfy two or more conditions are found by following these steps:

STEP 1: Identify the Given and the conditions.
STEP 2: Draw a diagram that reflects the given information.
STEP 3: Determine the locus that satisfies the first condition.
STEP 4: Determine the locus for each of the remaining conditions, using the same diagram.
STEP 5: Note the points of intersection of the loci (if any).

Two parallel lines are 8 inches apart. Point A is located on one of the lines. Find the number of points that are the same distance from each of the parallel lines *and* that are also at a distance from point A of:

a. 5 inches **b.** 4 inches **c.** 3 inches

SOLUTION

STEP 1: Identify the Given and the conditions.

GIVEN: Two parallel lines 8 inches apart and a point A located on one of them.

CONDITIONS
a. Points must be:
 1. Equidistant from the two lines.
 2. 5 inches from point A

STEP: 2: Draw a diagram that reflects the given information.

STEP 3: The locus of all points equidistant from two parallel lines is a line parallel to the original lines and midway between them.

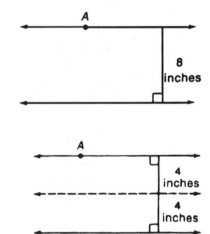

STEP 4: The locus of all points 5 inches from point A is a circle having point A as its center and a radius of 5 inches.

STEP 5: Since the length of the radius of the circle is greater than the distance of the middle line from the line that contains point A, the circle intersects this line in *two* distinct points

Thus, there are two points, P_1 and P_2, that are the same distance from each of the parallel lines and that are 5 inches from point A.

b. Since the circle has a radius of 4 inches, it is tangent to the line that is midway between the original pair of parallel lines. The loci intersect at point *P*. There is one point, *P*, that satisfies both conditions.

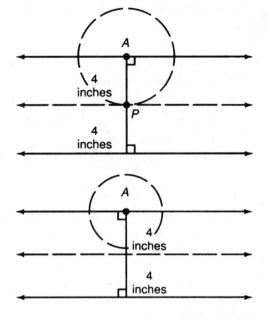

c. Since the circle has a radius of 3 inches, it does *not* intersect the line that is midway between the original pair of parallel lines. Since the loci do *not* intersect, there is no point that satisfies both conditions.

Point P on line ℓ. Find the loci of points that are 2 inches from line ℓ and also 2 inches from point *P*.

SOLUTION

- *Locus condition 1:* All points 2 inches from line ℓ. The desired locus is a pair of parallel lines on either side of line ℓ, each line at a distance of 2 inches from line ℓ.
- *Locus condition 2:* All points 2 inches from point *P*. The desired locus is a circle that has *P* as its center and has a radius of 2 inches. See the accompanying diagram.
- The required loci are the two points, *A* and *B*, that satisfy both conditions.

Locus and Coordinates

Points that satisfy one or more conditions can also be located in the coordinate plane.

Write the equation of the locus of all points whose ordinates exceed their abscissas by 3.

SOLUTION

The locus is a line. The *y*-coordinate (ordinate) of each point on this line is 3 more than the corresponding *x*-coordinate (abscissa). Therefore, the equation $y = x + 3$ completely describes the required locus.

Find the loci of all points 3 units from the *y*-axis.

SOLUTION

The loci of all points 3 units from the *y*-axis (a line)
may be determined by drawing two vertical lines
on either side of the *y*-axis and 3 units from the
y-axis. The loci of all points 3 units from the *y*-axis
are the lines whose equations are $x = -3$ and $x = 3$.

Find an equation of a line that describes the locus of points equidistant from the
lines whose equations are $y = 3x - 1$ and $y = 3x + 5$.

SOLUTION

The lines given by the equations $y = 3x - 1$ and $y = 3x + 5$ are parallel since the
slope of each line is 3. The line parallel to these lines and midway between them
must have a slope of 3 and a *y*-intercept of 2 since 2 is midway between (the aver-
age of) the *y*-intercepts of the original pair of lines. Hence, the locus of points
equidistant from the lines whose equations are $y = 3x - 1$ and $y = 3x + 5$ is the line
whose equation is $y = 3x + 2$.

How many points are 3 units from the origin and 2 units from the *y*-axis?

SOLUTION

- *Locus condition 1:* All points 3 units from the origin. The desired locus is a
 circle with the origin as its center and a radius of 3 units. Note in the accompa-
 nying diagram that the circle intersects each coordinate axis at 3 and –3.

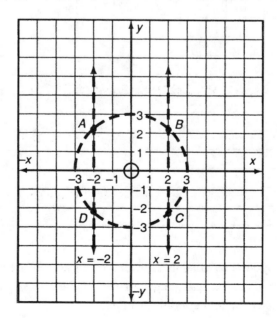

- *Locus condition 2*: All points 2 units from the y-axis. The desired loci are a pair of parallel lines; one line is 2 units to the right of the y-axis ($x = 2$), and the other line is 2 units to the left of the y-axis ($x = -2$). See the accompanying diagram.
- Since the loci intersect at points *A, B, C,* and *D*, there are four points that satisfy both conditions.

GIVEN: Points $A(2, 7)$ and $B(6, 7)$.
 a. Write an equation of $\overline{AB}$.
 b. Describe the locus or loci of points equidistant from:
 (i) points *A* and *B* **(ii)** the x- and y-axes
 c. How many points satisfy both conditions obtained in part **b**?

SOLUTIONS

a. Since the y-coordinates of points *A* and *B* are the same, points *A* and *B* determine a horizontal line, $\overline{AB}$, whose equation is $y = 7$.

b. **(i)** The locus of points equidistant from two points is the perpendicular bisector of the line segment determined by the given points. The midpoint of $\overline{AB}$ is $M(4, 7)$. Since $\overline{AB}$ is a horizontal line, a vertical line that contains $(4, 7)$ will be the perpendicular bisector of $\overline{AB}$. An equation of this line is $x = 4$.

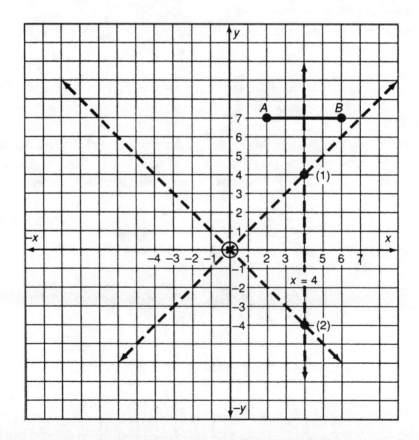

 (ii) The loci of points equidistant from the x- and y-axes are the pair of lines that bisect the pairs of vertical angles formed by the intersecting coordinate axes.

c. Two points satisfy the two conditions in part **b**.

Basic Constructions

Geometric constructions, unlike *drawings*, are made only with a straightedge (for example, an unmarked ruler) and compass. The point at which the pivot point of the compass is placed is sometimes referred to as the *center*, while the fixed compass setting that is used is called the *radius length*.

COPYING SEGMENTS AND ANGLES

Given a line segment or angle, it is possible to construct another line segment or angle that is congruent to the original segment or angle without using a ruler or protractor.

CONSTRUCTION 1 Given line segment $\overline{AB}$, construct a congruent segment.

STEP 1: Using a compass, measure $\overline{AB}$ by placing the compass point on A and the pencil point on B.

STEP 2: Draw any line, and choose any convenient point on it. Label the line as ℓ and the point as C.

STEP 3: Using the same compass setting, place the compass point on C and draw an arc that intersects line ℓ. Label the point of intersection as D.

Conclusion: $\overline{AB} \cong \overline{CD}$.

CONSTRUCTION 2 Given $\angle ABC$, construct a congruent angle.

STEP 1: Using any convenient compass setting, place the compass point on B and draw an arc intersecting $\overrightarrow{BC}$ at X and $\overrightarrow{BA}$ at Y.

STEP 2: Draw any line and choose any point on it. Label the line as ℓ and the point as S.

STEP 3: Using the same compass setting, place the compass point at S and draw arc WT, intersecting line ℓ at T.

STEP 4: Adjust the compass setting to measure the line segment determined by points X and Y by placing the compass point at X and the pencil at Y.

STEP 5: Using the same compass setting, place the compass point at T and construct an arc intersecting arc WT at point R.

STEP 6: Using a straightedge, draw $\overrightarrow{SR}$.

Conclusion: $\angle ABC \cong \angle RST$.

Why it works:
- The arcs are constructed in such a way that $\overline{BX} \cong \overline{ST}$, $\overline{BY} \cong \overline{SR}$, and $\overline{XY} \cong \overline{TR}$.
- Therefore, $\triangle XYB \cong \triangle TRS$ by the SSS postulate.
- By CPCTC, $\angle ABC \cong \angle RST$.

CONSTRUCTION 3 Given two points, construct the perpendicular bisector of the segment determined by the two points.

STEP 1: Label points A and B, and draw $\overline{AB}$. Choose any compass setting (radius length) that is more than one-half the length of $\overline{AB}$.

STEP 2: Using this compass setting, and points A and B as centers, construct a pair of arcs above and below $\overline{AB}$. Label the points at which the pairs of arcs intersect as P and Q.

STEP 3: Draw $\overleftrightarrow{PQ}$ and label the point of intersection of $\overleftrightarrow{PQ}$ and $\overline{AB}$ as M.

Conclusion: $\overleftrightarrow{PQ}$ is the perpendicular bisector of $\overline{AB}$.

Why it works:
- The arcs are constructed in such a way that $AP = BP = AQ = BQ$.
- Since quadrilateral $APBQ$ is equilateral, it is a rhombus.
- Since the diagonals of a rhombus are perpendicular bisectors, $\overline{AM} \cong \overline{BM}$ and $\overline{PQ} \perp \overline{AB}$.

CONSTRUCTION 4 Given an angle, construct the bisector of the angle.

STEP 1: Designate the angle as ∠*ABC*. Using *B* as a center, construct an arc, using any convenient radius length, that intersects $\overrightarrow{BA}$ at point *P* and $\overrightarrow{BC}$ at point *Q*.

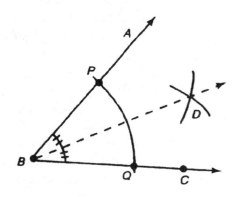

STEP 2: Using points *P* and *Q* as centers and the same radius length, draw a pair of arcs that intersect. Label the point at which the arcs intersect as *D*.

STEP 3: Draw $\overrightarrow{BD}$.

Conclusion: $\overrightarrow{BD}$ is the bisector of ∠*ABC*.

Why it works: Draw $\overline{PD}$ and $\overline{QD}$. Then:
* △*BPD* ≅ △*BQD* by SSS.
* By CPCTC, ∠*ABD* ≅ ∠*CBD*.

CONSTRUCTING PERPENDICULAR LINES

A line can be constructed perpendicular to a given line at a given point on the line or through a given point not on the line.

CONSTRUCTION 5 Given a line ℓ and a point *P* *not* on the line, construct a line through *P* and perpendicular to line ℓ.

STEP 1: Using *P* as a center and any convenient radius length, construct an arc that intersects line ℓ at two points. Label these points as *A* and *B*.

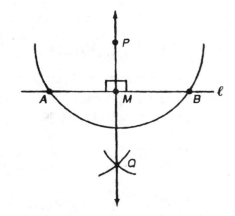

STEP 2: Choose a radius length greater than one-half the length of $\overline{AB}$. Using points *A* and *B* as centers, construct a pair of arcs that intersect at point *Q*.

STEP 3: Draw $\overleftrightarrow{PQ}$, intersecting line ℓ at point *M*.

Conclusion: $\overleftrightarrow{PQ}$ is perpendicular to line ℓ at point *M*.

Why it works: Draw $\overline{PA}$, $\overline{PB}$, $\overline{AQ}$, and $\overline{BQ}$. Then:

- $\triangle PAQ \cong \triangle PBQ$ by SSS.
- By CPCTC, $\angle APM \cong \angle BPM$.
- $\triangle PMA \cong \triangle PMB$ by SAS.
- By CPCTC, $\angle AMP \cong \angle BMP$ so $\overline{PQ} \perp \ell$.

CONSTRUCTION 6 Given a line ℓ and a point P on line ℓ, construct a line through P and perpendicular to line ℓ.

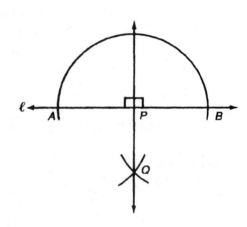

STEP 1: Using P as a center and any convenient radius length, construct an arc that intersects line ℓ at two points. Label these points as A and B.

STEP 2: Choose a radius length greater than one-half the length of $\overline{AB}$. Using points A and B as centers, construct a pair of arcs on either side of line ℓ that intersect at point Q.

STEP 3: Draw $\overleftrightarrow{PQ}$.

Conclusion: $\overleftrightarrow{PQ}$ is perpendicular to line ℓ at point P.

Why it works: Draw $\overline{AQ}$ and $\overline{BQ}$. Then:

- $\triangle PAQ \cong \triangle PBQ$ by SSS.
- By CPCTC, $\angle APQ \cong \angle BPQ$ so $PQ \perp \ell$.

CONSTRUCTING PARALLEL LINES

A line can be constructed parallel to a given line and through a given point not on the line by drawing any convenient transversal through the point and then constructing a congruent corresponding angle, using this point as a vertex.

CONSTRUCTION 7 Given a line $\overleftrightarrow{AB}$ and a point P *not* on the line, construct a line through P and parallel to $\overleftrightarrow{AB}$.

STEP 1: Through P draw any convenient line, extending it so that it intersects $\overleftrightarrow{AB}$. Label the point of intersection as Q.

STEP 2: Using P and Q as centers, draw arcs having the same radius length. Label the point at which the arc intersects the ray opposite $\overrightarrow{PQ}$ as R.

STEP 3: Construct at P an angle, one of whose sides is PR, congruent to $\angle PQB$.

STEP 4: Draw $\overleftrightarrow{PS}$.

Conclusion: $\overrightarrow{PS}$ is parallel to $\overleftrightarrow{AB}$.

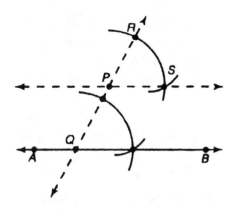

Why it works: $\angle PQB$ and $\angle RPS$ are congruent corresponding angles.

REVIEW EXERCISES FOR CHAPTER 16

In Exercises 1 to 4, 6, 8, and 9, find the loci.

1. The locus of all points 1 centimeter from a circle whose radius is 8 centimeters.

2. The locus of the centers of all circles tangent to each of two parallel lines that are 10 inches apart.

3. The locus of the center of a circle that rolls along a flat surface.

4. The locus of all points equidistant from two concentric circles having radii of 7 centimeters and 11 centimeters.

5. Two circles are externally tangent. One circle remains stationary while the other circle rolls around it so that the two circles remain externally tangent to each other. What is the locus of the center of the moving circle?

6. The locus of the vertices of all isosceles triangles having a common base.

7. Two circles are internally tangent. The inside circle rolls around the outside circle so that the two circles remain internally tangent. What is the locus of the center of the moving circle?

8. The loci of all points equidistant from points A and B and 4 inches from $\overleftrightarrow{AB}$.

9. The loci of all points equidistant from two intersecting lines and 3 inches from their point of intersection.

10. Point A is 4 inches from line k. What is the loci of all points 1 inch from line k and 2 inches from point A?

11. Point X is the midpoint of $\overline{PQ}$. If $PQ = 10$ centimeters, what is the loci of all points equidistant from points P and Q and 5 centimeters from point X?

12. Lines p and q are parallel and 10 centimeters apart. Point A is between lines p and q and 2 centimeters from line q. What is the loci of all points equidistant from lines p and q and d centimeters from point A, if d equals the given number of centimeters?
 (a) 6 cm (b) 1 cm (c) 3 cm

13. Points A and B are d units apart. What is the number of points equidistant from points A and B and r units from point A given the following conditions?

 (a) $r < \dfrac{d}{2}$ (b) $r = \dfrac{d}{2}$ (c) $r > \dfrac{d}{2}$

14. Write the equation(s) that describe the following loci.
 (a) The locus of all points that are 5 units from the x-axis.
 (b) The locus of all points that are 4 units from the y-axis.
 (c) The locus of all points such that their ordinates are twice their abscissas.
 (d) The locus of all points such that the sum of twice their ordinates and three times their abscissas is 6.
 (e) The locus of all points such that their ordinates exceed their abscissas by 5.

15. How many points are 3 inches from line ℓ and also 3 inches from a point on line ℓ?

16. Two parallel lines are 12 inches apart. Point P is located on one of the lines. How many points are equidistant from the parallel lines and are also at a distance of 5 inches from point P?

17. Find the number of points that are 3 inches from point A and 5 inches from point B if:
 (a) $AB = 8$
 (b) $AB = 6$
 (c) $AB = 10$

18. How many points are equidistant from two parallel lines and also equidistant from two points on one of the lines?

19. Give an equation or equations that describe the locus of points:
 (a) equidistant from the lines $x = -1$ and $x = 5$
 (b) equidistant from the lines $y = -3$ and $y = -7$
 (c) equidistant from the lines $y = 3x + 1$ and $y = 3x + 9$

20. Find the number of points that are 3 units from the origin and also 2 units from the x-axis.

21. Find the number of points that are equidistant from points $(-1, 0)$ and $(3, 0)$ and are also 2 units from the origin.

22. The coordinates of point P are $(3, 5)$.
 (a) Describe fully the locus of points at a distance of:
 1 d units from P **2** 1 unit from the y-axis
 (b) How many points satisfy the conditions in part **a** simultaneously for the
 following values of d?
 1 $d = 2$ **2** $d = 4$ **3** $d = 5$

23. The total number of points in a plane that are distance d from a given straight line
 and are also distance r from a certain point on the line can *not* be
 (1) 0 (2) 2 (3) 3 (4) 4

24. Draw obtuse triangle ABC, where $\angle B$ is obtuse. Construct:
 (a) the altitude to side $\overline{BC}$ (extended, if necessary)
 (b) the median to side $\overline{BC}$
 (c) a line through A and parallel to $\overline{BC}$

CUMULATIVE REVIEW EXERCISES: CHAPTERS 12–16

1. Given the points $A(-2, 4)$ and $B(4, -4)$. Find:
 (a) the slope of $\overleftrightarrow{AB}$. (b) the length of $\overline{AB}$.

2. If the diagonals of a rhombus have lengths 4 and 5, what is the area of the rhombus?

3. The length of $\overline{AB}$ is 5. The coordinates of A are $(2, -4)$, the coordinates of B are $(k, -4)$, and k is a positive integer. What is the value of k?

4. From the same external point, two tangents are drawn to a circle. If the tangents intercept arcs whose degree measures are 300° and 60°, what is the measure of the angle formed by the two tangents?

5. What is the perimeter of a regular hexagon inscribed in a circle whose radius is 5?

6. Tangents $\overline{PA}$ and $\overline{PB}$ are drawn to circle O from external point P, and chord $\overline{AB}$ is drawn. If m $\angle APB = 60$ and $AB = 8$, what is the length of $\overline{PB}$?

7. If the number of square centimeters in the area of a circle is equal to the number of centimeters in its circumference, what is the length (in centimeters) of the radius of the circle?

8. What is an equation for the locus of points that are equidistant from:
 (a) $(1, 0)$ and $(7, 0)$? (b) $(-2, -1)$ and $(-2, 5)$?

9. The degree measure of the central angle of a sector of a circle is 120°. If the area of the sector is 27π square units, what is the *circumference* of the circle?

10. In trapezoid $ABCD$, $\overline{AB} \parallel \overline{DC}$, $DC = 4$, $AB = 20$, $AD = 12$, and m $\angle A = 30$. What is the area of trapezoid $ABCD$?

11. In circle O, chords $\overline{PQ}$ and $\overline{RS}$ intersect at T. If $PT = 4$, $TQ = 4$, and $ST = 8$, what is TR?

12. What is the distance between the points whose coordinates are $(3, -2)$ and $(-4, 5)$?

13. In a circle, an arc measures 120°. If the length of the arc is 8π, find the circumference of the circle.

14. In the accompanying diagram, triangles AEB, EBD, and BDC are equilateral. If the perimeter of $ACDE$ is 20, then what is the area of $ACDE$?

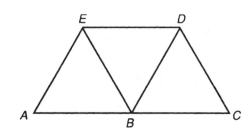

15. In circle, O, chords $\overline{AB}$ and $\overline{CD}$ intersect at point E. If $AE = EB$, $CE = 4$, and $ED = 9$, find the length of $\overline{AB}$.

16. Point M is the midpoint of $\overline{AB}$. If the coordinates of A are $(7, -3)$ and the coordinates of M are $(7, 7)$, what are the coordinates of B?

17. The area of a triangle is 42. If the length of the base is 14, what is the length of the altitude drawn to this base?

18. What is the slope of the line that is parallel to the line whose equation is $2y = 3x - 10$?

19. A regular hexagon with an apothem of 3 centimeters circumscribes a circle. What is the area of the circle in square centimeters?

20. What is an equation of the line that passes through the point $(0, 0)$ and has a slope of $-\dfrac{2}{3}$?

21. Secant $\overline{ADB}$ and tangent $\overline{AC}$ are drawn to circle O from external point A. If $AD = 3$ and $DB = 9$, what is the length of $\overline{AC}$?

22. If two circles with radii 2 and 5 are drawn so that the distance between their centers is 6, what is the maximum number of common tangents they can have?

23. As shown in the diagram, two semicircles of radius 2 are drawn in a square. If the length of a side of the square is 4, what is the area of the shaded portion of the figure?

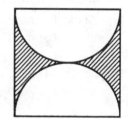

24. In the accompanying diagram, circle O is inscribed in quadrilateral $ABCD$ and $E, F, G,$ and H are the points of tangency of the sides. If $AH = 6$, $DG = 4$, $CF = 2$, and $BF = 3$, what is the perimeter of quadrilateral $ABCD$?

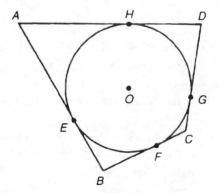

25. Quadrilateral $TRAP$ has vertices $T(0, 0)$, $R(0, 5)$, $A(9, 8)$, and $P(12, 4)$. Prove by coordinate geometry that quadrilateral $TRAP$ is an isosceles trapezoid.

26. In the accompanying figure, point $S(-3, 4)$ lies on circle O with center $(0, 0)$. Line $\overleftrightarrow{ASB}$ and radius $\overline{OS}$ are drawn.

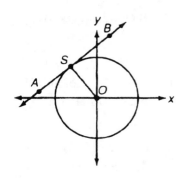

 (a) Find the length of $\overline{OS}$.
 (b) Write an equation of circle O.
 (c) If $\overleftrightarrow{AB} \perp \overline{OS}$, find the slope of $\overleftrightarrow{AB}$.
 (d) Write an equation of line $\overleftrightarrow{ASB}$.
 (e) Find the coordinates of any point on $\overleftrightarrow{AB}$ other than S.

27. Quadrilateral $ABCD$ has vertices $A(-1, 0)$, $B(3, 3)$, $C(6, -1)$, and $D(2, -4)$. Prove that quadrilateral $ABCD$ is a square.

28. The vertices of $\triangle ABC$ are $A(-4, -2)$, $B(2, 6)$, and $C(2, -2)$.
 (a) Write an equation of the locus of points equidistant from vertex B and vertex C.
 (b) Write an equation of the line parallel to $\overline{BC}$ and passing through vertex A.
 (c) Find the coordinates of the point of intersection of the locus in part **a** and the line determined in part **b**.
 (d) Write an equation of the locus of points that are 4 units from vertex C.
 (e) What is the total number of points that satisfy the loci described in parts **a** and **d**?

29. In a trapezoid, the length of one base is 5 times the length of the other base. The height of the trapezoid is 1 less than the length of the shorter base. If the area of the trapezoid is 90 square units, find the length of the *shorter* base.

30. Isosceles trapezoid $ABCD$ with bases $\overline{AB}$ and $\overline{DC}$ is inscribed in circle O. Secant $\overline{RCT}$ intersects $\overline{AB}$ at S, $\overline{RE}$ is tangent to the circle at B, chord $\overline{TB}$ is drawn, $m\angle AST = 50$, $m\angle ABR = 110$, and $m\widehat{AD} = 80$.

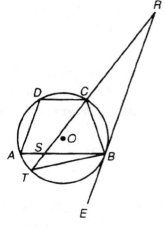

 Find: (a) $m\widehat{DC}$
 (b) $m\widehat{AT}$
 (c) $m\angle TBE$
 (d) $m\angle R$
 (e) $m\angle ADC$

31. The vertices of $\triangle ABC$ are $A(-2, 3)$, $B(0, -3)$, and $C(4, 1)$. Prove, by means of coordinate geometry, that:
 (a) $\triangle ABC$ is isosceles
 (b) The median to side $\overline{BC}$ is also the altitude to side $\overline{BC}$

32. The equation of line $\overleftrightarrow{AB}$ is $x = 2$.
 (a) Describe fully the locus of points d units from line $\overleftrightarrow{AB}$.
 (b) Describe fully the locus of points 1 unit from the origin.
 (c) How many points satisfy the conditions in parts **a** and **b** simultaneously for the following values of d?
 1 $d = 2$ **2** $d = 3$ **3** $d = 4$

33. GIVEN: $\overleftrightarrow{RN}$ tangent to circle T at N, and
diameter $\overline{AN}$.
 PROVE: $(AN)^2 = AR \times AB$

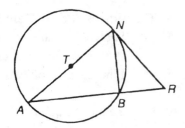

34. GIVEN: Secants $\overline{PAB}$ and $\overline{PCD}$ are drawn to
circle O, $\overline{PAB} \cong \overline{PCD}$, and $\overline{AE}$ and $\overline{CF}$ are
perpendicular to BD at points E and F,
respectively.
 PROVE: $\overline{AE} \cong \overline{CF}$

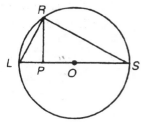

35. In the accompanying diagram, $\triangle LRS$ is inscribed in
circle O. The length of diameter $\overline{LOS}$ is 10. Altitude
$\overline{RP}$ is drawn so that $PO = 3$.
 (a) Find m∡LRS
 (b) Find RP
 (c) Using the result from part **b**, find
 1 m∡L to the *nearest degree*
 2 m∡PRS to the *nearest degree*

36. Regular pentagon $ABCDE$ is inscribed in circle O.
Diameter $\overline{AHJ} \perp \overline{DC}$. Segments $\overline{AHJ}$ and $\overline{ED}$ are
extended and intersect at external point G.
 Find: **(a)** m$\overset{\frown}{BC}$
 (b) m∡BDC
 (c) m∡AHB
 (d) m∡G
 (e) m∡CDG

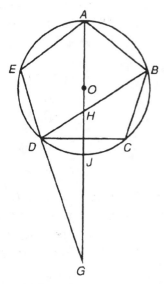

37. Lines $\overleftrightarrow{AB}$ and $\overleftrightarrow{CD}$ are parallel and 4 units apart. Point P is on $\overleftrightarrow{AB}$.
 (a) Describe fully the locus of points equidistant from $\overleftrightarrow{AB}$ and $\overleftrightarrow{CD}$.
 (b) Describe fully the locus of points at a distance of k units from P.
 (c) How many points will satisfy the conditions of both parts **a** and **b** if
 1 $k = 2$ **2** $k = 4$ **3** $k = 1$

38. The vertices of $\triangle ABC$ are $A(-1, -2)$, $B(3, 1)$, and $C(0, 5)$.
 (a) Show, by means of coordinate geometry, that $\triangle ABC$ is a right triangle and state a reason for your conclusion.
 (b) Find the area of $\triangle ABC$.

39. Find the area of quadrilateral $ABCD$ with vertices $A(5, 2)$, $B(0, 5)$, $C(-2, -2)$, and $D(0, 0)$.

40. GIVEN: $\overline{CEB}$ is tangent to circle O at B, diameter $\overline{BA} \parallel \overline{CD}$. Secants $\overline{DFB}$ and $\overline{EFA}$ intersect circle O at F.

 PROVE: $\dfrac{BD}{AE} = \dfrac{CD}{BE}$

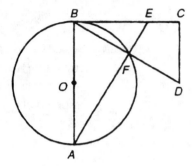

SOME GEOMETRIC FORMULAS WORTH REMEMBERING

RELATIONSHIPS BETWEEN ANGLES FORMED BY PARALLEL LINES

Type of Angle Pair	Angle Relationship	Line $\ell \parallel$ Line m
• Alternate interior angles	• $\angle 3 \cong \angle 6$ and $\angle 4 \cong \angle 5$	
• Corresponding angles	• $\angle 1 \cong \angle 5$ and $\angle 3 \cong \angle 7$; $\angle 2 \cong \angle 6$ and $\angle 4 \cong \angle 8$	
• Interior angles on the same side of the transversal	• $m \angle 3 + m \angle 5 = 180$ and $m \angle 4 + m \angle 6 = 180$	

ANGLE-SUM RELATIONSHIPS

Figure	Sum
• Triangle	• Sum of interior angles = 180
• Quadrilateral	• Sum of interior angles = 360
• n-Sided polygon	• Sum of interior angles = $(n - 2) \times 180$
• n-Sided polygon	• Sum of exterior angles = 360

ANGLE MEASUREMENT IN CIRCLES

Location of Vertex of Angle	Measure of Angle Equals. . .
• At center of circle	• the measure of the intercepted arc
• On circle	• one-half of the measure of the intercepted arc
• Inside circle	• one-half of the sum of the measures of the intercepted arcs
• Outside circle	• one-half of the difference of the measures of the intercepted arcs

CHORD, TANGENT, AND SECANT-SEGMENT RELATIONSHIPS

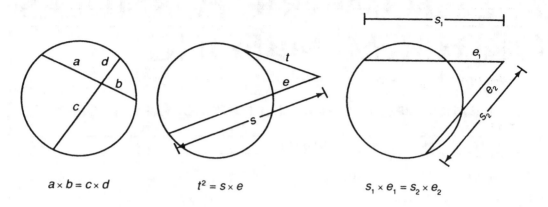

$a \times b = c \times d$ $t^2 = s \times e$ $s_1 \times e_1 = s_2 \times e_2$

AREAS OF POLYGONS

Figure	Area Formula
• Triangle	• $\frac{1}{2} \times$ base $\times$ height
• Equilateral triangle	• $\frac{(\text{side})^2}{4} \times \sqrt{3}$
• Rectangle	• length $\times$ width
• Square	• side $\times$ side or $\frac{1}{2} \times (\text{diagonal})^2$
• Parallelogram	• base $\times$ height
• Rhombus	• $\frac{1}{2} \times$ diagonal$_1 \times$ diagonal$_2$
• Trapezoid	• $\frac{1}{2} \times$ height $\times$ sum of bases
• Regular polygon	• $\frac{1}{2} \times$ apothem $\times$ perimeter

CIRCUMFERENCE AND AREA OF CIRCULAR REGIONS

Quantity	Formula
• Circumference of circle	• $2\pi \times$ radius
• Length of arc formed by central angle of $n°$	• $\frac{n}{360} \times$ circumference of circle
• Area of circle	• $\pi \times (\text{radius})^2$
• Area of sector formed by central angle of $n°$	• $\frac{n}{360} \times$ area of circle

COORDINATE FORMULAS

Description	Formula

Let $A = (x_A, y_A)$ and $B = (x_B, y_B)$

- Midpoint (x_m, y_m) of $\overline{AB}$.

- $x_m = \dfrac{x_A + x_B}{2}$; $y_m = \dfrac{y_A + y_B}{2}$

- Length of $\overline{AB}$.

- $AB = \sqrt{(x_B - x_A)^2 + (y_B - y_A)^2}$

- Slope m of $\overleftrightarrow{AB}$.

- $m = \dfrac{y_B - y_A}{x_B - x_A}$

- $\overleftrightarrow{AB}$ is a horizontal line.
- $\overleftrightarrow{AB}$ is a vertical line.
- $\overleftrightarrow{AB}$ is parallel to $\overleftrightarrow{CD}$.
- $\overleftrightarrow{AB}$ is perpendicular to $\overleftrightarrow{CD}$.

- Slope of $\overleftrightarrow{AB} = 0$.
- Slope of $\overleftrightarrow{AB}$ is undefined.
- Slope of $\overleftrightarrow{AB}$ = slope of $\overleftrightarrow{CD}$.
- Slope of $\overleftrightarrow{AB} = \dfrac{-1}{\text{slope of } \overleftrightarrow{CD}}$.

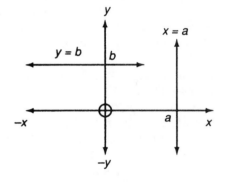

Equation of horizontal line:
$$y = b$$
Equation of vertical line:
$$x = a$$

Equation of oblique line:
$$y = mx + b$$
or
$$y - y_1 = m(x - x_1)$$

Equation of circle having radius r and center:

- at the origin

- $x^2 + y^2 = r^2$

- at (h, k)

- $(x - h)^2 + (y - k)^2 = r^2$

GLOSSARY

AA Theorem of Similarity Two triangles are similar if two pairs of corresponding angles are congruent.

AAS Theorem of Congruence Two triangles are congruent if two pairs of corresponding angles are congruent and a nonincluded pair of corresponding sides are congruent.

Abscissa The x-coordinate of a point in the coordinate plane.

Absolute value For a number x, denoted by $|x|$, its distance from 0 on the number line. Thus, $|x|$ always represents a nonnegative number.

Acute angle An angle whose degree measure is greater than 0 and less than 90.

Acute triangle A triangle with three acute angles.

Adjacent angles Two angles that have the same vertex and share one side, but do not have any interior points in common.

Alternate interior angles Pairs of angles formed when a transversal intersects two lines. The two angles in each pair are between the two lines, have different vertices, and lie on opposite sides of the transversal.

Altitude A segment that is perpendicular to the side of the figure to which it is drawn.

Angle The union of two rays that have the same end point.

Angle bisector A line or any part of a line that contains the vertex of an angle and that divides the angle into two congruent angles. An angle has exactly one angle bisector.

Angle of depression An angle formed by a horizontal ray of sight and the ray that is the line of sight to an object below the horizontal ray.

Angle of elevation An angle formed by a horizontal ray of sight and the ray that is the line of sight to an object above the horizontal ray.

Apothem For a regular polygon, the radius of its inscribed circle.

Arc of a chord The minor arc of a circle whose end points are the end points of the chord. If the chord is a diameter, then either semicircle is an arc of the diameter.

Area For a plane geometric figure, the number of square units it contains.

ASA Theorem of Congruence Two triangles are congruent if two pairs of corresponding angles are congruent and the sides included by these angles are congruent.

Base angles of an isosceles triangle The congruent angles that lie opposite the congruent sides of an isosceles triangle.

Base of an isosceles triangle The noncongruent side of the isosceles triangle.

Bases of a trapezoid The parallel sides of a trapezoid.

Betweenness of points A term that refers to the order of three collinear points. If A, B, and C are three different collinear points, point C is between points A and B if $AC + CB = AB$.

Bisect To divide into two equal parts.

Center of a regular polygon The common center of the circles inscribed and circumscribed in the polygon.

Central angle of a circle An angle whose vertex is at the center of a circle and whose sides are radii.

Central angle of a regular polygon An angle whose vertex is the center of the regular polygon and whose sides terminate at consecutive vertices of the polygon.

Centroid of a triangle The point at which the three medians of the triangle intersect.

Chord of a circle A segment whose end points are on the circle.

Circle The set of all points in a plane at a fixed distance from a given point called the *center*. The fixed distance is called the *radius* of the circle. An equation of a circle with center at point (h, k) and radius length r is $(x - h)^2 + (y - k)^2 = r^2$.

Circumference of a circle The distance around the circle.

Circumscribed circle about a polygon A circle that passes through each vertex of the polygon.

Circumscribed polygon about a circle A polygon that has all of its sides tangent to the circle.

Collinear points Points that lie on the same line.

Common external tangent to two circles A line that is tangent to both circles and does not intersect the line segment whose end points are the centers of the two circles.

Common internal tangent to two circles A line that is tangent to both circles and intersects the line segment whose end points are the centers of the two circles.

Complementary angles Two angles whose measures add up to 90.

Concentric circles Circles in the same plane that have the same center but have radii of different lengths.

Congruent angles Angles that have the same measure.

Congruent circles Circles with congruent radii.

Congruent line segments Line segments that have the same length.

Congruent polygons Polygons with the same number of sides that have the same size and the same shape. The symbol for congruence is $\cong$.

Congruent triangles Triangles whose vertices can be paired so that any one of the following conditions is true: (1) the sides of one triangle are congruent to the corresponding sides of the other triangle ($SSS \cong SSS$); (2) two sides and the included angle of one triangle are congruent to the corresponding parts of the other triangle ($SAS \cong SAS$); (3) two angles and the included side of one triangle are congruent to the corresponding parts of the other triangle ($ASA \cong ASA$); (4) two angles and the side opposite one of these angles of one triangle are congruent to the corresponding parts of the other triangle ($AAS \cong AAS$). Two right triangles are congruent if the hypotenuse and a leg of one right triangle are congruent to the corresponding parts of the other triangle ($Hy - Leg \cong Hy - Leg$).

Converse of a conditional statement Another conditional statement formed by interchanging the hypothesis ("Given") with the conclusion ("To Prove") of the original statement.

Convex polygon A polygon each of whose interior angles measures less than 180.

Coordinate plane A plane that is divided into four equal regions, called *quadrants*, by a horizontal number line and a vertical number line, called *axes*, intersecting at their zero points, called the *origin*. Each point in a coordinate plane is located by an ordered pair of numbers of the form (x, y). The first member, x, of the ordered pair gives the directed distance of the point from the zero point of the x-axis (horizontal). The second member, y, of the ordered pair gives the directed distance of the point from the zero point of the y-axis.

Corollary A theorem that can be easily proved by means of a closely related theorem.

Corresponding angles Pairs of angles formed when a transversal intersects two lines. The two angles in each pair lie on the same side of the transversal, but one angle is between the two lines and the other angle is exterior to the two lines.

Cosine of an acute angle of a right triangle The ratio of the length of the leg that is adjacent to the acute angle to the length of the hypotenuse.

Decagon A polygon with 10 sides.

Deductive reasoning A step-by-step process by which a set of accepted facts is used to arrive at a conclusion.

Degree A unit of angle measure. One degree is the measure of an angle formed by 1/360 of one complete rotation of a ray about its end point.

Diagonal of a polygon A line segment whose end points are nonconsecutive vertices of the polygon.

Diameter of a circle A chord of the circle that contains the center of the circle.

Distance from a point to a line The length of the perpendicular segment from the point to the line.

Distance formula A formula used to find the length of the segment determined by two points in the coordinate plane. The distance d between two points $A(x_A, y_A)$ and $B(x_B, y_B)$ is given by the formula $d = \sqrt{(x_B - x_A)^2 + (y_B - y_A)^2}$

Dodecagon A polygon with 12 sides.

Equiangular polygon A polygon in which all of the angles have the same measure.

Equiangular triangle A triangle in which all three angles have the same measure.

Equidistant having the same distance.

Equilateral polygon A polygon in which all of the sides have the same length.

Equilateral triangle A triangle whose three sides have the same length.

Exterior angle of a polygon An angle formed by a side of the polygon and the extension of an adjacent side of the polygon.

Externally tangent circles Tangent circles that lie on opposite sides of the common tangent.

Extremes The first and fourth terms of a proportion. In the proportion $\frac{a}{b} = \frac{c}{d}$, a and d are the extremes.

Geometric mean See **Means.**

Half-plane The set of points in a plane that lie on one side of a line.

Hexagon A polygon with six sides.

Hypotenuse The side of a right triangle that is opposite the right angle.

Incenter of a triangle The point at which the three angle bisectors of the triangle intersect.

Indirect proof A method of proof in which each possibility except the one that needs to be proved is eliminated by showing that it contradicts some known or given fact.

Inscribed angle of a circle An angle whose vertex lies on the circle and whose sides are chords of the circle.

Inscribed circle of a polygon A circle that is tangent to each side of the polygon.

Inscribed polygon A polygon that has all of its vertices on a circle.

Internally tangent circles Tangent circles that lie on the same side of the common tangent.

Isosceles trapezoid A trapezoid whose non-parallel sides, called *legs*, have the same length.

Isosceles triangle A triangle with two sides, called *legs*, that have the same length.

Leg of a right triangle Either of the two sides of the right triangle that are not opposite the right angle.

Line A term undefined in geometry; a line can be described as a continuous set of points forming a straight path that extends indefinitely in two opposite directions.

Line of centers of two circles The line segment whose end points are the centers of the circles.

Line segment Part of a line that consists of two different points on the line, called *end points*, and the set of all points on the line that are between them. The notation $\overline{AB}$ refers to the line segment with end points A and B, while AB refers to the length of $\overline{AB}$.

Locus (plural: **loci**) The set of all points, and only those points, that satisfy a given condition.

Major arc An arc of a circle whose degree measure is greater than 180.

Means The two middle terms of a proportion. In the proportion $\frac{a}{b} = \frac{c}{d}$, the terms b and c are

the means. If $b = c$, then either b or c is called the *mean proportional* between a and d, which are called the *extremes*.

Median of a trapezoid A line segment whose end points are the midpoints of the legs of the trapezoid.

Median of a triangle A line segment whose end points are a vertex of the triangle and the midpoint of the side opposite that vertex.

Midpoint formula A formula used to find the coordinates of the midpoint of a line segment in the coordinate plane. The midpoint of a line segment whose end points are $A(x_A, y_A)$ and $B(x_B, y_B)$ is $\left(\frac{x_A + x_B}{2}, \frac{y_A + y_B}{2} \right)$.

Midpoint of a line segment The point on the line segment that divides the segment into two segments that have the same length.

Minor arc An arc of a circle whose degree measure is less than 180.

Obtuse angle An angle whose degree measure is greater than 90 and less than 180.

Obtuse triangle A triangle that contains an obtuse angle.

Octagon A polygon with eight sides.

Opposite rays Two rays that have the same end point and form a line.

Ordinate The y-coordinate of a point in the coordinate plane.

Origin The zero point on a number line.

Orthocenter of a triangle The point at which the three altitudes of the triangle intersect.

Parallel lines Lines in the same plane that do not intersect.

Parallelogram A quadrilateral that has two pairs of parallel sides.

Pentagon A polygon with five sides.

Perimeter of a polygon The sum of the lengths of the sides of the polygon.

Perpendicular bisector of a segment A line, ray, or line segment that is perpendicular to the segment at its midpoint.

Perpendicular lines Two lines that intersect at 90° angles.

Plane A term undefined in geometry; a plane can be described as a flat surface that extends indefinitely in all directions.

Point A term undefined in geometry; a point can be described as a dot with no size that indicates location.

Polygon A closed figure in a plane whose sides are line segments that intersect at their end points.

Postulate A statement whose truth is accepted without proof.

Proportion An equation that states that two ratios are equal. In a proportion the product of the means equals the product of the extremes. Hence, if $\frac{a}{b} = \frac{c}{d}$, then $b \times c = a \times d$.

Pythagorean theorem In a right triangle, the sum of the squares of the lengths of the legs is equal to the square of the length of the hypotenuse.

Quadrant of a coordinate plane One of the four equal rectangular regions into which the coordinate plane is divided.

Quadrilateral A polygon with four sides.

Radius (plural: radii) **of a circle** A line segment whose end points are the center of the circle and any point on the circle.

Radius of a regular polygon The radius of its circumscribed circle.

Ratio A comparison of two numbers by division. The ratio of a to b can be represented by the fraction $\frac{a}{b}$, provided that $b \neq 0$.

Ratio of similitude of two similar polygons The constant ratio of the lengths of any two corresponding sides.

Ray The part of a line that consists of a fixed point, called an *end point*, and the set of all points on one side of the end point.

Reciprocal of a nonzero number The number that, when multiplied by the original number, gives 1. For example, the reciprocal of 5 is $\frac{1}{5}$ since $5 \times \frac{1}{5} = 1$.

Rectangle A parallelogram with four right angles.

Rhombus A parallelogram with four sides that have the same length.

Right angle An angle whose degree measure is 90.

Right triangle A triangle that contains a right angle.

SAS Theorem of Congruence Two triangles are congruent if two pairs of corresponding sides are congruent and the angles formed by these sides are congruent.

Scalene triangle A triangle in which no two sides have the same length.

Secant of a circle A line that intersect the circle in two different points.

Semicircle An arc whose end points are a diameter of the circle.

Similar polygons Figures that have the same shape but may have different sizes. Two polygons with the same number of sides are similar if corresponding angles are congruent and the lengths of corresponding sides are in proportion.

Sine of an acute angle of a right triangle The ratio of the length of the leg that is opposite the acute angle to the length of the hypotenuse.

Slope A numerical measure of the steepness of a nonvertical line. The slope of a line is the difference of the y-coordinates of any two different points on the line divided by the difference of the corresponding x-coordinates of the two points. The slope of a horizontal line is 0, and the slope of a vertical line is not defined.

Slope formula A formula used to calculate the slope of a nonvertical line when the coordinates of two points on the line are given. The slope m of a nonvertical line that contains points $A(x_A, y_A)$ and $B(x_B, y_B)$ is given by the formula $m = \frac{y_B - y_A}{x_B - x_A}$.

Slope-intercept form of an equation of a nonvertical line An equation that has the form $y = mx + b$, where m is the slope of the line and b is the y-coordinate of the point at which the line crosses the y-axis.

Square A rectangle all of whose sides have the same length.

SSS Theorem of Congruence Two triangles are congruent if three pairs of corresponding sides are congruent.

Supplementary angles Two angles whose measures add up to 180.

Tangent circles Circles in the same plane that are tangent to the same line at the same point.

Tangent ratio of an acute angle of a right triangle The ratio of the length of the leg that is opposite a given acute angle to the length of the leg that is adjacent to the same angle.

Tangent to a circle A line that intersects the circle in exactly one point, called the *point of tangency*.

Theorem A generalization that can be proved.

Transversal A line that intersects two lines at different points.

Trapezoid A quadrilateral with exactly one pair of parallel sides.

Triangle A polygon with three sides.

Undefined term A term that can be described but is so basic that it cannot be defined. The

terms *point, line,* and *plane* are undefined in geometry.

Vertex (plural: vertices) of a polygon The point at which two sides of the polygon intersect.

Vertex angle of an isosceles triangle The angle formed by the congruent sides of the isosceles triangle.

Vertical angles Pairs of nonadjacent (opposite) angles formed by two intersecting lines.

Volume of a solid figure The capacity of the solid figure as measured by the number of cubic units it contains.

***x*-axis** The horizontal number line in the coordinate plane.

***x*-coordinate** The first number in the ordered pair that represents the coordinates of a point in the coordinate plane. The x-coordinate gives the directed horizontal distance of the point from the origin.

***y*-axis** The vertical number line in the coordinate plane.

***y*-coordinate** The second number in the ordered pair that represents the coordinates of a point in the coordinate plane. The y-coordinate gives the directed vertical distance of the point from the origin.

***y*-intercept** The y-coordinate of the point at which a nonvertical line crosses the y-axis.

ANSWERS TO CHAPTER EXERCISES

CHAPTER 1

1. **(a)** $\overrightarrow{BA}$, $\overrightarrow{BE}$, $\overrightarrow{BC}$, $\overrightarrow{BD}$.
 (b) $\overleftrightarrow{BD}$, $\overleftrightarrow{BE}$, $\overleftrightarrow{DE}$
 (c) $\overleftrightarrow{AB}$, $\overleftrightarrow{BC}$, $\overleftrightarrow{AC}$.
 (d) Angles *ABE, EBC, CBD,* and *DBA.*
 (e) $\overrightarrow{BE}$ and $\overrightarrow{BD}$; $\overrightarrow{BA}$ and $\overrightarrow{BC}$.
2. **(a)** *D* **(b)** *A* **(c)** *B*
3. **(a)** $\angle AEF$ **(b)** $\angle BCA$ **(c)** $\angle EFA$
4. *B, N, R,* and *W.*
5. $\overline{NB}$ and $\overline{NW}$.
6. Triangles *ABN, NKR, TKR, TRW, NTR,* and *NTW.*
7. Angles *NRK, KRT, TRW, KRW, NRW,* and *NRT.*
8. Angle *ANT.*
9. $\overrightarrow{NB}$ and $\overrightarrow{NW}$; $\overrightarrow{RB}$ and $\overrightarrow{RW}$.
10. $\overline{TR}$ is a side of $\triangle TKR$ and $\triangle TWR$. Also, $\overline{KR}$ is a side of $\triangle NKR$ and $\triangle TKR$.
11. $\overline{RN}$ and $\overline{RW}$.
12. **(a)** An angle whose measure is less than 90° is an acute angle.
 (b) A triangle having three sides equal in length is an equilateral triangle.
 (c) A ray (or segment) that divides an angle into two congruent angles is the bisector of the angle.
13. **(a)** Inductive. **(b)** Deductive.
 (c) Inductive. **(d)** Deductive.
14. The length of the median drawn to the hypotenuse of a right triangle is equal to one-half the length of the hypotenuse
15. The medians intersect at the same point.
16. **(a)** Henry has green eyes.
 (b) The measure of the third angle of the triangle is 80.
17. **(a)** The formula generates prime numbers for integer values of *N* from $n = 0$ to $n = 15$.
 (b) For $n = 16$, the formula produces the value 289, which is not a prime number since it is divisible by 17.

CHAPTER 2

1. **(a)** Acute. **(b)** Straight. **(c)** Right.
 (d) Acute. **(e)** Right. **(f)** Obtuse.
 (g) Acute. **(h)** Obtuse.
2. 8
3. *Z.*
4. $m\angle RPL = 10$.
5. 12
6. 42
7. Right.
8. 20
9. $\overline{EY} \cong \overline{AS}$; $\angle PYS \cong \angle IAE$, $\angle YEP \cong \angle ASI$.
10. $\overline{AF} \cong \overline{CF}$.
11. $\angle STP \cong \angle OTP$.
12. $\overline{BE} \cong \overline{DE}$; $\angle ADB \cong \angle CDB$.
13. Transitive.
14. Reflexive.
15. Transitive.
16. Substitution.
17. Addition
18. Subtraction.
19. Addition.
20. Addition.
21. Halves of equals are equal.
22. Subtraction.
23. Subtraction.
24. Addition.
25. 3. Reflexive.
 4. Addition.
26. 1. Given.
 2. Congruent angles are equal in measure.
 3. Subtraction.
 4. Substitution.
 5. Angles equal in measure are congruent.
27. 1. Given
 2. Reflexive.
 3. Subtraction.
 5. Given.
 6. Congruent segments are equal in measure.
 7. Reflexive.
 8. Subtraction.
 10. Segments equal in measure are congruent.

28. 2. Congruent angles are equal in measure.
 3. Given.
 4. Definition of angle bisector.
 5. Halves of equals are equal.
 6. Angles equal in measure are congruent.

CHAPTER 3

1. Adjacent angle pairs 1 and 4; 2 and 3; 4 and 3; 5 and 6; 7 and 8. Vertical angle pairs 1 and 2; 9 and 10.
2. 75
3. 136
4. 45
5. The measure of the angle is 52 and its complement has measure 38.
6. 56
7. **(a)** 9 **(b)** 40 **(c)** 18
8. 143
9. True (provided the measures of the angles that form the vertical pair are less than 90).
10. True.
11. True.
12. True.
13. False. Consider triangle ABC in which $AC = BC$ Point C is equidistant from points A and B, but point C is not the midpoint of side $\overline{AB}$.
14. (1) $\angle 1 \cong \angle 4$ (Given);
 (2) $\angle 2 \cong \angle 1$ and $\angle 3 \cong \angle 4$ (Vertical angles are congruent);
 (3) $\angle 2 \cong \angle 3$ (Transitive property).
15. (1) BD bisects $\angle ABC$ (Given);
 (2) $\angle 3 \cong \angle 4$ (Definition of angle bisector);
 (3) $\angle 1 \cong \angle 2$ (Supplements of congruent angles are congruent).
16. (1) $\angle 3$ is complementary to $\angle 1$, $\angle 4$ is complementary to $\angle 2$ (Given);
 (2) $\angle 1 \cong \angle 2$ (Vertical angles are congruent);
 (3) $\angle 3 \cong \angle 4$ (Complements of congruent angles are congruent)
17. (1) $\overline{AB} \perp \overline{BD}$, $\overline{CD} \perp \overline{BD}$, $\angle 2 \cong \angle 4$ (Given);
 (2) $\angle 1$ is complementary to $\angle 2$ and $\angle 3$ is complementary to $\angle 4$ (Adjacent angles whose exterior sides are perpendicular are complementary);
 (3) $\angle 1 \cong \angle 3$ (Complements of congruent angles are congruent).
18. (1) $\overline{KL} \perp \overline{JM}$, $\overline{KL}$ bisects $\angle PLQ$ (Given);
 (2) $\angle 1$ is complementary to $\angle 2$ and $\angle 4$ is complementary to $\angle 3$ (Adjacent angles whose exterior sides are perpendicular are complementary);
 (3) $\angle 2 \cong \angle 3$ (Definition of angle bisector);

(4) $\angle 1 \cong \angle 4$ (Complements of congruent angles are congruent).

19. (1) $\overline{NW} \perp \overline{WT}$, $\overline{WB} \perp \overline{NT}$, $\angle 4 \cong \angle 6$ (Given);
 (2) $\angle 4 \cong \angle 3$ (Vertical angles are congruent);
 (3) $\angle 3 \cong \angle 6$ (Transitive);
 (4) $\angle 5$ is complementary to $\angle 6$ and $\angle 2$ is complementary to $\angle 3$ (Adjacent angles whose exterior sides are perpendicular are complementary);
 (5) $\angle 5 \cong \angle 2$ (Complements of congruent angles are congruent).
20. (1) $\overline{MT}$ bisects $\angle ETI$, $\overline{KI} \perp \overline{TI}$, $\overline{KE} \perp \overline{TE}$, $\angle 3 \cong \angle 1$, $\angle 5 \cong \angle 2$ (Given);
 (2) $\angle 1 \cong \angle 2$ (Definition of angle bisector);
 (3) $\angle 3 \cong \angle 5$ (Transitive);
 (4) $\angle 4$ is complementary to $\angle 3$ and $\angle 6$ is complementary to $\angle 5$ (Adjacent angles whose exterior sides are perpendicular are complementary);
 (5) $\angle 4 \cong \angle 6$ (Complements of congruent angles are congruent);

CHAPTER 4

1. Alternate interior.
2. Corresponding.
3. Corresponding.
4. **(a)** $x = 112, y = 68$. **(b)** $x = 104, y = 104$.
 (c) $x = 48, y = 80$. **(d)** $x = 45, y = 117$.
5. **(a)** 22 **(b)** 110 **(c)** 60
6. **(a)** 138 and 42 **(b)** 144 and 36
7. **(a)** 55 **(b)** 10
8. 30
9. **(a)** $m\angle 2 = 80$ $m\angle 8 = 100$
 $m\angle 3 = 100$ $m\angle 9 = 80$
 $m\angle 4 = 80$ $m\angle 10 = 130$
 $m\angle 5 = 80$ $m\angle 11 = 50$
 $m\angle 6 = 50$ $m\angle 12 = 130$
 $m\angle 7 = 50$ $m\angle 13 = 50$
 (b) $m\angle 1 = 127$ $m\angle 8 = 127$
 $m\angle 2 = 53$ $m\angle 9 = 53$
 $m\angle 3 = 127$ $m\angle 10 = 116.5$
 $m\angle 4 = 53$ $m\angle 11 = 63.5$
 $m\angle 5 = 53$ $m\angle 12 = 116.5$
 $m\angle 6 = 63.5$ $m\angle 13 = 63.5$
 $m\angle 7 = 63.5$
10. **(a)** If I live in the United States, then I live in New York. (False.)
 (b) If two angles are equal in measure, then the angles are congruent. (True.)
 (c) If two angles are congruent, then they are vertical angles. (False.)

(d) If the sum of the measures of two angles is 90, then the angles are complementary. (True.)

(e) If two angles have the same vertex, then they are adjacent. (False.)

(f) If two lines are parallel, then they are perpendicular to the same line. (False.)

11. (1) $\overline{BA} \parallel \overline{CF}$, $\overline{BC} \parallel \overline{ED}$ (Given);
 (2) $\angle 1 \cong \angle C$ (Postulate 4 1);
 (3) $\angle C \cong \angle 2$ (Theorem 4 2);
 (4) $\angle 1 \cong \angle 2$ (Transitive).

12. (1) $\overline{LT} \parallel \overline{WK}$, $\overline{AP}$, $\overline{PL} \parallel \overline{AG}$ (Given);
 (2) $\angle 1 \cong \angle KWP$ (Theorem 4.2);
 (3) $\angle KWP \cong \angle WPA$ (Postulate 4.1);
 (4) $\angle 1 \cong \angle WPA$ (Transitive);
 (5) $\angle WPA \cong \angle 2$ (Postulate 4.1);
 (6) $\angle 1 \cong \angle 2$ (Transitive).

13. (2)

14. (1) $\overline{QD} \parallel \overline{UA}$, $\overline{QU} \parallel \overline{DA}$ (Given);
 (2) $m\angle 1 = m\angle 4$, $m\angle 2 = m\angle 3$ (Postulate 4 1);
 (3) $m\angle 1 + m\angle 2 = m\angle 3 + m\angle 4$ (Addition);
 (4) $m\angle QUA = m\angle ADQ$ (Substitution);
 (5) $\angle QUA \cong \angle ADQ$ (Angles equal in measure are congruent).

15. (1) $\overline{AT} \parallel \overline{MH}$, $\angle M \cong \angle H$ (Given);
 (2) $\angle A$ is supplementary to $\angle M$ and $\angle T$ is supplementary to $\angle H$ (Theorem 4.3);
 (3) $\angle A \cong \angle T$ (Supplements of congruent angles are congruent).

16. (1) $\overline{IB} \parallel \overline{ET}$, IS bisects $\angle EIB$, $\overline{EC}$ bisects $\angle TEI$ (Given);
 (2) $m\angle BIS = \frac{1}{2} m\angle EIB$ and $m\angle TEC = \frac{1}{2} m\angle TEI$ (Definition of angle bisector);
 (3) $m\angle EIB = m\angle TEI$ (Postulate 4.1);
 (4) $m\angle BIS = m\angle TEC$ (Halves of equals are equal);
 (5) $\angle BIS \cong \angle TEC$ (Angles equal in measure are congruent).

17. (1) $\angle B \cong \angle D$, $\overline{BA} \parallel \overline{DC}$ (Given);
 (2) $\angle B \cong \angle C$ (Postulate 4.1);
 (3) $\angle C \cong \angle D$ (Transitive);
 (4) $\overline{BC} \parallel \overline{DE}$ (Postulate 4.2).

18. (1) $k \parallel \ell$, $\angle 5 \cong \angle 8$ (Given);
 (2) $\angle 5 \cong \angle 1$ (Theorem 4.2);
 (3) $\angle 1 \cong \angle 8$ (Transitive);
 (4) $j \parallel \ell$ (Postulate 4 2).

19. (1) $\angle K = \angle P$, $m\angle J + m\angle P = 180$ (Given);
 (2) $m\angle J + m\angle K = 180$ (Substitution);
 (3) $\overline{KL} \parallel \overline{JP}$ (Theorem 4.6).

20. (1) $\overline{AB} \perp \overline{BC}$, $\angle ACB$ is complementary to $\angle ABE$ (Given);

(2) $m\angle ABE + m\angle ABC + m\angle CBD = 180$ (A straight angle has measure 180);
(3) $m\angle ABE + 90 + m\angle CBD = 180$ (Substitution);
(4) $m\angle ABE + m\angle CBD = 90$ (Subtraction);
(5) $\angle CBD$ is complementary to $\angle ABE$ (Definition of complementary angles);
(6) $\angle ACB \cong \angle CBD$ (Complements of the same angle are congruent);
(7) $\overleftrightarrow{AC} \parallel \overleftrightarrow{EBD}$ (Postulate 4.2).

21. (1) $\overline{AG} \parallel \overline{BC}$, $\overline{KH} \parallel \overline{BC}$, $\angle 1 \cong \angle 2$ (Given);
 (2) $\overline{AG} \parallel \overline{KH}$ (Segments parallel to the same segment are parallel to each other—see Exercise 20);
 (3) Angles 1 and 2 are supplementary (Theorem 4.3);
 (4) Angles 1 and 2 are right angles (If two angles are supplementary and congruent, then each is a right angle);
 (5) $\overline{HK} \perp \overline{AB}$ (Segments that intersect to form a right angle are perpendicular).

22. GIVEN: $\ell \parallel p$ and $m \parallel p$.
 PROVE: $\ell \parallel m$
 PLAN: Show $\angle 1 \cong \angle 2$.
 $\angle 1 \cong \angle 3$ and $\angle 2 \cong \angle 3$. Since $\angle 1 \cong \angle 2$, $\ell \parallel m$ by Theorem 4.4.

23. GIVEN: $\overrightarrow{XL}$ bisects $\angle AXY$, $\angle YM$ bisects $\angle DYX$, $\overleftrightarrow{XL} \parallel \overleftrightarrow{YM}$.
 PROVE: $\overleftrightarrow{AXB} \parallel \overleftrightarrow{CYD}$
 PLAN: Show $m\angle AXY = m\angle DYX$
 $m\angle 1 = m\angle 2$; $m\angle AXY = 2m\angle 1$ and $m\angle DYX = 2m\angle 2$. Since doubles of equals are equal, $m\angle AXY = m\angle DYX$ and $\overleftrightarrow{AXB} \parallel \overleftrightarrow{CYD}$ by Postulate 4.2.

24. GIVEN: $\overleftrightarrow{RAS} \parallel \overleftrightarrow{PCQ}$,
 $\overrightarrow{AB}$ bisects $\angle HAS$,
 $\overrightarrow{CD}$ bisects $\angle QCA$.
 PROVE: $\overrightarrow{AB} \parallel \overrightarrow{CD}$
 PLAN: Show $m\angle 1 = m\angle 2$.
 $m\angle HAS = m\angle QCA$; $m\angle 1 = \frac{1}{2} m\angle HAS$ and $m\angle 2 = \frac{1}{2} m\angle QCA$. Hence, $m\angle 1 = m\angle 2$ and $\overrightarrow{AB} \parallel \overrightarrow{CD}$ by Theorem 4.4.

CHAPTER 5

1. **(a)** 52 **(b)** 67 **(c)** 47 **(d)** 22 **(e)** 127
2. $m\angle 1 = 70$ $m\angle 4 = 40$ $m\angle 7 = 70$
 $m\angle 3 = 70$ $m\angle 5 = 140$ $m\angle 8 = 40$

3. **(a)** $x = 18$. Right triangle since m∡$C = 90$.
 (b) $x = 22$. Obtuse triangle since m∡$B = 105$.
 (c) $x = 30$ Acute triangle.
4. 40
5. 50
6. 90
7. 113
8. **(a)** 360 **(b)** 720 **(c)** 1,260 **(d)** 1,980
9. **(a)** 12 **(b)** 17 **(c)** 5 **(d)** 14
10. **(a)** 125 **(b)** 104 **(c)** 170
11. **(a)** 108 **(b)** 165 **(c)** 135 **(d)** 156
12. **(a)** 20 **(b)** 10 **(c)** 9 **(d)** 30
13. (3)
14. 10
15. **(a)** 8 **(b)** 4 **(c)** 15
16–18 PLAN: Show two angles of one triangle are congruent to two angles of the other triangle. Then apply Corollary 5.2.4
16. ∡$B ≅$ ∡D and ∡$ACB ≅$ ∡ECD so that ∡$A ≅$ ∡E.
17. ∡$C ≅$ ∡DEB and ∡$B ≅$ ∡ B so that ∡$1 ≅$ ∡2.
18. ∡$P ≅$ ∡K and ∡$KML ≅$ ∡PRJ so that ∡$J ≅$ ∡L. Hence, $\overline{KL} ∥ \overline{PJ}$ by Postulate 4.2.
19. Angles 1 and 2 are each complementary to ∡DCE and are therefore congruent to each other.
20. Angles CAX and ACY are supplementary Hence, $\frac{1}{2}$ m∡ $CAX + \frac{1}{2}$ m∡ $ACY = \frac{1}{2} (180) = 90$. By substitution, the sum of the measures of angles BAC and BCA is equal to 90. This implies that m∡ $ABC = 90$ and ∡ ABC is a right angle.

9. Use AAS. ∡$ARF ≅$ ∡ARI; by subtraction, ∡$RFA ≅$ ∡RIA; $\overline{AR} ≅ \overline{AR}$.
10. Use SAS. $\overline{TS} ≅ \overline{RS}$; by subtraction, ∡$TSW ≅$ ∡RSP, $\overline{SW} ≅ \overline{SP}$.
11. Use AAS. Since $\overline{RP} ∥ \overline{SW}$, ∡ $TSW ≅$ ∡SRP (Corresponding angles). Also, since $\overline{SP} ∥ \overline{TW}$, ∡$STW ≅$ ∡RSP $\overline{TW} ≅ \overline{SP}$.
12. Use SSS. $\overline{AB} ≅ \overline{DE}$; $\overline{BM} ≅ \overline{EM}$; $\overline{AM} ≅ \overline{DM}$ by transitivity.
13. Use AAS ∡$1 ≅$ ∡2. ∡$A ≅$ ∡$D ≅$ ∡2, so that ∡$A ≅$ ∡D (Angle); since complements of congruent angles are congruent, ∡$AMB ≅$ ∡DME (Angle); $\overline{BM} ≅ \overline{EM}$ (Side).
14. Use ASA Right angle $AFC =$ right angle BDC, ∡$C ≅$ ∡C, $\overline{FC} ≅ \overline{DC}$.
15. Use SAS. $\overline{AD} ≅ \overline{BF}$; ∡$BAD ≅$ ∡ABF; $\overline{AB} ≅ \overline{AB}$.
16. Use Hy-Leg. $\overline{AB} ≅ \overline{DC}$ (Hyp); by subtraction, $\overline{AF} ≅ \overline{CE}$ (Leg).
17. Use AAS. Right angle $KLG ≅$ right angle JRG; ∡$G ≅$ ∡G; $\overline{KL} ≅ \overline{JR}$.
18. Use AAS. Right angle $KRO =$ right angle JLO; ∡$ROK ≅$ ∡LOJ; by halves of equals are equal, $\overline{KR} ≅ \overline{JL}$.
19. Use ASA. ∡$AEB ≅$ ∡DEC; $\overline{EB} ≅ \overline{EC}$; by halves of equals are equal, ∡$1 ≅$ ∡2.
20. Use SSS (all other methods, except Hy-Leg, can also be used).
 GIVEN: $\triangle ABC ≅ \triangle XYZ$ and $\triangle RST ≅ \triangle XYZ$.
 PROVE: $\triangle ABC ≅ \triangle RST$.
 PLAN: Using the reverse of the definition of congruent triangles, $\overline{AB} ≅ \overline{XY}$ and $\overline{RS} ≅ \overline{XY}$ so that $\overline{AB} ≅ \overline{RS}$. Similarly, show that each of the remaining pairs of corresponding sides are congruent.

CHAPTER 6

1. Use SAS. $\overline{BM} ≅ \overline{BM}$; right angle $BMA ≅$ right angle BMC; $\overline{AM} ≅ \overline{CM}$.
2. Use ASA. ∡$SRT ≅$ ∡WRT; $\overline{RT} ≅ \overline{RT}$; ∡$STR ≅$ ∡WTR.
3. Use ASA. ∡$F ≅$ ∡A; by addition, $\overline{AC} ≅ \overline{FD}$; ∡$EDF ≅$ ∡BCA.
4. Use ASA. ∡$R ≅$ ∡T; $\overline{SR} ≅ \overline{ST}$; ∡$S ≅$ ∡S.
5. Use ASA. Right angle $REW ≅$ right angle THW, $\overline{EW} ≅ \overline{HW}$; ∡$EWR ≅$ ∡HWT.
6. Use AAS. ∡$UQX ≅$ ∡DAX; ∡$UXQ ≅$ ∡DXA; $\overline{QU} ≅ \overline{DA}$.
7. Use Hy-Leg. $\overline{JL} ≅ \overline{EV}$ (Hyp) and by addition, $\overline{KL} ≅ \overline{TV}$ (Leg).
8. Use ASA. Right angle $JKL ≅$ right angle ETV; $\overline{KL} ≅ \overline{TV}$; by taking supplements of congruent angles, ∡$EVT ≅$ ∡JLK.

CHAPTER 7

1. **(a)** 76 **(b)** 20 **(c)** 40
 (d) 32 **(e)** 16 **(f)** 36
2. (2)
3. Show $\triangle HGF ≅ \triangle HJF$ by SAS. By CPCTC, ∡$GFH ≅$ ∡JFH. ∡$1 ≅$ ∡2 since supplements of congruent angles are congruent.
4. Show $\triangle ABC ≅ \triangle DCB$ by Hy-Leg. ∡$1 ≅$ ∡2 by CPCTC.
5. Show $\triangle JPY ≅ \triangle KPX$ by ASA. $\overline{JY} ≅ \overline{KX}$ by CPCTC.
6. $\overline{PJ} ≅ \overline{PK}$ by addition. Show $\triangle KXP ≅ \triangle JYP$ by SAS where $\overline{PK} ≅ \overline{PJ}$, ∡$P ≅$ ∡P, and ∡$PX ≅$ ∡PY. By CPCTC, $\overline{KX} ≅ \overline{JY}$.
7. Show $\triangle UTQ ≅ \triangle DWA$ by SAS where $\overline{UT} ≅ \overline{DW}$, ∡$QTU ≅$ ∡ AWD (Alternate exterior

angles); $\overline{QT} \cong \overline{AW}$ (Subtraction) By CPCTC, $\measuredangle UQT \cong \measuredangle DAW$ which implies $\overline{UQ} \parallel \overline{AD}$.

8. Show $\triangle SRT \cong \triangle HWT$ by AAS. By CPCTC, $\overline{RT} \cong \overline{WT}$ and T is the midpoint of $\overline{RW}$.

9. Show $\triangle ADB \cong \triangle CDB$ by SSS. By CPCTC, $\measuredangle ADB \cong \measuredangle CDB$ and $\overline{DB}$ bisects $\measuredangle ADC$.

10. Show $\triangle BAM \cong \triangle CDM$ by AAS. By CPCTC, $\overline{AM} \cong \overline{DM}$ and $\overline{BC}$ bisects $\overline{AD}$.

11. $\measuredangle RHK \cong \measuredangle NHK$ and $\measuredangle HRK \cong \measuredangle HNK$ (Supplements of congruent angles are congruent). If two angles of one triangle are congruent to corresponding angles of another triangle, then the third pair of angles are congruent. Hence, $\measuredangle HKR \cong \measuredangle HKN$ which implies $\overline{HK} \perp \overline{RN}$.

12. Show $\triangle RTS \cong \triangle ACB$ by SSS. By CPCTC, $\measuredangle 1 \cong \measuredangle 2$. If two angles are supplementary (see Given) and congruent, then each is a right angle. Hence, $\overline{ST} \perp \overline{TR}$ and $\overline{BC} \perp \overline{AC}$.

13. (1) Show $\triangle RLS \cong \triangle RLT$ by Hy-Leg where $\overline{RL} \cong \overline{RL}$ (Hyp) and $\overline{RS} \cong \overline{RT}$ (Leg)
 (2) By CPCTC, $\overline{SL} \cong \overline{TL}$ and $RLS \cong \measuredangle RLT$. Taking supplements of congruent angles, $\measuredangle SLW \cong \measuredangle TLW$. $\triangle SLW \cong \triangle TLW$ by SAS: $\overline{SL} \cong \overline{TL}$ (Side), $\measuredangle SLW \cong \measuredangle TLW$ (Angle), and $\overline{LW} \cong \overline{LW}$ (Side). By CPCTC, $\measuredangle SWL \cong \measuredangle TWL$ so that $\overline{WL}$ bisects $\measuredangle SWT$.

14. Show $\triangle PMK \cong \triangle PML$ by Hy-Leg. By CPCTC, $\overline{KM} \cong \overline{LM}$ and M is the midpoint of $\overline{KL}$.

15. Show $\triangle PMK \cong \triangle PML$ by SSS. By CPCTC, $\measuredangle KMP \cong \measuredangle LMP$ which implies $\overline{PM} \perp \overline{KL}$.

16. Show $\triangle PSL \cong \triangle LTP$ by AAS. By CPCTC, $\overline{PS} \cong \overline{LT}$.

17. Show $\triangle XLR \cong \triangle XPS$ by Hy-Leg where $\overline{XR} \cong \overline{XS}$ (Hyp) and $\overline{XL} \cong \overline{XP}$ (Leg). By CPCTC, $\measuredangle R \cong \measuredangle S$ which implies $\triangle RTS$ is isosceles.

18. $\measuredangle 1 \cong \measuredangle 3$. Since parallel lines form congruent corresponding angles, $\measuredangle 1 \cong \measuredangle 2$ and $\measuredangle 3 \cong \measuredangle 4$. By transitivity, $\measuredangle 2 \cong \measuredangle 4$ which implies $\triangle PQR$ is isosceles.

19. Show $\triangle TWL \cong \triangle PXF$ by Hy-Leg where by addition $\overline{TL} \cong \overline{PF}$ (Hyp) and $\overline{WL} \cong \overline{XF}$ (Leg). By CPCTC, $\measuredangle 1 \cong \measuredangle 2$ which implies $\triangle FML$ is isosceles.

20. Show $\triangle KLE \cong \triangle ABW$ by AAS: $\measuredangle E \cong \measuredangle W$ (Theorem 7.1); $\measuredangle BAW \cong \measuredangle LKE$ (Supplements of congruent angles are congruent), $\overline{EL} \cong \overline{WB}$. By CPCTC, $\overline{KL} \cong \overline{AB}$.

21. By Theorem 7.1, $\measuredangle 1 \cong \measuredangle 2$ and $\measuredangle 3 \cong \measuredangle 4$ since supplements of congruent angles are congruent. $\triangle VOK \cong \triangle VLZ$ by SAS. By

CPCTC, $VK \cong VZ$, which implies that $\triangle KVZ$ is isosceles.

22. Show $\triangle BJH \cong \triangle KJL$ by ASA where right $\measuredangle BHJ$ is congruent to right $\measuredangle KLJ$, $\overline{HJ} \cong \overline{LJ}$ (Theorem 7.2), and $\measuredangle HJB \cong \measuredangle LJK$. By CPCTC, $\overline{JB} \cong \overline{JK}$.

23–27. These exercises involve double congruence proofs requiring that one pair of triangles be proved congruent in order to obtain congruent pairs of parts. These pairs may then be used to prove a second pair of triangles congruent.

23. First show $\triangle ABC \cong \triangle ADC$ by SAS in order to obtain $\overline{BC} \cong \overline{DC}$ and $\measuredangle ACB \cong \measuredangle ACD$. Taking supplements of congruent angles yields $\measuredangle BCE \cong \measuredangle DCE$. Since $\overline{CE} \cong \overline{CE}$, $\triangle BCE \cong \triangle DCE$ by SAS.

24. First show $\triangle BEC \cong \triangle DEC$ by SSS in order to obtain $\measuredangle BCE \cong \measuredangle DCE$ so that $\measuredangle BCA \cong \measuredangle DCA$. Show $\triangle ABC \cong \triangle ADC$ by SAS. By CPCTC, $\measuredangle BAC \cong \measuredangle DAC$ which implies that $\overline{CA}$ bisects $\measuredangle DAB$.

25. First show $\triangle ABL \cong \triangle CDM$ by SAS in order to obtain $\overline{BL} \cong \overline{DM}$ and $\measuredangle ALB \cong \measuredangle CMD$. Taking supplements of congruent angles gives $\measuredangle BLC \cong \measuredangle DMA$. Using the addition property, $\overline{AM} \cong \overline{CL}$. Hence, $\triangle CLB \cong \triangle AMD$ by SAS By CPCTC, $\measuredangle CBL \cong \measuredangle ADM$.

26. Show $\triangle AFD \cong \triangle CFE$ by ASA where right $\measuredangle ADF =$ right $\measuredangle CEF$, $\overline{AF} \cong \overline{CF}$ (Theorem 7.2), $\measuredangle AFD \cong \measuredangle CFE$ By CPCTC, $\overline{FD} \cong \overline{FE}$. $\triangle BDF \cong \triangle BEF$ by Hy-Leg. By CPCTC, $\measuredangle DBF \cong \measuredangle EBF$ which implies that $\overline{BF}$ bisects $\measuredangle DBE$.

27. Show $\triangle BDF \cong \triangle BEF$ by SSS so that $\measuredangle BDF \cong \measuredangle BEF$. Show $\triangle AFD \cong \triangle CFE$ by ASA where $\measuredangle DFA \cong \measuredangle EFC$; $\overline{DF} \cong \overline{EF}$; $\measuredangle ADF \cong \measuredangle CEF$. By CPCTC, $AF \cong CF$ which implies that $\triangle AFC$ is isosceles.

28. GIVEN: $\triangle ABC \cong \triangle RST$ and $\overline{BH}$ and $\overline{SK}$ are altitudes.
 PROVE: $\overline{BH} \cong \overline{SK}$.
 PLAN: Show $\triangle BAH \cong \triangle SRK$ by AAS where right $\measuredangle BHA =$ right $\measuredangle SKR$, and using the reverse of the definition of congruent triangles, $\measuredangle A \cong \measuredangle S$ and $AB \cong RS$ By CPCTC, $BH \cong SK$.

29. GIVEN: Equilateral $\triangle ABC$
 PROVE: $\measuredangle A \cong \measuredangle B \cong \measuredangle C$.
 PLAN: Since $\overline{AB} \cong \overline{BC}$, $\measuredangle A \cong \measuredangle C$ Since $\overline{AC} \cong \overline{BC}$, $\measuredangle A \cong \measuredangle B$. Hence, $\measuredangle B \cong \measuredangle C$ so that $\measuredangle A \cong \measuredangle B \cong \measuredangle C$.

30. GIVEN: $\triangle PEG$, $\overline{PE} \cong \overline{GE}$, altitudes $\overline{GH}$ and $\overline{PK}$ are drawn.
 PROVE: $\overline{GH} \cong \overline{PK}$.

PLAN: Show $\triangle GHP \cong \triangle PKG$ by AAS where right $\angle PHG \cong$ right $\angle GKP$ $\angle HPG \cong \angle KGP$ (Base Angles Theorem); $\overline{PG} \cong \overline{PG}$. By CPCTC, $\overline{GH} \cong \overline{PK}$.

31. GIVEN: $\triangle ART$, $\overline{AR} \cong \overline{TR}$, medians $\overline{AY}$ and $\overline{TX}$ are drawn.
 PROVE: $\overline{AY} \cong \overline{TX}$.
 PLAN: Show $\triangle AXT \cong \triangle TYA$ by SAS where $\overline{AT} \cong \overline{AT}$; $\angle XAT \cong \angle YTA$ (Theorem 7.1); $\overline{AX} \cong \overline{TY}$ (Halves of equals are equal) By CPCTC, $\overline{AY} \cong \overline{TX}$.

32. GIVEN: M is the midpoint of $\overline{AB}$, $\overline{PM} \perp \overline{AB}$
 PROVE: $\overline{PA} = \overline{PB}$
 PLAN: Show $\triangle PMA \cong \triangle PMB$ by SAS since $\overline{PM} \cong \overline{PM}$, rt. $\angle PMA \cong$ rt. $\angle PMB$, and $\overline{AM} \cong \overline{BM}$.

33. GIVEN: $\overline{AB}$ and points P and Q such that $PA = PB$ and $QA = QB$. $\overleftrightarrow{PQ}$ intersects $\overline{AB}$ at point M.
 PROVE: $\overline{AM} \cong \overline{BM}$ and $\overleftrightarrow{PQ} \perp \overline{AB}$
 PLAN: $\triangle PAQ \cong \triangle PBQ$ by SSS so that $\angle APM \cong \angle BPM$. Therefore, $\triangle APM \cong \triangle BPM$ by SAS. By CPCTC, $\overline{AM} \cong \overline{BM}$ and $\angle PMA \cong \angle PMB$. Since angles PMA and PMB are adjacent, $\overleftrightarrow{PQ} \perp \overline{AB}$.

CHAPTER 8

1. $\overline{AB}$. 3. $\overline{BC}$. 5. $\overline{KL}$.
2. $\overline{AB}$. 4. $\overline{AB}$. 6. $\overline{BC}$.
7. $\angle S$.
8. (a) True. (b) False. (c) True.
 (d) True. (e) True. (f) False.
9. $\angle B$.
10. (a) $\angle R$. (b) $\angle S$.
11. (a) Yes. (b) No. (c) No. (d) Yes.
12. $m\angle 1 = m\angle 2$ and $m\angle 3 > m\angle 2$. By substitution, $m\angle 3 > m\angle 1$ which implies $AB > BD$.
13. (4)
14. $m\angle 1 > m\angle A$. By substitution, $m\angle 2 > m\angle A$ which implies $AD > ED$.
15. $m\angle 4 > m\angle 3$ and $m\angle 3 > m\angle AEC$. Hence, $m\angle 4 > m\angle AEC$.
16. Since $AC > BC$, $m\angle 3 > m\angle EAC$. Since $m\angle 2 < m\angle EAC$, $m\angle 3 > m\angle 2$ which implies $AD > BD$.
17. Since $AD > BD$, $m\angle 3 > m\angle 2$ m. $\angle 1 = m\angle 2$ and by substitution $m\angle 3 > m\angle 1$. But $m\angle 4 > m\angle 3$. Hence, $m\angle 4 > m\angle 1$ which implies $AC > DC$.
18–23 Use an indirect proof.

18. Assume $\angle 1 \cong \angle 2$. By transitivity, $\angle 2 \cong \angle 3$. But this contradicts Theorem 5.4. Hence, $\angle 1 \not\cong \angle 2$.
19. Assume $\overline{AB} \cong \overline{BC}$. Then $\angle 2 \cong \angle C$ and, by transitivity, $\angle 1 \cong \angle C$. But this contradicts Theorem 5.4. Hence, $\overline{AB} \not\cong \overline{BC}$.
20. Assume $RW = WL$ Then $m\angle WRL = m\angle WLR$ $m\angle WLR > m\angle T$. By substitution, $m\angle WRL > m\angle T$. Since $m\angle SRT > m\angle WRL$ $m\angle SRT > m\angle T$ which implies that $TS > RS$. But this contradicts the Given ($RS = TS$). Hence, $RW \not\cong WL$.
21. Assume $\triangle ADC$ is isosceles. Then $AD = CD$ $\triangle ADB \cong \triangle CDB$ by SAS by CPCTC, $\overline{AB} \cong \overline{CB}$, but this contradicts the Given ($\triangle ABC$ is *not* isosceles) Hence, $\triangle ADC$ is *not* isosceles.
22. Assume $\overline{AB} \parallel \overline{DE}$. Then $\angle B \cong \angle D$. Since $\overline{DE} \cong \overline{CE}$, $\angle D \cong \angle DCE$. By transitivity, $\angle B \cong \angle DCE$ Since $\angle ACB \cong \angle DCE$, $\angle B \cong \angle ACB$ which implies $\overline{AC} \cong \overline{AB}$. But this contradicts the Given ($AC > AB$). Hence $\overline{AB}$ is *not* parallel to $\overline{DE}$.
23. Assume $\overline{BD} \perp \overline{AC}$. $\triangle ABD \cong \triangle CBD$ by ASA. By CPCTC, $\overline{AB} \cong \overline{CB}$ which contradicts the assumption that $\triangle ABC$ is scalene Hence, $\overline{BD}$ is *not* perpendicular to $\overline{AC}$.
24. Since $\overline{BE}$ bisects $\angle ABD$, $m\angle ABE = m\angle DBE$. By substitution, $m\angle E > m\angle DBE$. Therefore, $BD > ED$. In $\triangle BCD$, since $\angle BDC \cong \angle C$, $BC = BD$. By substitution, $BC > ED$.
25. (a) By subtracting the measures of pairs of congruent base angles, $m\angle CAD = m\angle CBD$.
 (b) In $\triangle EBC$, the measure of exterior angle AED is greater than the measure of angle CBD. By substitution, $m\angle AED > m\angle CAD$. Therefore, $AD > DE$.
26. GIVEN: Equilateral $\triangle RST$. Point X is any point on $\overline{RT}$ and $\overline{SX}$ is drawn.
 PROVE: $SX < RS$.
 PROOF: $m\angle SXR > m\angle T$. Since an equilateral triangle is also equiangular, $m\angle R = m\angle T$, then $m\angle SXR > m\angle R$, which implies that $RS > SX$ or, equivalently, $SX < RS$. Since $RS = ST = RT$. $SX < ST$ and $SX < RT$.
27. GIVEN: $\triangle BET$ with $\overline{BE} \cong \overline{TE}$. Angle E is obtuse.
 PROVE: $BT > BE$.
 PROOF: $m\angle 1 < 90$ since a triangle may have almost one nonacute angle. Hence, $m\angle E > m\angle T$ which implies $BT > BE$. Since $BE = TE$, $BT > TE$.

28. Suppose point P is any point not on line ℓ. Draw a segment from point P perpendicular to ℓ, intersecting ℓ at point A. Choose any other point on ℓ, say point X, so that A and X are different points. Draw $\overline{PX}$. $\overline{PX}$ is the hypotenuse of right triangle PAX. Since $\overline{PX}$ lies opposite the greatest angle of the triangle, it must be the longest side of the triangle. Hence, $PA < PX$.

29. GIVEN: Acute scalene $\triangle ABC$. $\overline{BH}$ is the altitude drawn to side $\overline{AC}$.
PROVE: $\overline{BH}$ does not bisect $\angle ABC$.
PROOF: Assume $\overline{BH}$ bisects $\angle ABC$. Then $\triangle ABH \cong \triangle CBH$ by ASA so that $\overline{AB} \cong CB$ by CPCTC. This contradicts the assumption that $\triangle ABC$ is scalene. Hence, $\overline{BH}$ does not bisect $\angle ABC$.

CHAPTER 9

1. 36
2. $m\angle T = m\angle M = 105$. $m\angle H = m\angle A = 75$.
3. $m\angle R = m\angle G = 55$. $m\angle T = m\angle I = 125$.
4. (1)
5. 23
6. $EF = 19$, $RT = 38$
7. (a) $YT = 28.5$ and $BE = 41.5$.
 (b) $YT = 35$, $LM = 41$, $BE = 47$.
8. 22
9. 45
10. In $\square ABCD$, $\overline{AB} \cong \overline{CD}$ and $\overline{AB} \perp \overline{CD}$ so that $\angle BAE \cong \angle DCF$. Show $\triangle AEB \cong \triangle CFD$ by SAS. By CPCTC, $\angle ABE \cong \angle CDF$
11. In $\square ABCD$, $\overline{AD} \parallel \overline{BC}$ so that $\overline{BH} \parallel \overline{DE}$ and $\angle H \cong \angle E$. $\angle BAD \cong \angle BCD$. Taking supplements of congruent angles, $\angle FAE \cong \angle GCH$. $\triangle FAE \cong \triangle GCH$ by AAS. By CPCTC, $\overline{AF} \cong \overline{CG}$.
12. In $\square ABCD$, $\overline{AB} \cong \overline{CD}$ and $\overline{AB} \parallel \overline{DC}$. $\overline{AB} \cong \overline{BE}$ (Definition of midpoint). By transitivity, $\overline{CD} \cong \overline{BE}$; $\angle EBF \cong \angle DCF$, $\angle EFB \cong \angle DFC$; $\triangle BEF \cong \triangle CDF$ by AAS. By CPCTC, $\overline{EF} \cong \overline{FD}$.
13. Right $\triangle ABM \cong$ right $\triangle DCM$ by SAS. By CPCTC, $\overline{AM} \cong \overline{DM}$, which implies that $\triangle AMD$ is isosceles.
14. Since the diagonals of a rhombus are the perpendicular bisectors of each other, $\overline{AR} \cong \overline{CR}$ and angles ARS and CRS are right angles. Show $\triangle SRA \cong \triangle SRC$ by SAS. By

CPCTC, $\overline{SA} \cong \overline{SC}$ which implies that $\triangle ASC$ is isosceles.
15. $\angle EBC \cong \angle ECB$. Since $\overline{AD} \parallel \overline{BC}$, $\angle AFB \cong \angle EBC$ and $\angle DGC \cong \angle ECB$. By transitivity, $\angle AFB \cong \angle DGC$. $\angle A \cong \angle D$ and $\overline{AB} \cong \overline{DC}$ so that $\triangle FAB \cong \triangle GDC$ by AAS. By CPCTC, $\overline{AF} \cong \overline{DG}$.
16. Since $\overline{AD} > \overline{DC}$, $m\angle ACD > m\angle DAC$. Since $\overline{AB} \parallel \overline{DC}$, $m\angle BAC = m\angle ACD$. By substitution, $m\angle BAC > m\angle DAC$.
17. Since $ABCD$ is a parallelogram, $m\angle ABC = m\angle ADC$. $m\angle RBS = m\angle SDR$ since halves of equals are equal. In triangles BAR and DCS, $\angle A \cong \angle C$ and $\angle ABR \cong \angle CDS$. This implies that angles CSD and ARB, the third pair of angles, are congruent. Since supplements of congruent angles are congruent, $\angle BRD \cong \angle DSB$. $BRDS$ is a parallelogram since both pairs of opposite angles are congruent.
18. $\overline{BL} \cong \overline{DM}$ and $\overline{BL} \parallel \overline{DM}$. Taking supplements of congruent angles, $\triangle ALB \cong \triangle CMD$. $\triangle ALB \cong \triangle CMD$ by SAS. By CPCTC, $AB \cong CD$ and $\angle BAL \cong \angle DCM$ so that $\overline{AB} \parallel \overline{CD}$. Hence, $ABCD$ is a parallelogram by Theorem 9.9
19. $\overline{AB} \cong \overline{CD}$ and $\overline{AB} \parallel \overline{CD}$ so that $\angle BAL \cong \angle DCM$. $\triangle ALB \cong \triangle CMD$ by ASA, $\overline{BL} \cong \overline{DM}$ and $\angle ALB \cong \angle CMD$ by CPCTC. Taking supplements of congruent angles, $\angle BLM \cong \angle DML$ so that $\overline{BL} \parallel \overline{DM}$. Hence, $BLDM$ is a parallelogram by Theorem 9.9.
20. $\overline{AB} \cong \overline{BC}$ since a rhombus is equilateral. $\triangle ABL \cong \triangle BCM$ by SSS. By CPCTC, $\angle B \cong \angle C$. Since $\overline{AB} \parallel \overline{DC}$, angles B and C are supplementary. Hence, each is a right angle and $ABCD$ is a square.
21. Use an indirect proof. Assume $ABCD$ is a rectangle. Then $\triangle BAD \cong \triangle CDA$ by SAS. By CPCTC, $\angle 1 \cong \angle 2$ which contradicts the Given ($m\angle 2 \cong m\angle 1$).
22. $DE = \frac{1}{2}BC$ and $DF = \frac{1}{2}AB$. Since $DE = DF$, $\frac{1}{2}BC = \frac{1}{2}AB$ implies that $BC = AB$. Hence, $\triangle ABC$ is isosceles.
23. If $\overline{BC} \parallel \overline{WT}$, then $\overline{AC} \parallel \overline{WT}$ (Extensions of parallel segments are parallel). $\overline{AW} \parallel \overline{TC}$ (Segments of parallel lines are parallel). Hence, $WACT$ is a parallelogram.
24. Show $\triangle RSW \cong \triangle WTR$ by SAS. By CPCTC, $\angle TRW \cong \angle SWR$ so that $\overline{RP} \cong \overline{WP}$ (Theorem 7.2).
25. By addition, $AG = DF$, $\overline{EF} \cong \overline{EG}$ (Given) and by Theorem 7.2 $\angle EGF \cong \angle EFG$.

$\overline{BG} \cong \overline{CF}$ (Given). $\triangle BAG \cong \triangle CDF$ by SAS. By CPCTC, $\overline{AB} \cong \overline{DC}$. Hence, trapezoid $ABCD$ is isosceles.

26. By Theorem 7.1, $\angle PLM \cong \angle PML$. Since LM is a median, $\overline{LM} \parallel \overline{AD}$ so that $\angle PLM \cong \angle APL$ and $\angle PML \cong \angle DPM$. By transitivity, $\angle APL \cong \angle DPM$. Show $\triangle LAP \cong \triangle MDP$ by SAS. By CPCTC, $\angle A \cong \angle D$ which implies trapezoid $ABCD$ is isosceles.

27. Since $ABCD$ is a trapezoid, $\overline{BC} \parallel \overline{KD}$. Show $\overline{BK} \parallel \overline{DC}$ as follows: $\angle BAK \cong \angle CDA$ and by transitivity $\angle BKA \cong \angle CDA$; since corresponding angles are congruent, $\overline{BK} \parallel \overline{DC}$. $BKDC$ is a parallelogram since both pairs of opposite sides are parallel.

28. **(a)** Since $\angle B \cong \angle C$, $\overline{AB} \cong \overline{AC}$ (Theorem 7.2). $\overline{BD} \cong \overline{CF}$ (Halves of congruent segments are congruent). Therefore, $\triangle DEB \cong \triangle FGC$ by SAS.

 (b) $\angle 1 \cong \angle 2$ (CPCTC). Since angles 1 and 2 are also supplementary (Given), $\overline{FG}$ and $\overline{DE}$ are perpendicular to $\overline{BC}$, $\overline{DE} \parallel \overline{FG}$ (lines perpendicular to the same line are parallel) and $\overline{DE} \cong \overline{FG}$ (by CPCTC). Hence, $DEGF$ is a parallelogram. Parallelogram $DEGF$ is a square since it contains a right angle (for example, $\angle DEG$ is a right angle since perpendicular lines meet to form right angles) and a pair of congruent adjacent sides ($\overline{FD} \cong \overline{FG}$ is given).

29. See the hint in the statement of the problem. Since the diagonals of a rhombus bisect its angles, $m\angle EDA > m\angle EAD$. This implies that $AE > DE$. Equivalently, $2AE > 2DE$. Since the diagonals of a rhombus bisect each other, AC may be substituted for $2AE$ and BD may be substituted for $2DE$ so that $AC > BD$.

30. See the hint in the statement of the problem. By dropping perpendiculars, a parallelogram is formed. Since the opposite sides of a parallelogram are equal in length, the perpendicular segments have the same length.

31. GIVEN: $\square ABCD$. Points X, Y, Z, and W are the midpoints of sides $\overline{AB}$, $\overline{BC}$, $\overline{CD}$, and $\overline{DA}$, respectively.
 PROVE: Quadrilateral $WXYZ$ is a parallelogram.
 PLAN: Draw diagonal $\overline{BD}$. In $\triangle BAD$, $\overline{XW}$ is parallel to $\overline{BD}$. In $\triangle BCD$, $\overline{YZ}$ is parallel to $\overline{BD}$. Hence, $\overline{XW} \parallel \overline{YZ}$. Similarly, by drawing diagonal $\overline{AC}$, show that $\overline{XY} \parallel \overline{WZ}$. Since both pairs of opposite sides are parallel, quadrilateral $WXYZ$ is a parallelogram.

32. GIVEN: Rectangle $ABCD$. Points E, F, G, and H are the midpoints of sides $\overline{AB}$, $\overline{BC}$, $\overline{CD}$, and $\overline{DA}$, respectively.
 PROVE: Quadrilateral $EFGH$ is a rhombus.
 PLAN: First prove $EFGH$ is a parallelogram (see Exercise 29). Next, show that an adjacent pair of sides are congruent. Show $\overline{HE} \cong \overline{HG}$ by proving $\triangle EAH \cong \triangle GDH$ by SAS.

33. In trapezoid $ABCD$ with congruent diagonals $\overline{BD}$ and $\overline{CA}$, altitudes $\overline{BE}$ and $\overline{CF}$ form right triangles BED and CFA which are congruent by Hy-Leg. Since $\angle CAD \cong \angle BDA$ by CPCTC, $\triangle ABD \cong \triangle DCA$ by SAS. By CPCTC, $\overline{AB} \cong \overline{DC}$.

CHAPTER 10

1. 84
2. 72
3. 75, 105, 75, 105
4. **(a)** 24 **(b)** 10 **(c)** $\dfrac{47}{8}$
 (d) 5 **(e)** −3 or 7
5. **(a)** 8 **(b)** $6e^2$ **(c)** $\dfrac{1}{4}$ **(d)** $\sqrt{54}$
6. **(a)** Yes. **(b)** Yes. **(c)** No. **(d)** No.
7. $\dfrac{AL}{AB} = \dfrac{AM}{AG}$; $\dfrac{LB}{AB} = \dfrac{MG}{AG}$; $\dfrac{AL}{LB} = \dfrac{AM}{MG}$.
8. **(a)** Yes. **(b)** Yes. **(c)** No.
 (d) No. **(e)** Yes.
9. **(a)** 15 **(b)** 4 **(c)** 7
 (d) 24 **(e)** $KE = 9$, $KH = 3$.
10. $\angle G \cong \angle S$, $\angle A \cong \angle H$, $\angle L \cong \angle E$
 $\dfrac{GA}{SH} = \dfrac{AL}{HE} = \dfrac{GL}{SE}$
11. 40
12. $MY = 5$, $MX = 35$.
13. 44
14. 12 and 27
15. **(a)** $JB = 12$ and $RA = 8$
 (b) $LK\ 17.5 =$ and $TS = 14$
 (c) $ST = 10$ and $KL = 16$
 (d) 5
16. 2. Parallel Postulate.
 3. Postulate 10.1.
 4. Substitution.
 5. If two lines are parallel, then the corresponding angles are congruent.
 6. Given.
 7. Transitive property.
 8. If two angles of one triangle are congruent to two angles of another triangle, then the third pair of angles are congruent.

9. ASA Postulate

10. CPCTC.

11. Substitution.

17–29. are based on the application of Theorem 10.3.

17. $\angle W \cong \angle Y$ and right $\angle HAW \cong$ right $\angle KBY$ implies $\triangle HWA \sim \triangle KYB$.

18. $\angle XBC \cong \angle YST$ and $\angle C \cong \angle T$ (Reverse of the definition of similar triangles). Hence, $\triangle BXC \sim \triangle SYT$.

19. $\angle FCD \cong \angle AEF$ and $\angle AFE \cong \angle CFD$. Therefore, triangles EFA and CFD are similar by Theorem 10.3.

20. $\angle AEF \cong \angle AFE$ and $\angle DFC \cong \angle FCD$. Since $\angle AFE \cong \angle DFC$, by the transitive property, $\angle AEF \cong \angle FCD$. Therefore, $\triangle AFE \sim \triangle DFC$ by Similarity 10.3 and $\dfrac{AF}{EF} = \dfrac{DF}{FC}$.

21. $\angle CTM \cong \angle AWB$ (Reverse of the definition of similar triangles). Since $\angle AWB \cong \angle CWB$ (Definition of angle bisector), $\angle CTM \cong \angle CWB$ by transitivity. $\angle MCT \cong \angle BCW$. Hence, $\triangle MCT \sim \triangle BCW$.

22. $\angle S \cong \angle MWB$ (Angle) and $\angle A \cong \angle WCA$ (Theorem 7.1). Since $\overline{AW} \parallel \overline{ST}$, $\angle A \cong \angle BTS$ and by transitivity, $\angle BTS \cong \angle WCA$ (Angle). Hence, $\triangle BCW \sim \triangle BTS$.

23. $\angle WBC \cong \angle WCB$ and taking supplements of these angles, $\angle ABW \cong \angle TCW$ (Angle). Since $\overline{AT}$ bisects $\angle STW$, $\angle STA \cong \angle WTC$. $\angle A \cong \angle STA$ since $\overline{ST} \parallel \overline{AW}$. By transivity, $\angle A \cong \angle WTC$ (Angle). Hence, $\triangle ABW \sim \triangle TCW$.

24. Show $\triangle TAX \sim \triangle WHY$, $\angle W \cong \angle ATX$ and $\angle Y \cong \angle AXT$ since $\parallel$ line segments have congruent corresponding angles.

25. Show $\triangle RMN \sim \triangle RAT$. $\angle RMN \cong \angle A$ and $\angle RNM \cong \angle T$. Write $\dfrac{MN}{AT} = \dfrac{RN}{RT}$. By Theorem 7.2 $NT = MN$ Substitute NT for MN in the proportion.

26. Show $\triangle SQP \sim \triangle WRP$, $\angle SPQ \cong \angle WPR$ (Angle). Since $\overline{SR} \cong \overline{SQ}$, $\angle SRQ \cong \angle SQR$. $\angle SRQ \cong \angle WRP$ (Definition of angle bisector). By transitivity, $\angle SQR \cong \angle WRP$ (Angle).

27. Show $\triangle PMQ \sim \triangle MKC$. Right $\angle MCK \cong$ right $\angle PMQ$. Since $\overline{TP} \cong \overline{TM}$, $\angle TPM \cong \angle TMP$.

28. Show $\triangle PMT \sim \triangle JKT$. Since $\overline{MP} \cong \overline{MQ}$, $\angle MPQ \cong \angle MQP$. Since $\overline{JK} \parallel \overline{MQ}$, $\angle J \cong \angle MQP$ so that by transitivity, $\angle J \cong \angle MPQ$ (Angle) $\angle K \cong \angle QMT$ (Congruent alternate interior angles). $\angle QMT \cong \angle PMT$ (since $\triangle MTP \cong \triangle MTQ$ by SSS). By transitivity,

$\angle K \cong \angle PMT$ (Angle). Write the proportion $\dfrac{PM}{JK} = \dfrac{PT}{JT}$. Substitute TQ for PT (see Given) in the proportion.

29. Show $\triangle EIF \sim \triangle HIG$. Since $\overline{EF}$ is a median, $\overline{EF} \parallel \overline{AD}$ so that $\angle FEI \cong \angle GHI$ and $\angle EFI \cong \angle HGI$.

30. Rewrite the product as $ST \times ST = TW \times RT$. Show $\triangle SWT \sim \triangle RST$. Right $\angle RST =$ right $\angle SWT$ and $\angle T \cong \angle T$.

31. Show $\triangle AEH \sim \triangle BEF$. $\angle BEF \cong \angle HEA$ and $\angle EAH \cong \angle EBF$ (Halves of equals are equal).

32. Apply Postulate 10.1. Since $\overline{XY} \parallel \overline{LK}$.
$\dfrac{JY}{YK} = \dfrac{JX}{XL}$. Since $\overline{XZ} \parallel \overline{JK}$, $\dfrac{JX}{XL} = \dfrac{KZ}{ZL}$. By transitivity, $\dfrac{JY}{YK} = \dfrac{KZ}{ZL}$. $YK = XZ$ since $YKZX$ is a parallelogram. Using substitution, $\dfrac{JY}{XZ} = \dfrac{KZ}{ZL}$. Cross-multiplying yields the desired product.

33. Rt. $\angle B \cong$ Rt. $\angle DCA$. Angles BAC and CDA are complementary to $\angle CAD$ and are, therefore, congruent to each other. Triangles ABC and DCA are similar which means $\dfrac{AD}{AC} = \dfrac{AC}{BC}$, and $BC \times AD = (AC)^2$.

34. (a) $\overline{EF} \parallel \overline{AC}$, $\angle ACF \cong \angle GFE$ since parallel lines form congruent alternate interior angles. Similarly, since $\overline{DE} \parallel \overline{AB}$, $\angle AFC \cong \angle EGF$. Hence, $\triangle CAF \sim \triangle FEG$.

(b) $\triangle CAF \sim \triangle CDG$. By the transitive property, $\triangle CDG \sim \triangle FEG$ so that $\dfrac{DG}{EG} = \dfrac{GC}{GF}$ which implies $DG \times GF = EG \times GC$.

CHAPTER 11

1. $r = 8$, $s = 8\sqrt{3}$, $t = 4\sqrt{3}$.

2. $r = 25$, $s = 5\sqrt{5}$, $t = 10\sqrt{5}$.

3. $r = 16$, $s = 9$, $t = 15$.

4. 8

5. 6

6. $4\sqrt{6}$

7. $9\sqrt{5}$

8. 7

9. $4\sqrt{10}$

10. (a) 4 and 16

(b) $8\sqrt{5}$

11. 8

12. $6\sqrt{2}$

13. 24

14. $100\sqrt{7}$

15. $\dfrac{20}{3}$

16. 80

17. 18

18. $5\sqrt{3}$

19. $\sqrt{74}$

20. **(a)** 4 **(b)** $8\sqrt{3}$

21. $6\sqrt{2}$

22. $36\sqrt{2}$

23. **(a)** 4 **(b)** $4\sqrt{3}$ **(c)** $4\sqrt{2}$

24. **(a)** Altitude = $8/\sqrt{3}$, leg = $16/\sqrt{3}$.

 (b) Altitude = 8, leg = $8\sqrt{2}$.

 (c) Altitude = $8\sqrt{3}$, leg = 16.

25. $x = 4\sqrt{3}$, $y = 4$, $z = 2\sqrt{37}$.

26. $2\sqrt{37}$

27. $10\sqrt{3}$

28. $\dfrac{24}{25}$

29. $\cos R = \dfrac{40}{41}$, $\tan R = \dfrac{9}{40}$.

30. 32°

31. Base = 15, leg = 12.

32. **(a)** 57.7 **(b)** 112.9

33. $x = 12.9$, $y = 3.6$.

34. $AD = 6.2$, $CD = 21.9$.

CHAPTER 12

1. 66 6. 38

2. 52.5 7. 66

3. 25 8. 61

4. 66 9. 24

5. 12.5 10. 75

11. **(a)** 36 **(b)** 30

12. 26 13. 106 14. 25

15. **(a)** 74 **(d)** 16 **(g)** 16

 (b) 16 **(e)** 32 **(h)** 16

 (c) 74 **(f)** 90 **(i)** 74

16. **(a)** m$\angle DEB$ = 90, m$\angle ABC$ = 31, m$\angle AED$ = 90.

 (b) m$\overarc{AC}$ = 56, m$\angle DEB$ = 79, m$\angle AED$ = 101.

 (c) m$\overarc{AC}$ = 51, m$\angle ABC$ = 25.5, m$\angle AED$ = 97.

 (d) m$\overarc{BD}$ = 150, m$\overarc{AC}$ = 82, m$\angle DEB$ = 116.

17. $x = 82$, $y = 59$, $z = 39$.

18. **(a)** m$\angle CAQ$ = 60, m$\angle DBC$ = 40, m$\angle AEB$ = 60, m$\angle CPQ$ = 40.

 (b) m$\overarc{AB}$ = 58, m$\angle CAQ$ = 66, m$\angle DBC$ = 37, m$\angle AEB$ = 66.

 (c) m$\overarc{ADC}$ = 164, m$\angle CAQ$ = 82, m$\angle DBC$ = 50, m$\angle CPQ$ = 50

 (d) m$\overarc{ADC}$ = 142, m$\overarc{AB}$ = 44, m$\angle DBC$ = 49, (m$\angle AEB$ = 71.

19. **(a)** m$\overarc{WA}$ = 30, m$\overarc{AT}$ = 90, m$\overarc{ST}$ = 150, m$\overarc{SW}$ = 90.

 (b) 45. **(c)** 90. **(d)** 120. **(e)** 15.

20. **(a)** m$\overarc{KF}$ = 76, m$\overarc{FM}$ = 104, m$\overarc{JK}$ = 100, m$\overarc{JM}$ = 80, m$\overarc{HM}$ = 28.

 (b) 76. **(c)** 76. **(d)** 12.

21. Show $\triangle AOM \cong \triangle BOM$ by Hy-Leg. By CPCTC, central $\angle AOX \cong$ central $\angle BOX$ which implies $\overarc{AX} \cong \overarc{BX}$.

22. By subtraction, $\overarc{RS} \cong \overarc{WT}$. $\angle RTS \cong \angle WST$ since inscribed angles that intercept congruent arcs are congruent.

23. Since $OA > AC$, m$\angle OCA >$ m$\angle AOC$, m$\angle BOC >$ m$\angle OCA$ so that m$\angle BOC >$ m$\angle AOC$. Since m$\angle BOC =$ m$\overarc{BC}$ and m $\angle AOC =$ m$\overarc{AC}$, by substitution, m$\overarc{BC} >$ m$\overarc{AC}$.

24. Since m$\overarc{BC} >$ m$\overarc{AC}$, m$\angle BOD >$ m$\angle AOD$. m$\angle ADO >$ m$\angle BOD$ so that m$\angle ADO >$ m$\angle AOD$. This implies that $OA > AD$.

25. Show $\triangle XPM \cong \triangle YQM$ by AAS where right $\angle XPM \cong$ right $\angle YQM$, $\angle XMP \cong \angle YMQ$, $\overline{XP} \cong \overline{YQ}$.

26. Since $\overline{FE} \cong \overline{FG}$, m$\angle G =$ m$\angle E$. m$\overarc{FG} = 2$m$\angle E$ so that m$\angle BFG =$ m$\angle E$. By transitivity, m$\angle G =$ m$\angle BFG$ so that $\overline{AB} \parallel \overline{EG}$.

27. Show $\triangle AMB \cong \triangle AMD$ by SAS. $\overline{BM} \cong \overline{DM}$; $\angle M$ is inscribed in a semicircle so that right $\angle AMB \cong$ right $\angle AMD$; $\overline{AM} \cong \overline{AM}$. By CPCTC, $\angle BAM = \angle DAM$ which implies that $\overarc{BM} \cong \overarc{CM}$.

28. **(a)** $\angle W \cong \angle RSW$ since inscribed angles that intercept congruent arcs are congruent. $\angle RSW \cong \angle NSW$ (Definition of angle bisector). By transitivity, $\angle W \cong \angle NSW$ which implies that $\triangle NSW$ is isosceles.

 (b) $\overarc{WT} \cong \overarc{RW}$ since congruent inscribed angles intercept congruent arcs. Since $\overarc{RW} \cong \overarc{SK}$ (Given), $\overarc{WT} \cong \overarc{SK}$. $\angle STK \cong \angle TKW$ (Inscribed angles that intercept congruent arcs are congruent). Since the base angles of $\triangle NTK$ are congruent, the triangle is isosceles.

29. m$\widehat{LM}$ ≅ m$\widehat{KL}$ implies that m∠MKL >
m∠LMK. Since $\overline{JK}$ ∥ $\overline{LM}$, m∠JKM =
m∠LMK. By substitution, m∠MKL >
m∠JKM.
30. Since $\overline{MA}$ ≅ $\overline{MT}$, ∠MAT ≅ ∠ATM. ∠SRA ≅
∠ATM (Inscribed angles that intercept the
same arc are congruent). By transitivity,
∠SRA ≅ ∠MAT which implies that $\overline{SR}$ ∥
$\overline{ATW}$. Quadrilateral $RSTW$ is a parallelogram
since $\overline{SR}$ ≅ $\overline{TW}$ (Given) and $\overline{SR}$ ≅ $\overline{TW}$.

CHAPTER 13

1. (a) 2 common internal tangents.
 2 common external tangents.
 (b) No common tangent.
 (c) 1 common external tangent.
2. (a) 11 (b) 3
3. 3 7. 18 11. 5
4. 9 8. 2 12. 27
5. 14 9. 20 13. 5
6. 1 or 21 10. 1 14. 12
15. 12
16. (a) 13
 (b) JW = 48, OA = 7.
 (c) 12
17. 2√5 18. 6
19. (a) 66 20. (a 16
 (b) 314 (b) 75
 (c) 34π (c) $\frac{11.5}{\pi}$
21. 9π 22. 9π√2
23. (a) 6π (b) 45 (c) 10
24. 3π√2
25. (a) 120° (b) 12π (c) 36π + 36√3
26. Triangles OAR and OBR are right triangles
since angles S and T are right angles and
$\overline{OA}$ ∥ $\overline{MS}$ and $\overline{OB}$ ∥ $\overline{MT}$ making correspond-
ing angles congruent. △OAR ≅ △OBR by
Hy-Leg. By CPCTC, OA = OB. $\overline{SR}$ ≅ $\overline{TR}$ by
Theorem 13.2.
27. $\overline{OA}$ ⊥ $\overline{SR}$ and $\overline{OB}$ ⊥ $\overline{TR}$ (See Exercise 26).
By Theorem 13.1, OA = OB so that △OAR
≅ △OBR by Hy-Leg. By CPCTC, ∠AOR ≅
∠BOR. Taking supplements of these congru-
ent angles gives ∠AOM ≅ ∠BOM.
28. Since a square contains 4 right angles and is
equilateral, $\overline{OX}$ ⊥ $\overline{QT}$, $\overline{OY}$ ⊥ $\overline{PJ}$, and QX = OY.
By Theorem 13.2, $\overline{QT}$ ≅ $\overline{PJ}$ so that $\overline{QT}$ ≅
$\overline{PJ}$. By arc subtraction, $\widehat{QP}$ ≅ $\widehat{JT}$.
29. Show △HBW ~ △MBL. ∠ABW ≅ ∠H since
they are measured by one-half of the same
arc measure. Since $ABLM$ is a parallelo-

gram, $\overline{AB}$ ∥ $\overline{ML}$ so that ∠ABW ≅ ∠BML. By
transitivity, ∠H ≅ ∠BML (Angle) ∠MBL ≅
∠HBW (Angle).
30. Show CD : BC = BC : AB by proving △ABC ~
△BCD. Right ∠ACB ≅ right ∠CDB. ∠CAB
≅ ∠CBD since they are measured by one-half
the same arc measure.
31. Show △KLP ~ △KJM. ∠LKP ≅ ∠JKM and
right ∠KLP ≅ right ∠KJM.
32. Show △KLP ~ △KJM. Right ∠KLP ≅ right
∠KJM (Angle). Since $\overline{JP}$ ≅ $\overline{JM}$, ∠JPM ≅
∠JMK ∠KPL ≅ ∠JPM. By transitivity,
∠KPL ≅ ∠JMK (Angle).
33. (a) ∠NTK ≅ ∠WTK Since $\overline{KW}$ ≅ $\overline{TW}$.
 ∠WKT ≅ ∠WTK. By transitivity, ∠NTK
 ≅
 ∠WKT which implies that $\overleftrightarrow{NTP}$ ∥ $\overleftrightarrow{KW}$
 (b) Rewrite product as TW · TW = JT · TK.
 Show △JTW ~ △WTK. ∠T ≅ ∠T
 (Angle). ∠JWT ≅ ∠NTK and ∠K ≅
 NTK so that ∠K ≅ ∠JWT (Angle).
34. GIVEN: Circles A and B are tangent externally
 at point P. $\overleftrightarrow{XY}$ is tangent to ⊙A at
 point X and tangent to ⊙B at point Y.
 $\overleftrightarrow{PQ}$ intersects $\overleftrightarrow{XY}$ at point Q.
 PROVE: $\overline{QX}$ ≅ $\overline{QY}$.
 PLAN: By Theorem 13.3, $\overline{QX}$ ≅ $\overline{QP}$ and
 $\overline{QP}$ ≅ $\overline{QY}$. By transitivity, $\overline{QX}$ ≅ $\overline{QY}$.
35. GIVEN: $\overleftrightarrow{XY}$ is tangent to ⊙A at point X and
 tangent to ⊙B at point Y. $\overleftrightarrow{PQ}$ is tan-
 gent to ⊙A at point P and tangent to
 ⊙B at point Q.
 PROVE: $\overline{XY}$ ≅ $\overline{PQ}$.
 PLAN: *Case 1.* Assume circles A and B are
 not congruent. $\overleftrightarrow{XY}$ cannot be parallel
 to $\overleftrightarrow{PQ}$. Therefore, $\overleftrightarrow{XY}$ and $\overleftrightarrow{PQ}$ will
 intersect, extended if necessary, say
 at point C. By Theorem 13.3, $\overline{CX}$ ≅
 $\overline{CP}$ and $\overline{CY}$ ≅ $\overline{CQ}$. By subtraction,
 $\overline{XY}$ ≅ $\overline{PQ}$.
 Case 2. Assume circles A and B are
 congruent. Then $\overleftrightarrow{XY}$ ∥ $\overleftrightarrow{PQ}$. Draw the
 line of centers $\overline{AB}$ and radii $\overline{AX}$, $\overline{BY}$,
 $\overline{AP}$, and $\overline{BQ}$. Quadrilateral $AXYB$ is
 a parallelogram since $\overline{AX}$ ≅ $\overline{BY}$
 (Congruent circles have congruent
 radii) and $\overline{AX}$ ∥ $\overline{BY}$ (Segments perpen-
 dicular to the same line are parallel).
 Similarly, $APQB$ can be shown to be
 a parallelogram. Hence, $\overline{XY}$ ≅ $\overline{AB}$
 and $\overline{AB}$ ≅ $\overline{PQ}$ so that $\overline{XY}$ ≅ $\overline{PQ}$.
36. GIVEN: Lines ℓ, m, and k are tangent to
 circle P at points C, D, and E,

respectively, so that $\ell \parallel m$ and line k intersects line ℓ at point A and line m at point B.

PROVE: $\angle APB$ is a right angle.

PLAN: Since $\ell \parallel m$, angles CAB and DBA are supplementary. $\overline{PA}$ bisects $\angle CAB$ and PB bisects $\angle DBA$ (refer to the discussion of the proof of Theorem 13.3). Hence, the sum of the measures of angles PAB and PBA is 90, which implies that the remaining angle of $\triangle APB$, namely, $\angle APB$, must have a measure of $180 - 90$ or 90.

CHAPTER 14

1. (a) 48 (b) 32 (c) $90\sqrt{3}$
2. (a) Side = 12 cm.
 (b) Altitude = 5, base = 15.
 (c) Altitude = 15, base = 9.
3. 60 7. 104
4. $35\sqrt{3}$ 8. 52
5. $10\sqrt{2}$ 9. 17.5
6. $9\sqrt{3}$ 10. 162
11. $54 + 36\sqrt{3}$
12. (a) 8 (b) $2\sqrt{11}$ (c) $\dfrac{4}{\sqrt[4]{3}}$
13. (a) 12 (b) 14
14. (a) 4 (b) 5 (c) 6
15. 240
16. (a) 42 (b) 21 (c) $21\sqrt{3}$
17. $\dfrac{16}{9}$
18. (a) $\dfrac{9}{11}$ (b) 55
19. 192 21. 30
20. 45 22. $150\sqrt{3}$
23. (a) Radius = $3\sqrt{2}$, apothem = 3, area = 36.
 (b) Side = 7, radius = $3.5\sqrt{2}$, apothem = 3.5.
 (c) Side = 10, radius = $5\sqrt{2}$, area = 100.
 (d) Side = $8\sqrt{2}$, apothem = $4\sqrt{2}$, area = 128.
24. $\dfrac{3}{4}$
25. (a) 30.8 cm
 (b) 3,080 cm^2
26. (a) $D = 10$, $C = 10\pi$, $A = 25\pi$
 (b) $r = 4.5$, $C = 9\pi$, $A = \dfrac{81}{4}\pi$
 (c) $r = 7$, $D = 14$, $C = 14\pi$
 (d) $r = 9$, $D = 18$, $A = 81\pi$

(e) $r = \dfrac{8}{5}$, $D = \dfrac{16}{5}$, $C = \dfrac{16}{5}\pi$

27. 15 cm
28. 312.5 square units
29. 20π square units
30. (a) Area of circle = 144π, area of sector = 18π.
 (b) Radius = 6, degree measure of sector arc = $120°$.
 (c) Radius = 10, area of circle = 100π.
 (d) Area of circle = 81π, area of sector = 36π.
31. (a) 24π square units
 (b) $24\pi - 36\sqrt{3}$ square units
32. $8\pi - 16$ cm^2
33. $\dfrac{1}{2}$
34. $\dfrac{625}{8}\pi - 84$ square units
35. $256 - 64\pi$ square units
36. 80π cm^2
37. $16\pi - 32$ square units
38. (a) 64π square units
 (b) $96\sqrt{3}$ square units
 (c) $\dfrac{32}{3}\pi - 16\sqrt{3}$ square units
39. (a) From point L draw the altitude to side $\overline{CV}$, intersecting $\overline{CV}$ at point H. Area $\triangle CLG = \dfrac{1}{2} CG \cdot LH$ and area $\triangle VLG = \dfrac{1}{2}$

 $VG \cdot LH$. Since $CG = VG$, area $\triangle CLG$ area $\triangle VLG$. (*Note: In $\triangle CLV$, $\overline{LG}$ is a median. We have just proved that a median divides a triangle into two triangles having the same area.*)

 (b) In $\triangle BCV$, $\overline{BG}$ is a median so that area $\triangle BCG$ = area $\triangle BVG$ (see part **a** of this exercise). Next, subtract:

 $\text{Area } \triangle BCG = \text{area} \triangle BVG$
 $-\text{Area } \triangle CLG = \text{area} \triangle VLG$
 $\overline{\text{Area} \triangle BLC = \text{area} \triangle BLV}$

40. (a) 11 (b) 396 square units
41. (a) By subtraction, $\angle ABC \cong \angle EBD$ so that $\triangle ABC \cong \triangle EBD$ by SAS.
 (b) Since congruent triangles have equal areas, by addition, area $\triangle ABC$ + area $\triangle BCD$ = area $\triangle EBD$ + area $\triangle BCD$. Therefore, area polygon $ABDC$ = area polygon $BEDC$.
42. The area of $\square ABCD$ is equal to the sum of the areas of triangles AEB, CEB and ADC. Since $\triangle CED \cong \triangle AEB$ and $\triangle AED \cong \triangle CEB$, the area of $\square ABCD$ is equal to the

sum of the areas of triangles CED, AED, and ADC. $\overline{BE} \cong \overline{ED}$ so that $\overline{ED} \cong \overline{FD}$. Since $\overline{CD}$ and $\overline{AD}$ are medians of triangles ECF and EAF, respectively, area$\triangle CED$ = area $\triangle CDF$ and area $\triangle AED$ = area $\triangle FAD$. By substitution, the area of $\square ABCD$ is equal to the sum of the areas of triangles CDF, FAD, and ADC which is equivalent to $\triangle FAC$.

43. **(a)** Circle O is inscribed in equilateral $\triangle ABC$ where apothem $\overline{OX} \perp \overline{AC}$. In right $\triangle AOX$, $\overline{OA}$ represents the radius of the circumscribed circle, while OX is a radius of the inscribed circle. Since

$m\angle OAX = 30$ and $AX = \frac{1}{2}AC$, $OX = \frac{1}{\sqrt{3}}$

$AX = \frac{1}{2\sqrt{3}}AC = \frac{\sqrt{3}}{6}AC$. The altitude to

side $AC = \frac{\sqrt{3}}{2}AB = \frac{\sqrt{3}}{2}AC$. Hence, OX is

one-third the length of the altitude.

(b) $AX = \frac{\sqrt{3}}{2}OA$ or $OA = \frac{2}{\sqrt{3}}AX = \frac{2}{\sqrt{3}}$

$\left(\frac{1}{2}AC\right)$. OA is therefore equal to $\frac{1}{\sqrt{3}}$.

Since OX may be represented by $\frac{1}{2\sqrt{3}}AC$

(see part **a**), $OX = \frac{1}{2}OA$. That is, the

radius of the inscribed circle is one-half the length of the radius of the circumscribed circle. Since their areas have the same ratio as the square of the ratio of their radii, the ratio of the areas of the circle is 1:4.

44. **(a)** 24 **(b)** 36 square units
45. 27 46. 87.5π cm³
47. **(a)** 150 square units
 (b) 94 square units
 (c) 64 cubic units
48. 1800π cm³ 49. 88 square units
50. $4\sqrt{3}$

CHAPTER 15

1. 20 square units
2. 12 square units
3. 17 square units
4. 24 square units
5. **(a)** $M(2, 5)$, $AB = \sqrt{40}$.
 (b) $M(-3.5, -3)$, $AB = 13$.

 (c) $B(4, -8)$, $AB = \sqrt{148}$.
 (d) $A(3, -9)$, $AB = \sqrt{164}$.
 (e) $M(b, a)$, $AB = 2\sqrt{a^2 + b^2}$.
 (f) $M\left(\frac{a+c}{2}, \frac{b+d}{2}\right)$, $AB =$

 $\sqrt{(a-c)^2 + (b-d)^2}$.

6. **(a)** $O(-1, 1)$.
 (b) 5
 (c) Yes, since $OK = 5$.
7. $y = -1$.
8. $\sqrt{125}$
9. First show $JKLM$ is a parallelogram. Since slope of $JK = \frac{2}{3}$ and the slope of $KL = -\frac{3}{2}$, $\overline{JK} \perp \overline{KL}$. Since $\overline{JK} = \overline{KL} = \sqrt{13}$, $JKLM$ is a square.
10. **(a)** Slope of $JK = \frac{6}{5}$ and slope of $KL = -\frac{5}{6}$ so that $\overline{JK} \perp \overline{KL}$.
 (b) $JL = \sqrt{305}$, $KL = \sqrt{61}$, and $JK = \sqrt{244}$ so that $(JL)^2 = (KL)^2 + (JK)^2$.
 (c) 61 square units
11. **(a)** $\frac{8}{3}$ **(b)** $-\frac{3}{2}$ **(c)** $\frac{7}{6}$
12. **(a)** Collinear.
 (b) Noncollinear.
 (c) Collinear.
13. Slope of $\overline{ST} = \frac{8}{5}$, slope of $\overline{TA} = -\frac{5}{8}$, slope of $\overline{AR} = \frac{8}{5}$ and slope of $\overline{SR} = -\frac{5}{8}$. Hence, $\overline{ST} \parallel \overline{AR}$ and $\overline{TA} \parallel \overline{SR}$. Also, $\overline{ST} \perp \overline{TA}$ and so on.
14. $y = \frac{1}{2}x + 5$.
15. $y = \frac{3}{2}x - \frac{7}{2}$.
16. **(a)** $y = 7$.
 (b) $y = -x + 3$
 (c) $x = 6$.
 (d) $y = -4x + 1$.
17. **(a)** $y = 2$.
 (b) $x = -5$.
18. **(a)** $y = x + 1$.
 (b) $y = \frac{11}{9}x - \frac{25}{9}$.
 (c) $y = -\frac{2}{3}x + 12$.
19. $y = -\frac{1}{4}x - 2$.

20. $k = -25, t = -7$.
21. (a) $x^2 + y^2 = 16$.
 (b) $(x - 3)^2 + (y - 4)^2 = 36$.
 (c) Center = (1, 0), radius = 9.
 (d) Center = (0, 0), radius = 7.
 (e) Center = (2, -5), radius = $\sqrt{51}$.
22. $x^2 + y^2 = 25$.
23. $k = \dfrac{5}{3}, h = \dfrac{36}{5}$.
24. $x^2 + (y - 9)^2 = 25$.
25. $\sqrt{41}$
26. (a) 36.5 square units
 (b) 90 square units
 (c) 76 square units
 (d) 77 square units
27. (a) $k = 10$
 (b) $y = 2x + 4$
 (c) 24.5 square units
28. (a) $y = -x + 4$ (c) $x = 0$
 (b) $y = 4$ (d) $x^2 + y^2 = 16$
29. Show slope of $\overline{JA}$ = slope of $\overline{EK} = 0 \cong$ slope of $\overline{JE}$. Also show $JE = AK = a\sqrt{10}$.
30. Let the coordinates of $\square ABCD$ be represented by $A(0, 0)$, $B(s, t)$, $C(r + s, t)$, and $D(r, O)$. Use the distance formula to verify $AB = DC = \sqrt{s^2 + t^2}$ and $BC = AD = r$.
31. Let the coordinates of right $\triangle ABC$ be represented by $A(0, 0)$, $B(0, 2r)$, and $C(2s, 0)$. If M is the midpoint of $\overline{BC}$, then its coordinates are $M(s, r)$. $AM = \sqrt{s^2 + r^2}$ and $BC = 2\sqrt{s^2 + r^2}$ so that $AM = \dfrac{1}{2} BC$.
32. Let the coordinates of $\triangle ABC$ be represented by $A(0, 0)$, $B(2r, 2s)$, and $C(2t, 0)$. Let L and M be the midpoints of $\overline{AB}$ and $\overline{BC}$, respectively. Then $L(r, s)$ and $M(r + t, s)$. Show slope of $\overline{LM}$ = slope of $\overline{AC}$ and that $LM = \dfrac{1}{2} AC$.
33. Let the coordinates of isosceles trapezoid $ABCD$ be represented by $A(0, 0)$, $B(r, s)$, $C(t, s)$, and $D(r + t, 0)$. $AC = \sqrt{t^2 + s^2} = BD$.

CHAPTER 16

1. Two circles concentric with the original circle having radii of 7 cm and 9 cm.

2. A line midway between the 2 parallel lines, parallel to each line, and 5 in. from each line.
3. A line parallel to the surface at a distance equal to the radius length.
4. A concentric circle having a radius of 9 cm.
5. A circle concentric with the stationary circle whose radius length is the sum of the lengths of the radii of the two given circles.
6. The perpendicular bisector of the base.
7. A circle concentric with the stationary circle whose radius length is difference of the lengths of the radii of the two given circles.
8. Two points determined by the intersection of the perpendicular bisector of $\overline{AB}$ and the two lines that are parallel to and on either side of $\overleftrightarrow{AB}$ and 4 in. from $\overleftrightarrow{AB}$.
9. Four points determined by the angle bisector of each pair of vertical angles formed by the intersecting lines and a circle whose center is their point of intersection.
10. No points.
11. The two points determined by the intersection of the perpendicular bisector of $\overline{PQ}$ and the circle whose center is at point X and which has a radius of 5 cm.
12. (a) 2 (b) 0 (c) 1
13. (a) 0 (b) 1 (c) 2
14. (a) $y = 5, y = -5$.
 (b) $x = 4, x = -4$.
 (c) $y = 2x$.
 (d) $2x + 3y = 6$.
 (e) $y = x + 5$.
15. 2
16. 0
17. (a) 1 (b) 2 (c) 0
18. 1
19. (a) $x = 2$ (b) $y = -5$ (c) $y = 3x + 5$
20. 4
21. 2
22. (a) **1** A circle having point P as its center and a radius d units in length.
 2 The parallel lines $x = +1$ and $x = -1$.
 (b) **1** 1 **2** 3 **3** 4
23. (3)
24. Construction

SOLUTIONS TO CUMULATIVE REVIEW EXERCISES

CHAPTERS 1–7

1. 18 2. 130 3. 20 4. 144 5. 40 6. 40
7. $\triangle DAF \cong \triangle ECG$ by ASA since $\angle A \cong \angle C$, $\overline{AF} \cong \overline{CG}$ (halves of congruent segments are congruent) and $\angle AFD \cong \angle CGE$ (supplements of congruent angles are congruent). By CPCTC, $\overline{FD} \cong \overline{GE}$.
8. $\triangle DAF \cong \triangle ECG$ by AAS since $\angle FDA \cong \angle GEC$, $\angle AFD \cong \angle CGE$, and $\overline{AD} \cong \overline{CE}$ (using the subtraction property). By CPCTC, $\angle A \cong \angle C$. Therefore, $\triangle ABC$ is isosceles.
9. (a) $\angle STR \cong \angle BRT \cong \angle TRS$. By the transitive property, $\angle STR \cong \angle TRS$. Therefore, $\overline{RS} \cong \overline{ST}$ (Theorem 7.2).
 (b) $\triangle RSM \cong \triangle TSM$ by SSS. By CPCTC, $\angle RSM \cong \angle TSM$. Therefore $\overline{SM}$ bisects $\angle RST$.
10. $\overline{AM} \cong \overline{BM}$ (definition of median), $\angle CMB \cong \angle PMA$, and $\overline{CM} \cong \overline{PM}$ so $\triangle AMP \cong \triangle BMC$ by SAS. By CPCTC, $\angle P \cong \angle BCM$. Since alternate interior angles are congruent, $\overline{AP} \parallel \overline{CB}$.
11. (a) $\angle A \cong \angle B$, $\overline{AF} \cong \overline{BD}$ (addition property), $\angle x \cong \angle y$. Therefore, $\triangle AFG \cong \triangle BDE$ by ASA.
 (b) $\overline{AC} \cong \overline{BC}$ (Theorem 7.2), $\overline{AG} \cong \overline{BE}$ by CPCTC. Therefore, $\overline{GC} \cong \overline{EC}$ (subtraction property).
12. Show $\triangle KQM \cong \triangle LQN$ by SAS. By CPCTC, $\angle K \cong \angle L$ and $\overline{MK} \cong \overline{NL}$ so that $\triangle PMK \cong \triangle RNL$ by ASA. Therefore, $\overline{PM} \cong \overline{NR}$ by CPCTC.
13. By Theorem 7.1. $\angle CAB \cong \angle CBA$. By the subtraction property, $\angle EAP \cong \angle DBP$. In $\triangle APB$, $\overline{AP} \cong \overline{BP}$ by Theorem 7.2. Since $\angle EPA \cong \angle DPB$, $\triangle EPA \cong \triangle DPB$ by ASA. By CPCTC, $\overline{PE} \cong \overline{PD}$.
14. First show $\triangle CAE \cong \triangle CBD$ by AAS since $\angle A \cong \angle B$, $\angle CEA \cong \angle CDB$, and $\overline{AC} \cong \overline{BC}$. Next, show $\triangle GAE \cong \triangle FBD$ by SAS since $\overline{AE} \cong \overline{BD}$ (CPCTC), $\angle A \cong \angle B$, and $\overline{AG} \cong \overline{BF}$ (halves of congruent segments are congruent). By CPCTC, $\overline{EG} \cong \overline{DF}$.

CHAPTERS 8–11

1. (2) 2. (4) 3. 60 4. (4) 5. 24 6. 1
7. 40 8. (1) 9. (1) 10. (1) 11. $6\sqrt{2}$
12. 6 13. (3) 14. (4) 15. $2\sqrt{3}$ 16. $\dfrac{5}{13}$

17. Use Theorem 10.3. $\angle CMB \cong \angle E$ (all right angles are congruent) and $\angle B \cong \angle D$ since complements of congruent (vertical) angles are congruent. Therefore, $\triangle CMB \sim \triangle DEA$.
18. Since the measure of an exterior angle of a triangle is greater than the measure of a non-adjacent interior angle of the triangle, $m\angle ADC > m\angle BCD$. Since $m\angle BCD = m\angle ACD$, by substitution $m\angle ADC > m\angle ACD$ which implies $CA > DA$.
19. Assume $\overline{AC}$ bisects $\angle BAD$. Therefore, $\angle BAC \cong \angle DAC$. Since $\overline{BC} \parallel \overline{AD}$, $\angle BCA \cong \angle DAC$. By the transitive property, $\angle BAC \cong \angle BCA$, but this contradicts the Given, ($\triangle ABC$ is *not* isosceles). Therefore, the assumption $\overline{AC}$ bisects $\angle BAD$ is false, and the statement $\overline{AC}$ does *not* bisect $\angle BAD$ is true.
20. (a) $\angle CEF \cong \angle EDF$. $\angle CFE \cong \angle DEF$ since each angle is complementary to the same angle (angle DFE). $\triangle FEC \sim \triangle EDF$ by Theorem 10.3.
 (b) Since $\angle A \cong \angle C$ (Theorem 7.1) and $\angle DFA \cong \angle CEF$ (right angles are congruent), $\triangle DFA \sim \triangle FEC$. Since $\triangle FEC \sim \triangle EDF$, by the transitive property of similarity, $\triangle DFA \sim \triangle EDF$.
21. (a) $\overline{EB} \parallel \overline{CD}$ (extensions of parallel segments are parallel) and $\overline{CE} \parallel \overline{DB}$ (given) so that quadrilateral $BECD$ is a parallelogram.
 (b) Since the diagonals of the trapezoid are congruent, the trapezoid is isosceles so that triangles DAB and CBA are congruent. Thus, $\angle CAB \cong \angle DBA$. Since $\angle E \cong \angle DBA$ (parallel lines form congruent alternate interior angles), by the transitive property. $\angle CAB \cong \angle E$. By Theorem 7.2 $\overline{AC} \cong \overline{CE}$.

(c) Since $\angle CAB \cong \angle DBA$ (see part **b**), $\triangle AFB$ is isosceles.

22. (a) Since $ABCD$ is a rectangle, $\overline{AD} \cong \overline{BC}$, $\angle D \cong \angle C$, and $\overline{DE} \cong \overline{CF}$ (by addition). Therefore, $\triangle ADE \cong \triangle BCF$ by SAS and $\angle 1 \cong \angle 2$ by CPCTC.
(b) Angles A and B are right angles, $\angle 3 \cong \angle 4$ (complements of congruent angles are congruent).
(c) $\overline{AG} \cong \overline{GB}$ by Theorem 7.2.

23. $\overline{BC} \parallel \overline{AD}$ since alternate interior angles are congruent. $\triangle CEB \cong \triangle AED$ by ASA since $\angle 1 \cong \angle 2$, $\overline{AE} \cong \overline{CE}$, and $\angle AED \cong \angle CEB$. Therefore, $\overline{BC} \cong \overline{AD}$ by CPCTC. Quadrilateral $ABCD$ is a parallelogram since the same pair of sides are both parallel and congruent.

24. First show $\triangle RWT \sim \triangle RVS$. Since $\overline{TW} \cong \overline{TV}$, $\angle W \cong \angle TVW \cong \angle SVR$. By the transitive property, $\angle W \cong \angle SVR$. Since $\overline{RVW}$ bisects $\angle SRT$, $\angle TRW \cong \angle SRV$, and $\triangle RWT \sim \triangle RVS$ by Theorem 10.3. Therefore, $\dfrac{RW}{RV} = \dfrac{TW}{SV}$ so that $RW \times SV = RV \times TW$.

CHAPTERS 12–16

1. (a) $-\dfrac{4}{3}$ (b) 10
2. 10 square units 3. 7 4. 120°
5. 30 6. 8 7. 2
8. (a) $x = 4$ (b) $y = 2$
9. 18π 10. 72 square units 11. 2
12. $\sqrt{98}$ 13. 24π 14. $12\sqrt{3}$ square units
15. 12 16. (7, 17) 17. 6 18. $\dfrac{3}{2}$
19. 9π 20. $y = -\dfrac{2}{3}x$ 21. 6 22. 2
23. $16 - 4\pi$ square units 24. 30
25. Show slope of $\overline{RA}$ = slope of $\overline{PT} = \dfrac{1}{3}$ and slope of $\overline{RT} \neq$ slope of $\overline{AP}$. $RT = AP = 5$ so the trapezoid is isosceles.
26. (a) 5 (b) $x^2 + y^2 = 25$
(c) $\dfrac{3}{4}$ (d) $y = \dfrac{3}{4}x + \dfrac{25}{4}$ (e) $(0, \dfrac{25}{4})$
27. Show $AB = BC = CD = DA = 5$ and slope

$\overline{AB} \times$ slope $\overline{AD} = -1$.

28. (a) $y = 2$ (b) $x = -4$ (c) $(-4, 2)$
(d) $(x - 2)^2 + (y + 2)^2 = 16$ (e) 1
29. 6
30. (a) 60 (b) 20 (c) 60
(d) 20 (e) 110
31. (a) Show $AB = AC = \sqrt{40}$.
(b) Midpoint of $\overline{BC}$ is $M(2, -1)$. Show slope $\overline{AM} \times$ slope $\overline{BC} = -1$.
32. (a) A pair of parallel lines whose equations are $x = 2 - d$ and $x = 2 + d$.
(b) The circle whose equation is $x^2 + y^2 = 1$
(c) **1** 2 **2** 1 **3** 0
33. First show $\triangle ANR \sim \triangle ABN$. $\angle ANR \cong \angle ABN$ (all right angles are congruent) and $\angle A \cong \angle A$. Therefore, $\dfrac{AN}{AB} = \dfrac{AR}{AN}$ so that $(AN)^2 = AR \times AB$.
34. Show $\triangle AEB \cong \triangle CFD$. $\angle B \cong \angle D$ (Theorem 7.1) so $ACD \cong CAB$. By arc subtraction, $AB \cong CD$ which implies $\overline{AB} \cong \overline{CD}$. Since $\angle E \cong \angle F$, $\triangle AEB \cong \triangle CFD$ by AAS, $\overline{AE} \cong \overline{CF}$ by CPCTC.
35. (a) 90 (b) 4 (c) **1** 63° **2** 63°
36. (a) 72 (b) 36 (c) 54
(d) 18 (e) 72
37. (a) Line parallel to and midway between the given lines.
(b) Circle having P as its center and a radius of k units.
(c) **1** 1 **2** 2 **3** 0
38. (a) Show $(AB)^2 + (BC)^2 = (AC)^2$
(b) 12.5 square units
39. 17.5 square units
40. First show $\triangle BCD \sim \triangle ABE$. $\angle DBC \cong \angle A$ since the measure of each angle is equal to one-half the measure of the same arc (BF). $\angle ABE$ is a right angle. Since $\overline{BA} \parallel \overline{CD}$ and interior angles on the same side of the transversal are supplementary, $\angle DCB$ is also a right angle so $\angle DCB \cong \angle ABC$. Therefore, $\triangle BCD \sim \triangle ABE$ by Theorem 10.3, so $\dfrac{BD}{AE} = \dfrac{CD}{BE}$.

INDEX